ADVANCES IN
Applied Microbiology
VOLUME 11

CONTRIBUTORS TO THIS VOLUME

R. G. Board
Victor H. Edwards
G. O. Gale
William Gold
I. L. Hepner
William J. Kelleher
G. A. Kemp
J. S. Kiser
George Luedemann
K. E. Price
Selman A. Waksman

ADVANCES IN

Applied Microbiology

Edited by D. Perlman
School of Pharmacy
The University of Wisconsin
Madison, Wisconsin

VOLUME 11

 1969

ACADEMIC PRESS, New York and London

COPYRIGHT © 1969, BY ACADEMIC PRESS, INC.
ALL RIGHTS RESERVED.
NO PART OF THIS BOOK MAY BE REPRODUCED IN ANY FORM,
BY PHOTOSTAT, MICROFILM, RETRIEVAL SYSTEM OR ANY
OTHER MEANS, WITHOUT WRITTEN PERMISSION FROM
THE PUBLISHERS.

ACADEMIC PRESS, INC.
111 Fifth Avenue, New York, New York 10003

United Kingdom Edition published by
ACADEMIC PRESS, INC. (LONDON) LTD.
Berkeley Square House, London W1X 6BA

LIBRARY OF CONGRESS CATALOG CARD NUMBER: 59-13823

PRINTED IN THE UNITED STATES OF AMERICA

LIST OF CONTRIBUTORS

Numbers in parentheses indicate the pages on which the authors' contributions begin.

R. G. BOARD, *School of Biological Sciences, Bath University of Technology, Bath, Somerset, England* (245)

VICTOR H. EDWARDS, *School of Chemical Engineering, Cornell University, Ithaca, New York* (159)

G. O. GALE, *Agricultural Division, American Cyanamid Company, Princeton, New Jersey* (77)

WILLIAM GOLD, *Department of Microbiology, College of Dentistry, New York University, New York* (135)

I. L. HEPNER, *Process Biochemistry, London, England* (283)

WILLIAM J. KELLEHER, *Division of Pharmacognosy, Pharmacy Research Institute, University of Connecticut, Storrs, Connecticut* (211)

G. A. KEMP, *Agricultural Division, American Cyanamid Company, Princeton, New Jersey* (77)

J. S. KISER, *Agricultural Division, American Cyanamid Company, Princeton, New Jersey* (77)

GEORGE LUEDEMANN, *Department of Microbiology, Schering Corporation, Bloomfield, New Jersey* (101)

K. E. PRICE, *Research Division, Bristol Laboratories, Division of Bristol-Myers Company, Syracuse, New York* (17)

SELMAN A. WAKSMAN, *Institute of Microbiology, Rutgers, The State University, New Brunswick, New Jersey* (1)

PREFACE

The scope of this volume is somewhat restricted compared with previous volumes in this serial publication in that we will concentrate on one particular topic—the current status of antibiotic research. This is a good year for a review of this area of applied microbiology since it marks 25 years of intensive use of penicillin, 25 years of research on and use of streptomycin, 10 years of study of semisynthetic penicillins, and 10 years of study of antibiotic resistance.

We are fortunate in having an introspective paper by the Nobel Laureate S. A. Waksman on his successes and failures in antibiotic research. The chapter by K. E. Price on semisynthetic penicillin is a very timely summary of the objectives of programs researching new penicillins. Dr. Price also proposes some systematic arrangements to summarize the current information on more than 10,000 of these compounds. Kiser *et al.* provide a timely evaluation of the problem of antibiotic resistance, and Gold's study of periodontal microbiology indicates new areas in which antibiotic chemotherapy in conjunction with a better understanding of the disease in question will result in successful treatment. As Dr. Waksman indicates, the search for new antibiotics continues, and Luedemann provides an exhaustive study of the very important *Micromonospora* species which will be the source of some of these new compounds. Professor Edwards' review of the technology involved in recovery of microbial metabolites summarizes progress in industry and highlights new approaches. Kelleher, in his contribution, has included much of his own experimentation on ergot alkaloid-producing fermentations, and because of the interest in these compounds as intermediates in the preparation of new pharmacologically active agents the realistic picture given of the availability of the starting materials is invaluable.

As the scope of applied microbiology increases, the requirements for training of qualified practitioners will be broadened. Hepner's suggestions of what the modern microbiologist should study is quite a timely challenge to those who are responsible for arranging curricula for the training of undergraduate and graduate students. The trend toward interesting microbiology students in molecular studies has limited to some extent the progress of the fermentation industries in the United States, whereas in Japan, where the emphasis has been on fermentation technology, the industrial applications have been more successful. The future of microbiology is still bright, and promising and interesting developments will be highlighted in future volumes of this serial publication.

Madison, Wisconsin   D. PERLMAN
October, 1969

CONTENTS

CONTRIBUTORS v
PREFACE vii
CONTENTS OF PREVIOUS VOLUMES xiii

Successes and Failures in the Search for Antibiotics
SELMAN A. WAKSMAN

	Introductory Remarks	1
I.	Antibiotics Produced by Bacteria	2
II.	Antibiotics Produced by Fungi	4
III.	Antibiotics Produced by Actinomycetes	5
IV.	Future Potential Developments in the Field of Antibiotics	7
V.	The Search for Antiviral and Antitumor Agents	9
VI.	Recognition of Antibiotics as Therapeutic Agents	12
VII.	Summary	14
	References	16

Structure-Activity Relationships of Semisynthetic Penicillins
K. E. PRICE

I.	Introduction	17
II.	History	18
III.	Structure–Activity Relationships	23
IV.	Future Objectives	66
	References	68

Resistance to Antimicrobial Agents
J. S. KISER, G. O. GALE, AND G. A. KEMP

I.	Introduction	77
II.	Evolution of Drug Resistance	78
III.	Biochemical Basis of Drug Resistance	80
IV.	Genetic Basis of Drug Resistance	83
V.	The Hospital Environment and Drug Resistance	91
VI.	The Agricultural Environment and Drug Resistance	93
VII.	Summary	98
	Note Added in Proof	98
	References	99

Micromonospora Taxonomy
GEORGE LUEDEMANN

I.	Introduction	101
II.	Review of Taxonomic Literature	103

III.	Relationships	122
IV.	Impressions and Reflections	126
	References	131

Dental Caries and Periodontal Disease Considered as Infectious Diseases

William Gold

I.	Introduction	135
II.	Caries in Man and Animals	136
III.	Periodontal Disease in Man and Animals	148
IV.	Perspectives	152
V.	Summary	153
	References	153

The Recovery and Purification of Biochemicals

Victor H. Edwards

I.	Introduction	159
II.	Removal of Suspended Solids	163
III.	Liquid Ion Exchange	168
IV.	Gel Filtration	173
V.	Membrane Separations	190
VI.	Conclusions	207
	References	207

Ergot Alkaloid Fermentations

William J. Kelleher

I.	Introduction	211
II.	Mycology	212
III.	Chemistry	214
IV.	Pharmacology	218
V.	Occurrence	218
VI.	Extraction and Analysis	219
VII.	Clavine Alkaloid Fermentations	221
VIII.	Lysergic Acid Alkaloid Fermentations	222
IX.	Biosynthesis	236
X.	Summary	241
	References	242

The Microbiology of the Hen's Egg

R. G. Board

I.	Introduction	245
II.	The Egg	246

III. The Course of Infection	261
Addendum	274
References	274

Training for the Biochemical Industries

I. L. HEPNER

Text	283
AUTHOR INDEX	289
SUBJECT INDEX	307

CONTENTS OF PREVIOUS VOLUMES

Volume 1

Protected Fermentation
Miloš Herold and Jan Nečásek

The Mechanism of Penicillin Biosynthesis
Arnold L. Demain

Preservation of Foods and Drugs by Ionizing Radiations
W. Dexter Bellamy

The State of Antibiotics in Plant Disease Control
David Pramer

Microbial Synthesis of Cobamides
D. Perlman

Factors Affecting the Antimicrobial Activity of Phenols
E. O. Bennett

Germfree Animal Techniques and Their Applications
Arthur W. Phillips and James E. Smith

Insect Microbiology
S. R. Dutky

The Production of Amino Acids by Fermentation Processes
Shukuo Kinoshita

Continuous Industrial Fermentations
Philip Gerhardt and M. C. Bartlett

The Large-Scale Growth of Higher Fungi
Radcliffe F. Robinson and R. S. Davidson

AUTHOR INDEX – SUBJECT INDEX

Volume 2

Newer Aspects of Waste Treatment
Nandor Porges

Aerosol Samplers
Harold W. Batchelor

A Commentary on Microbiological Assaying
F. Kavanagh

Application of Membrane Filters
Richard Ehrlich

Microbiol Control Methods in the Brewery
Gerhard J. Hass

Newer Development in Vinegar Manufactures
Rudolph J. Allgeier and Frank M. Hildebrandt

The Microbiological Transformation of Steroids
T. H. Stoudt

Biological Transformation of Solar Energy
William J. Oswald and Clarence G. Golueke

SYMPOSIUM ON ENGINEERING ADVANCES IN FERMENTATION PRACTICE

Rheological Properties of Fermentation Broths
Fred H. Deindoerfer and John M. West

Fluid Mixing in Fermentation Processes
J. Y. Oldshue

Scale-up of Submerged Fermentations
W. H. Bartholemew

Air Sterilization
Arthur E. Humphrey

Sterilization of Media for Biochemical Processes
Lloyd L. Kempe

Fermentation Kinetics and Model Processes
Fred H. Deindoerfer

Continuous Fermentation
W. D. Maxon

Control Applications in Fermentation
George J. Fuld

AUTHOR INDEX — SUBJECT INDEX

Volume 3

Preservation of Bacteria by Lyophilization
Robert J. Heckly

Sphaerotilus, Its Nature and Economic Significance
Norman C. Dondero

Large-Scale Use of Animal Cell Cultures
Donald J. Merchant and C. Richard Eidam

Protection against Infection in the Microbiological Laboratory: Devices and Procedures
Mark A. Chatigny

Oxidation of Aromatic Compounds by Bacteria
Martin H. Rogoff

Screening for and Biological Characterizations of Antitumor Agents Using Microorganisms
Frank M. Schabel, Jr., and Robert F. Pittillo

The Classification of Actinomycetes in Relation to Their Antibiotic Activity
Elio Baldacci

The Metabolism of Cardiac Lactones by Microorganisms
Elwood Titus

Intermediary Metabolism and Antibiotic Synthesis
J. D. Bu'Lock

Methods for the Determination of Organic Acids
A. C. Hulme

AUTHOR INDEX — SUBJECT INDEX

Volume 4

Induced Mutagenesis in the Selection of Microorganisms
S. I. Alikhanian

The Importance of Bacterial Viruses in Industrial Processes, Especially in the Dairy Industry
F. J. Babel

Applied Microbiology in Animal Nutrition
Harlow H. Hall

Biological Aspects of Continuous Cultivation of Microorganisms
T. Holme

Maintenance and Loss in Tissue Culture of Specific Cell Characteristics
Charles C. Morris

Submerged Growth of Plant Cells
L. G. Nickell

AUTHOR INDEX — SUBJECT INDEX

Volume 5

Correlations between Microbiological Morphology and the Chemistry of Biocides
Adrien Albert

Generation of Electricity by Microbial Action
J. B. Davis

Microorganisms and the Molecular Biology of Cancer
G. F. Gause

Rapid Microbiological Determinations with Radioisotopes
Gilbert V. Levin

The Present Status of the 2,3-Butylene Glycol Fermentation
Sterling K. Long and Roger Patrick

Aeration in the Laboratory
  W. R. Lockhart and R. W. Squires

Stability and Degeneration of Microbial Cultures on Repeated Transfer
  Fritz Reusser

Microbiology of Paint Films
  Richard T. Ross

The Actinomycetes and Their Antibiotics
  Selman A. Waksman

Fusel Oil
  A. Dinsmoor Webb and John L. Ingraham

AUTHOR INDEX—SUBJECT INDEX

Volume 6

Global Impacts of Applied Microbiology: An Appraisal
  Carl-Göran Hedén and Mortimer P. Starr

Microbial Processes for Preparation of Radioactive Compounds
  D. Perlman, Aris P. Bayan, and Nancy A. Giuffre

Secondary Factors in Fermentation Processes
  P. Margalith

Nonmedical Uses of Antibiotics
  Herbert S. Goldberg

Microbial Aspects of Water Pollution Control
  K. Wuhrmann

Microbial Formation and Degradation of Minerals
  Melvin P. Silverman and Henry L. Ehrlich

Enzymes and Their Applications
  Irwin W. Sizer

A Discussion of the Training of Applied Microbiologists
  B. W. Koft and Wayne W. Umbreit

AUTHOR INDEX—SUBJECT INDEX

Volume 7

Microbial Carotenogenesis
  Alex Ciegler

Biodegradation: Problems of Molecular Recalcitrance and Microbial Fallibility
  M. Alexander

Cold Sterlization Techniques
  John B. Opfell and Curtis E. Miller

Microbial Production of Metal-Organic Compounds and Complexes
  D. Perlman

Development of Coding Schemes for Microbial Taxonomy
  S. T. Cowan

Effects of Microbes on Germfree Animals
  Thomas D. Luckey

Uses and Products of Yeasts and Yeast-like Fungi
  Walter J. Nickerson and Robert G. Brown

Microbial Amylases
  Walter W. Windish and Nagesh S. Mhatre

The Microbiology of Freeze-Dried Foods
  Gerald J. Silverman and Samuel A. Goldblith

Low-Temperature Microbiology
  Judith Farrell and A. H. Rose

AUTHOR INDEX—SUBJECT INDEX

Volume 8

Industrial Fermentations and Their Relations to Regulatory Mechanisms
  Arnold L. Demain

Genetics in Applied Microbiology
  S. G. Bradley

Microbial Ecology and Applied Microbiology
Thomas D. Brock

The Ecological Approach to the Study of Activated Sludge
Wesley O. Pipes

Control of Bacteria in Nondomestic Water Supplies
Cecil W. Chambers and Norman A. Clarke

The Presence of Human Enteric Viruses in Sewage and Their Removal by Conventional Sewage Treatment Methods
Stephen Alan Kollins

Oral Microbiology
Heiner Hoffman

Media and Methods for Isolation and Enumeration of the Enterococci
Paul A. Hartman, George W. Reinbold, and Devi S. Saraswat

Crystal-Forming Bacteria as Insect Pathogens
Martin H. Rogoff

Mycotoxins in Feeds and Foods
Emanuel Borker, Nino F. Insalata, Colette P. Levi, and John S. Witzeman

AUTHOR INDEX—SUBJECT INDEX

Volume 9

The Inclusion of Antimicrobial Agents in Pharmaceutical Products
A. D. Russell, June Jenkins, and I. H. Harrison

Antiserum Production in Experimental Animals
Richard M. Hyde

Microbial Models of Tumor Metabolism
G. F. Gause

Cellulose and Cellulolysis
Brigitta Norkrans

Microbiological Aspects of the Formation and Degradation of Cellulosic Fibers
L. Jurášek, J. Ross Colvin, and D. R. Whitaker

The Biotransformation of Lignin to Humus—Facts and Postulates
R. T. Oglesby, R. F. Christman, and C. H. Driver

Bulking of Activated Sludge
Wesley O. Pipes

Malo-lactic Fermentation
Ralph E. Kunkee

AUTHOR INDEX—SUBJECT INDEX

Volume 10

Detection of Life in Soil on Earth and Other Planets. Introductory Remarks
Robert L. Starkey

For What Shall We Search?
Allan H. Brown

Relevance of Soil Microbiology to Search for Life on Other Planets
G. Stotzky

Experiments and Instrumentation for Extraterrestrial Life Detection
Gilbert V. Levin

Halophilic Bacteria
D. J. Kushner

Applied Significance of Polyvalent Bacteriophages
S. G. Bradley

Proteins and Enzymes as Taxonomic Tools
Edward D. Garber and John W. Rippon

Micotoxins
Alex Ciegler and Eivind B. Lillehoj

Transformation of Organic Compounds by Fungal Spores
Claude Vézina, S. N. Sehgal, and Kartar Singh

Microbial Interactions in Continuous Culture
Henry R. Bungay, III and Mary Lou Bungay

Chemical Sterilizers (Chemosterilizers)
Paul M. Borick

Antibiotics in the Control of Plant Pathogens
M. J. Thirumalachar

AUTHOR INDEX—SUBJECT INDEX

CUMULATIVE AUTHOR INDEX—CUMULATIVE TITLE INDEX

Successes and Failures in the Search for Antibiotics[1]

SELMAN A. WAKSMAN

*Institute of Microbiology,
Rutgers, The State University,
New Brunswick, New Jersey*

	Introductory Remarks	1
I.	Antibiotics Produced by Bacteria	2
	A. Gram-Negative Bacteria	2
	B. Spore-Forming Bacteria	3
II.	Antibiotics Produced by Fungi	4
	A. Green Penicillia and Green or Brown Aspergilli	4
	B. Other Preparations Produced by Penicillia	4
	C. Other Fungal Products	5
	D. Semisynthetic Penicillins	5
III.	Antibiotics Produced by Actinomycetes	5
	A. Early Reports	5
	B. Isolation of Actinomycin A	5
	C. Extensive Screening Programs	6
IV.	Future Potential Developments in the Field of Antibiotics	7
V.	The Search for Antiviral and Antitumor Agents	9
VI.	Recognition of Antibiotics as Therapeutic Agents	12
VII.	Summary	14
	References	16

Introductory Remarks

The period of 1939 to 1960 has often been referred to as "The Golden Age of Chemotherapy." This is due, in no small measure, to the discovery and utilization of antibiotics, or those chemical substances produced by microorganisms which have the capacity to inhibit the growth and even to destroy other microorganisms in dilute solution. Antibiotics have found extensive applications in the treatment of numerous infectious diseases of man, animals, and, to a lesser degree, plants. They are also being used extensively in animal nutrition, in food conservation, and in the preservation of biological materials, such as animal semen and virus preparations. No wonder that antibiotics have been popularly dubbed "miracle drugs," and

---

[1] Observations made during my association, for three decades, with the problems involving formation by microorganisms of antimicrobial substances that came to be known as antibiotics and the practical utilization of these substances.

1

that the very period under consideration has also been spoken of as "The Antibiotic Era."

How did all this come about in such a short period of time? What were the successes and what were the failures that led to such fantastic developments? Any attempt to trace earlier observations made through the centuries concerning the effect of moldy products upon bacterial infections may be left out of present consideration. I shall also abstain from quoting here the more recent (prior to 1939) scientific reports made by various bona fide investigators on the destructive effects of fungi upon the growth of bacteria and of bacteria upon fungi. I shall instead limit myself mainly to our present knowledge of the production of antibiotics by three of the major groups of microorganisms: the bacteria, fungi, and actinomycetes.

The beginnings of my own association with the antagonistic effects of soil-inhabiting microbes, especially the actinomycetes, and the production of antimicrobial substances may be said to have begun in 1937 with the publication of a comprehensive review on "Associative and Antagonistic Relations of Microorganisms." Three years later (1940), in addressing the local Sigma Xi Society, I said, "We are finally approaching a new field of domestication of microorganisms for combating the microbial enemies of man and of his domesticated plants and animals. Surely, microbiology is entering a new phase of development." Some of the following observations comprise successes and failures, including many of my own. They served, however, both as stages in the development of a new and important science that has achieved spectacular practical results and has led to one of the most remarkable periods in the history of microbiology, as well as in the history of medical and veterinary science.

The observations that I am about to record are not presented in either historical or logical sequence, but are based upon the results obtained on the formation of antibiotics by those groups of microorganisms that proved to be the major producers of antimicrobial substances.[2]

I. Antibiotics Produced by Bacteria

A. GRAM-NEGATIVE BACTERIA

These comprise largely the *Pseudomonas aeruginosa* and the *Escherichi coli* groups.

---

[2] References to earlier literature are found in my book on "Microbial Antagonisms and Antibiotic Substances" (Waksman, 1945, 1947) and in the monumental treatise by Florey et al. (1949) on "Antibiotics."

The recognition of the remarkable antibacterial properties of *Pseudomonas aeruginosa* (spoken of originally as *Bacillus pyocyaneus*) began with the work of Bouchard and Emmerich and Low, carried out between 1890 and 1900. The active substance was at first considered to be an enzyme and was designated as pyocyanase. The organism was also found to produce a pigment known as pyocyanin, which possessed antimicrobial properties. The numerous contributions made to the subject of pyocyanase culminated in the excellent work of Doisy and his collaborators (Hays *et al.*, 1945) on the pyo compounds. Unfortunately, these substances, for various reasons, have found no practical applications in chemotherapy.

The antimicrobial products of *E. coli* fall into a somewhat different category. They are largely active upon members of the group producing them, and have come to be known as colicins. They appear to be related more closely to the bacteriophages rather than to the antibiotics. At the present time, the practical potentialities of most, if not all, of these preparations are limited.

B. Spore-Forming Bacteria

Here, as well, the story is a long and often confusing one. One need only begin with the fundamental studies on the antimicrobial properties of *Bacillus subtilis* by M. Nicolle in 1907 and the observations of E. Pringsheim on *Bacillus mycoides* in 1920, leading to the work of Dubos (1939) on tyrothricin produced by *Bacillus brevis*, and later of Meleney and Johnson on bacitracin, also a product of *B. subtilis*. Finally, one must add here the isolation of the polymyxins, carried out independently in three different laboratories, from strains of *Bacillus polymyxa*. Polymyxin B and polymyxin E are of considerable importance as chemotherapeutic agents in the treatment of infections caused by gram-negative bacteria.

Numerous other products were isolated from spore-forming bacteria, notably strains of *B. subtilis*. More than 25 names have been used to designate preparations derived from this group of organisms. It is sufficient to mention, in addition to bacitracin, also aterrimin, bacilipins, bacillin, bacillocin, bacillomycins, bacilysin, bulbiformin, datemycin, edeine, eumycin, fluvomycin, fungistatin, globicin, iturin, mycobacillin, mycosubtilin, neocidin, petrin, subtenolin, subtilin, kantellin, yeastcidin, aspergillus factor, and the rhizoctonia factor. Some of these are undoubtedly synonyms. The preparation "ayfivin," for example, named for the number of the culture in the type collection, was isolated by Florey's groups at Oxford and first believed to be a "new antibiotic"; it later proved to be synonymous with bacitracin. Only the latter has been used clinically and for animal feeding.

II. Antibiotics Produced by Fungi

A. GREEN PENICILLIA AND GREEN OR BROWN ASPERGILLI

The most important of these antibiotics are the penicillins. Active preparations have been isolated from cultures of penicillia and aspergilli by Gosio (1896), Duchesne (1897), Vaudremer (1913), and Gratia (1924). These led to the observations of Alexander Fleming in 1928 on the antibacterial properties of *Penicillium notatum*, and finally resulted in the isolation by Chain, Florey, and their associates of one of the most important groups of antibiotics, namely, the penicillins (1940). These studies were followed by extensive investigations during the second World War in the United States, under the sponsorship of the Committee on Medical Research of the Office of Scientific Research and Development, and in Great Britain by government and industrial laboratories.

It may be of interest to record here two personal observations made in 1939 and in 1940, bearing upon the isolation of penicillin, to prove that it all began that year. At a symposium on "Chemotherapy of Bacterial Infections," held at the Third International Congress for Microbiology in New York, in 1939, or just on the eve of Chain and Florey's classical report, penicillin was not mentioned, even in the paper by Fleming "On the Testing of Chemotherapeutic Drugs." The second observation deals with the meeting of the Society of American Bacteriologists held in St. Louis, during Christmas week of 1940. At a round table discussion on antimicrobial drugs, two of the preparations known at that time and produced by bacteria and actinomycetes, namely, tyrothricin and actinomycin, were discussed. Someone in the audience of 200 asked whether anything further had been heard about penicillin recently reported by the Oxford group in Great Britain. Nobody had any comments. The word "antibiotic" was not used at either of these conferences. The audience left the meeting not much wiser about the whole field, but vitally interested in it.

B. OTHER PREPARATIONS PRODUCED BY PENICILLIA
(AT FIRST BELIEVED TO BE PENICILLINS)

Only four preparations that proved to be quite distinct from the true penicillins need be mentioned here: penicillin B, penicidin, notatin, and a substance isolated by various Soviet investigators from cultures of *Penicillium crustosum* (Utkin) and designated as "penicillin-crustosin." The first three are enzymatic preparations. None of them proved to be of any practical significance.

C. OTHER FUNGAL PRODUCTS

Numerous other antibiotic preparations have been isolated from cultures of fungi. The more important ones are: the cephalosporins from *Cephalosporium;* griseofulvin from *Penicillium griseofulvum;* fusidin from *Fusarium;* various substances produced by *Aspergillus clavatus* and by certain penicillia: clavacin, claviformin, and patulin (this was supposed to combat the common cold); fumigatin, fumigacin, and fumagillin from *Aspergillus fumigatus.* More than 100 different antibiotic preparations from fungi have now been reported. Only the first two and fumagillin were of therapeutic value in 1968.

D. SEMISYNTHETIC PENICILLINS

A new field for the study of antibiotics has been opened through the enzymatic breakdown of the penicillins to 6-aminopenicillinic acid and chemical conversion of this compound to new penicillins. The question still remains how much these are really superior to the natural penicillins as chemotherapeutic agents. These new penicillins are said to have acid stability, activity against gram-negative bacteria, activity against penicillin-resistant bacteria, and give higher blood levels than found with penicillin.

III. Antibiotics Produced by Actinomycetes

A. EARLY REPORTS

One must begin here, as well, with the numerous observations concerning the capacity of this group of organisms to produce antimicrobial substances. It is sufficient to mention those of Greig-Smith in 1917, of Lieske in 1920, and of a group of Russian investigators, notably Borodulina, Koreniako, Kriss, and Krassilnikov in the 1930's. These did not result in the isolation of any active compound possessing antimicrobial properties. An interesting contribution was made in this field by Gratia and his collaborators in Belgium on mycolysate in 1924, and later by Welsch on actinomycetin, an enzyme complex.

B. ISOLATION OF ACTINOMYCIN A

The first crystalline antibiotic isolated from a culture of an actinomycete in 1940 was actinomycin. It proved to be extremely toxic to animals and was at first facetiously designated as "rat poison." The isolation of this antibiotic was followed by the isolation of streptothricin in 1942, which possessed a delayed toxicity in animal tests, and finally to streptomycin in 1943. A new era was now begun in the search for antibiotics of actinomycetes. In our own laboratories, it resulted in the isolation of grisein, micromonosporin, the neomycins (neomycin complex), candicidin, and other antifungal antibiotics.

C. EXTENSIVE SCREENING PROGRAMS

The isolation of the above antibiotics led to extensive screening programs, first in this country, and later in numerous laboratories throughout the world. Numerous new antibiotics, with new and important therapeutic properties, were discovered. They included the antibacterial chloramphenicol, tetracyclines, and erythromycins, and the antifungal nystatin, trichomycin, and other polyenes. Much confusion has developed in the naming of antibiotics and of antibiotic-producing organisms. Insufficient comparative studies were often made of the new isolates; a substance or preparation was often given a "new" name, with an eye to the naming of a "new" organism for patenting purposes. This has reached such an extreme that occasionally an organism has been named as "new," when, on further study, it turned out to be a strain of another organism. The new name was discarded and the older name substituted; the new culture was reduced to synonymy with the old one.

Among the large screening programs, it is sufficient to mention those carried out in the U.S.S.R. and in Japan. Unfortunately, the former have often resulted in the rediscovery of old compounds to which new names were given. Examples include the discovery of (a) "albomycin," which is similar to, if not identical with, grisein, discovered several years previously in our own laboratories and discarded because of certain undesirable properties that make it an unsuitable therapeutic agent; (b) "colimycin" and "meserin," which are identical with neomycin; and (c) "levomycetin," which is identical with chloramphenicol. These rediscoveries are largely a result of the previous screening programs carried out in this country. The first antibiotic usually isolated in such programs is some form of actinomycin; the next is some form of streptothricin; others then follow. In other words, "ontogeny recapitulates phylogeny," namely, the results of the first screening program carried out in our laboratory, beginning in 1939, were repeated with certain variations elsewhere. The Japanese screening programs, however, have yielded a number of highly desirable antibiotics, such as cellocidin, kanamycin, pyrrolnitrin, trichomycin, blasticidin S, and a number of others of clinical and agricultural importance.

IV. Future Potential Developments in the Field of Antibiotics

As the numbers of newly discovered antibiotics that possess desirable chemotherapeutic properties are gradually diminishing, although 50 or 60 have been reported every year since 1967 in extensive screening programs carried out throughout the world, and as the numbers of freshly isolated preparations are found to consist either of compounds that are already known or such that possess serious limitations, one begins to wonder whether such programs should be continued further or whether the methods that have commonly been used in the previous programs should be reconsidered. Perhaps some of the microbial groups that have yielded desirable antibiotics should be reexamined. What are these groups and what are their further potentials? The following should, perhaps, be given further consideration:

1. In addition to the genus *Streptomyces* (and other constituents or closely related groups often designated as new genera, such as *Streptoverticillium*), among the actinomycetes, only *Micromonospora* and *Thermoactinomyces* and other thermophilic genera appear to offer promise. The first antibiotic isolated from a member of the genus *Micromonospora* was micromonosporin. It was a labile, orange-colored substance active largely upon gram-positive bacteria. More recently, a new chemotherapeutically useful basic, water-soluble antibiotic complex, belonging to the 2-deoxystreptamine family, was isolated from several species of *Micromonospora* and designated as *gentamicin*.

Thermophilic actinomycetes have yielded several compounds that, at first, appeared to offer some definite promise. It is sufficient to mention thermomycin, thermoviridin, and refuin (anthramycin). In view of the number of thermophilic genera now known, their wide occurrence in composts and in soils, and their characteristic conditions of growth, they certainly deserve reexamination.

2. Among the fungi, both the filamentous and the higher or mushroom groups deserve further consideration. One should only keep in mind the large number of genera among the former; only three or four of which have been studied to any great extent, and which have

yielded such desirable compounds as the penicillins, cephalosporins, and griseofulvin. What about numerous other fungus genera, such as *Trichoderma, Verticillium,* and *Chaetomium?* What about the Mucorales? As to the higher fungi, we know definitely that some of them are capable of producing antimicrobial and even antineoplastic agents, such as clitocybin and polysporin. One wonders whether these, as well, have been studied sufficiently.

3. The yeasts have been given only cursory consideration. It is most interesting to note that while numerous yeasts produce various alcohols, which possess certain antimicrobial properties, these cannot be called antibiotics, however, since the very definition of the word implies their activity in dilute solutions, which obviously is not true of the alcohols.

4. Further consideration should also be given to some of the bacterial groups. Aside from the aerobic spore formers (genus *Bacillus*), certain gram-negative forms *(Pseudomonas aeruginosa)*, and certain micrococci (production of nisin), the vast majority of bacteria have hardly been studied sufficiently. Some appear to offer little promise; the anaerobes, for example, in spite of the ability to form alcohols (butyl), organic acids (butyric), and certain other products (acetone), have so far not yielded any true antibiotics.

5. One may also consider the chlorophyll-bearing lower forms of life. In the early days of antibiotic research, several compounds were isolated both from the free-living algae (chlorellin) and the symbiotic forms, notably the lichens (lichenin), but little further work has been done in more recent years.

6. I have often been asked about the possibility of finding antibiotic-producing organisms in the sea. In my own early work (1940–1942) on antibiotics, I looked into this matter rather carefully. I was at that time still under the impression, due to my earlier experience with soil enrichment procedures, that antibiotic-producing organisms could be obtained by feeding a natural substrate, such as soil, with cultures of certain pathogens. I tried that method with both dead and living cells of the tuberculosis organism on soil, and with *E. coli* cells on seawater. Both experiments failed. The streptomycin-producing *S. griseus* was isolated in other ways than by the enrichment method. The *E. coli* cells died when added to seawater, but such enriched water failed to yield any antibiotic-producing organism. The microbial population of the sea consists largely of diatoms and other microscopic plant and animal forms, as well as of bacteria. Fungi and actinomycetes are lacking entirely, except close to shore or upon

floating algal material. Neither the diatoms nor the bacteria appear to offer much promise as antibiotic producers, although some further investigation may be justified.

V. The Search for Antiviral and Antitumor Agents

The successful introduction of antibiotics in the treatment of infectious diseases caused by bacteria and other pathogenic microorganisms aroused great hopes that some of the more important non-microbial diseases would also soon become subject to chemotherapy. Among such diseases, the viral and the neoplastic diseases came first into consideration. The searches begun for chemical agents that had the capacity to combat them were based largely upon methods developed in the discovery of antimicrobial substances. Extensive screening programs were initiated. Numerous laboratories throughout the world were soon announcing the discovery of various antibiotics or other products of microbial metabolism that appeared to have some effect against viruses and neoplasms. Unfortunately, the therapeutic results obtained with these preparations did not fully support the hopeful announcements. Disappointments soon followed.

The reasons for these disappointments are obvious, especially when one attempts to analyze the reasons for the failure of the antibiotic approach to the search for antiviral and antineoplastic agents. Antibiotics are active primarily upon living microbes, whether they are bacteria, fungi, or protozoa, which infect the human or animal body. These pathogens are able to grow and divide, metabolize, and produce enzyme systems in a manner quite distinct from the growth and metabolism of the cells of higher forms of life which they infect. Any antimicrobial substance, especially the antibiotics, which can interfere with the growth, nutrition, synthesis, or enzyme mechanisms of the cells of the pathogenic microbes, without interfering with the growth and nutrition of the host cells, will automatically lead to the destruction of the infecting microbes without seriously affecting the host.

This is not true, however, for infections caused by viruses, nor is it true of the development of abnormal neoplastic cells within the human or animal body. Whether these cells are initiated by viruses or by chemical carcinogens, they are not subject, except to a limited degree, to the action of therapeutic agents, such as antibiotics, whereby the abnormal or neoplastic cell would be destroyed without injury to the normal body cells.

The earliest observations on the effect of microorganisms on tumors were concerned with certain acute bacterial infections, notably streptococcal organisms causing erysipelas, which were found to influence tumor growth in some way. Products of various bacteria, notably streptococci and *Serratia marcescens*, also had the capacity to suppress neoplasms. Patients suffering from various forms of cancer, when treated with these products, frequently showed hemorrhages and necroses of the tumors. Such preparations, including "Coley's toxins," were manufactured and used with a certain degree of success. Unfortunately, most of the cases treated successfully with bacterial preparations showed only partial or temporary regression. Only very few cases underwent complete and permanent regression. The criticism directed against most of these studies was that the preparations were not sufficiently standardized and that no suitable animal bioassay had been resorted to.

Roskin and Kluyeva found that certain toxic substances of protozoa (*Trypanosoma cruzi*) had a marked antitumor action; these claims were not generally substantiated. More recently, "krebiozen," an extract of the serum of horses previously infected with *Actinomyces bovis*[3] which was said to have a pronounced effect on tumors, also attracted attention but was without scientific merit. Other claims concerning the ability of various crude cultures of fungi or bacteria to suppress tumor growth have been presented, but they too lacked experimental verification. The ability of certain viruses to dissolve tumors has also been suggested. The true significance of this phenomenon is still difficult to evaluate.

Microbial products closely related to the antibiotics have also been receiving considerable attention in recent years. Some of them have given definite indications of favorable potentialities in the treatment of experimental and human tumors. This is true particularly of certain specific antibiotics produced by actinomycetes.

Attention has already been directed to the antitumor potentialities of actinomycin. This was first reported, in 1950, by Chester Stock.[4] In 1952, Hackmann found that actinomycin C exerted a favorable action upon contact with Ehrlich carcinoma for 3 hours at 0°C.; the tumors thus treated showed greatly reduced capacity to initiate fresh growth. Walker carcinoma in rats was repressed by treatment with 1:20,000 to 1:100,000 dilutions. Therapeutic use of this antibiotic in cases of

---

[3] It is highly doubtful whether this was a true culture of the anaerobe, an organism rather difficult to isolate in pure culture.

[4] For a detailed review of the literature, see the recent book on "Actinomycin," edited by the writer (Waksman, 1968).

malignant diseases of the lymphatic system was therefore suggested. On the basis of these studies, Schulte treated a number of clinical cases of lymphogranulomatosis with this antibiotic. No detrimental effects were observed in connection with the employed doses. In some patients, the glandular tumors disappeared without the necessity for any supplementary treatment; in others, it was possible to reduce considerably the supplementary roentgen doses.

Another form, actinomycin D, isolated in 1954 proved to be a markedly homogeneous substance and highly efficacious in the treatment of experimental tumors. (It was later approved by the Food and Drug Administration.) A comparative study of the several actinomycins brought out the fact that they are very similar in their toxicities and antibacterial effects. They were, therefore, also expected to be very similar in their antitumor effects.

Among the other antibiotics that were found to exert a marked effect upon experimental tumors, and in a few cases also upon human tumors, was azaserine and a closely related compound, diazo-5-oxo-L-norleucine, known as DON. They inhibited growth of a variety of tumors and leukemias. The treatment of a number of patients has so far not shown any substantial therapeutic results, except for some transient benefits in a few cases.

Sarkomycin[5] also possesses marked antitumor activity, especially against Ehrlich carcinoma of mice. This was accomplished by daily intravenous injection of 1 mg., intraperitoneal injection of 2.5 mg., or oral administration of 5 mg. per mouse. Yoshida sarcoma in rats was also inhibited. It has shown clinical response in stomach cancer, malignant chorioepithelioma, reticulosarcoma, and malignant struma, but it had no effect in Hodgkin's disease, osteosarcoma, or hypernephroma. Carzinophilin[5] was shown to exert a specific activity against Yoshida sarcoma, Ehrlich carcinoma, and ascitic hepatoma. Its use brought about the destruction of tumor cells and the prolongation of the patients' survival time. In addition to these antibiotics, several other preparations were reported to possess antitumor properties. These range from crude culture broths to specific chemical substances, such as amicetin, puromycin, diromycin, rubomycin, daunomycin, and mitomycin C.

The action of microbial products upon tumors, like that of antibiotics on bacteria and fungi, is selective in nature. Since a tumor shows a type of growth and general metabolic pattern which are distinct in many respects from those of normal tissues, one could easily

---

[5] These antibiotics are used for cancer treatment in Japan.

conceive of antimetabolic agents that would affect the growth of tumors without influencing that of healthy tissues. A similar phenomenon may be observed in the action of antibiotics and other chemotherapeutic agents on bacterial cells as compared to those of the host which the bacteria attack. Another striking comparison can be observed between the action of antimicrobial agents on rapidly growing and actively dividing bacteria, as compared to the action upon resting bacterial cells, and action of antineoplastic agents upon the rapidly multiplying abnormal tumor cells, as compared to the slow growing normal cells. The development of resistance to antitumor agents has also its parallel in the effect of antibacterial agents. The possible use of combinations of two or more substances active against tumors offers a further basis of comparison between the antitumor and antibacterial activities of microbial or chemical agents.

The mechanism of action of the microbial products as antitumor agents is still very little understood. There has been such a great urgency for discovering "cancer cures" that the investigators have not had sufficient time to evaluate the results calmly and come to definite rational conclusions. The whole field of cancer chemotherapy is still so recent, the problems involved are so complex, and the practical need is so great that one does not wonder at the "hit and miss" type of research commonly resorted to.

Another problem worthy of mention is the tendency to compare antitumor agents with antibiotics. The latter have brought about so great a revolution in the therapy of infectious diseases that they have proved to be in a true sense "miracle drugs." The public has, therefore, come to expect miracles from antitumor agents as well. Many clinicians, for example, have frequently injected actinomycin into two or three patients with terminal cancers already previously subjected to every possible treatment, ranging from radiation therapy and mustard gas to administration of aminopterin and other agents; when the patients were not "cured" a few days after treatment with actinomycin, the conclusion was reached that actinomycin is not an "effective cure."

While it is impossible, therefore, to state at present whether antibiotics will contribute to the solution of the tumor problem, there remains hope that some contributions will be made.

VI. Recognition of Antibiotics as Therapeutic Agents

Antibiotics did not easily acquire universal recognition and establishment as therapeutic agents, especially in some countries. The

Germans in North Africa are said to have referred to the discovery of penicillin as "an American bluff." Even in Great Britain, the "home of penicillin," the use of antibiotics for the treatment of infectious diseases was not always looked upon with great favor. In 1942, Almroth Wright (Cope, 1966), of St. Mary's Hospital in London, wrote an indignant letter to the London Times concerning their omission (in a discussion of Florey and Chain's remarkable work on penicillin) of Fleming's name, insisting that the latter was the real "discoverer of penicillin." Shortly afterward, however, he stated, "I see good reason to fear that as Chemotherapy is more and more widely resorted to Medical Practice—at any rate the Treatment of Bacterial Infections—will degenerate into a system of pure empiricism. And everyone who reflects will appreciate that treatment becomes pure empiricism, and indeed it becomes sometimes very like quackery, when the only questions that a Medical Man can be persuaded to take interest in are these: What is the drug that I should prescribe? And what is the dose I should administer and how often in the day must it be repeated?"

Streptomycin, the first antibiotic that proved to be effective in the treatment of tuberculosis, was severely criticized at the beginning, especially in certain countries, as in Great Britain, because of its toxic reactions and the rapid development of bacterial resistance. Later, however, it received its due recognition, as cited in a recent British editorial (1964):

> Our great success story in modern medicine is the control of tuberculosis. Half a century ago, it was an enigma wrapped up in tragedy. Some forty or fifty thousand people died in a single year, and it was generally assumed that for each death there were ten actual cases. Today we can confidently say to a patient: by skillful use of our three powerful drugs you can be cured. A large city may record no deaths from tuberculosis in a year, and the chest physician can say with truth that all the known cases are under his supervision.

In concluding these rambling remarks on the development of antibiotics, I may be permitted to mention a recent address by H. F. Dowling, the eminent medical microbiologist of the University of Illinois School of Medicine, before the 1966 Annual Conference on Antimicrobial Agents and Chemotherapy:

> If the progress of a drug from conception to consumption is viewed as a long production line, then the segments of this production line are managed by four different professional groups . . . (a) the executives of the pharmaceutical companies; (b) the basic and clinical scientists in the medical schools; (c) the administrators in the Federal Food and Drug Administration; and (d) the practicing physicians.

Dowling should have also given consideration to certain scientific groups that make this "production line" possible; the microbiologist, who isolates the antibiotic-producing cultures and discovers the new drugs; the chemist who purifies the compounds and determines their chemical structures; and the pharmacologist who evaluates the antibiotics *in vitro* and *in vivo*, before their potential usefulness is decided upon.

I should like to draw your attention to the recent comment by Chain (1966), that the discovery of biodynamic substances has had "in the vast majority of cases [as] the starting point a biologic phenomenon which was subsequently followed up by chemical methods ... with very few exceptions, the successful search for new biodynamic substances has followed a definite characteristic pattern in three phases: the biologic observations (followed by) the isolation of the active principle and elucidation of its structure (followed by) the synthesis (of the biodynamic substance), and the synthesis of substances with related or suitably modified structures." This became true with the recognition that microorganisms can be used both as biological test systems and as producers of biodynamic substances.

VII. Summary

The last three decades have often been designated, as far as biomedical research and practical applications in the control of infectious and other diseases are concerned, as the "golden age of chemotherapy," even as the "antibiotic era." Millions of human lives have been saved annually throughout the world. Numerous infectious diseases that have afflicted the human race since time immemorial have been brought under control. Diseases of childhood have practically disappeared. The average life span of man has been prolonged by nearly two decades. The ancient disease tuberculosis, or, as it was popularly called, "consumption," has been reduced from the number one killer of the human race to a position of thirteenth or fifteenth place among the diseases that are now afflicting mankind.

The discoveries that have led largely to the development of the antibiotics program were made in two small laboratories. No expenditures of large sums of money were involved. With only one exception, the United States federal government itself has contributed very little toward the solution of the vast problems in the search for new antibiotics, their evaluation, and practical utilization. There was no organized, government supported large-scale planned "basic" or biomedical research, as it is now commonly designated.

Of the first two most important antibiotics, one (penicillin) was discovered in a small laboratory in a British medical school; the other (streptomycin) was a result of a long-term program of a study of soil-inhabiting microorganisms, carried out in one of the American agricultural experiment stations. No fanfare and no front page publicity accompanied these discoveries. As soon as their promising potentialities were recognized, the practical applications followed immediately. As soon as it was shown that these antibiotics might prove to be effective in the treatment of infectious diseases not only in small experimental animals, but also in human beings, industry undertook at once their practical developments, by initiating an extensive research program, soon followed by large-scale manufacturing. These were followed by comprehensive screening programs throughout the world, consisting of a search for new and more effective antibiotics, their isolation, purification, and characterization, and by a pharmaceutical program for their practical evaluation. A multimillion dollar industry thus resulted, and a tremendous revolution was initiated in medical and veterinary science and application, in animal feeding, and in food preservation.

What remains to be done in the future? Certainly new methods should be developed. Any new method for the isolation of antibiotic-producing organisms will tend to yield new types of antibiotics. Usually *in vitro* procedures have been used. The introduction of *in vivo* methods, especially in search for antineoplastic agents is highly desirable. As indicated previously, the common methods now in use have tended to yield largely the antibiotics already known.

When young microbiologists are encouraged to use the screening program for the isolation of antibiotics and are given a free hand to modify existing procedures as they see best, they are bound to come up with new ideas, if not with new antibiotics. I have seen this happen time and again. One illustration will suffice. When in the early days of screening programs, the streaking of several test bacteria against the freshly isolated culture was replaced by the inoculation of the whole plate with a test culture after various dilutions of soil had been suspended in the same plate, zones of inhibition were produced around some of the developing organisms present in the soil suspension. Such cultures were isolated and tested further. This in itself opened a new approach to the isolation of antibiotic-producing organisms. In a similar manner, new ideas and new approaches are bound to lead to the discovery of new antibiotics. Certainly a young microbiologist, starting on a program of antibiotic research, should begin by screening for new cultures. New antibiotics are bound to come.

References

Chain, E. B. (1966). The quest for new biodynamic substances. *In* "Reflections on Research and the Future of Medicine. A Symposium" (C. E. Lyght, ed.). McGraw-Hill, New York, New York.

Chain, E., Florey, H. W., Gardner, A. D., Heatley, N. G., Jennings, M. A., Orr-Ewing, J., and Sanders, A. G. (1940). Penicillin as a chemotherapeutic agent. *Lancet* **ii**, 226–228.

Cope, Z. (1966). "Almroth Wright, Founder of Modern Vaccine-Therapy." Nelson, London.

Dowling, H. F. (1967). Drugs and the four cultures. *Actinomicrob. Agents Chemother.* pp. 1–5.

Editorial. (1964). Success story—uninspired! *Chest Heart Assoc. J. Health*, p. 163.

Florey, H. W., *et al.* (1949). "Antibiotics." Oxford Univ. Press, London.

Hays, E. E., Wells, I. C., Katsman, P. A., Cain, C. K., Jacobs, F. A., Thayer, S. A., Doisy, E. A., Gaby, W. L., Roberts, E. C., Muir, R. D., Carroll, C. J., Jones, L. R., and Wade, N. J. (1945). Antibiotic substances produced by *Pseudomonas aeruginosa*. *J. Biol. Chem.* **159**, 725–750.

Utkin, L. M. (1946). Physiology of penicillin-forming molds. I. Penicillin formation by *Penicillium crustosum* in media containing glucose. *Mikrobiologiya* **15**, 211–221.

Waksman, S. A. (1937). Associative and antagonistic effects of microorganisms. I. Historical review of antagonistic relationships. *Soil Sci.* **43**, 51–68.

Waksman, S. A. (1940). Microbes in a changing world. *Sci. Monthly* **51**, 422–427.

Waksman, S. A. (1945–1947). "Microbial Antagonisms and Antibiotic Substances." Commonwealth Fund, New York.

Waksman, S. A. (ed.) (1968). "Actinomycin." Wiley, New York.

Structure-Activity Relationships of Semisynthetic Penicillins

K. E. PRICE

Research Division, Bristol Laboratories,
Division of Bristol-Myers Company,
Syracuse, New York

I.	Introduction	17
II.	History	18
	A. Development of "Naturally" Produced Penicillins	18
	B. Development of Synthetic and Semisynthetic Penicillins	22
III.	Structure–Activity Relationships	23
	A. Effect of Nuclear Modifications on Intrinsic Activity	23
	B. Intrinsic Activity of 6-Substituted Penicillins	25
	C. Antibacterial Spectrum	33
	D. Acid Stability and Oral Absorbability	42
	E. Serum Binding	46
	F. Staphylococcal β-Lactamase Resistance	51
	G. Gram-Negative β-Lactamase Resistance	55
	H. Inhibition of β-Lactamases	60
	I. Hypersensitivity	64
IV.	Future Objectives	66
	References	68

I. Introduction

The gateway to preparation of a virtually limitless list of "semisynthetic" penicillins was not fully opened until isolation of the penicillin "nucleus," 6-aminopenicillanic acid (6-APA), was achieved by Batchelor and his co-workers in 1959 (18). As this fermentation-produced chemical became available in quantity, efforts were directed toward producing structurally modified penicillins having (a) a greater degree of intrinsic activity and a wider spectrum than that possessed by penicillin G, along with acid stability and oral absorbability comparable to that of penicillin V, (b) a low degree of deleterious binding to serum proteins, (c) reduced allergenicity, and (d), perhaps most important, resistance to microbial β-lactamases. The present report, which extends and updates a previous review by the author and his colleagues (155), considers the extent of success achieved in each of these areas. Also presented, in an effort to better emphasize the significance of some of the improvements attained, is a brief review of penicillin history over the more than 30-year period from its discovery by Alexander Fleming up to the point where production of new semisynthetic penicillins has become a routine procedure.

II. History

A. Development of "Naturally" Produced Penicillins

Fleming, although unable to purify the relatively unstable antibacterial agent present in *Penicillium notatum* fermentation broths, nevertheless demonstrated its inhibitory effects against a wide variety of gram-positive bacteria, as well as *Neisseria* and *Haemophilus* species *(82)*. He found over the next few years that the penicillin preparation could facilitate isolation of *Haemophilus influenzae* when incorporated into agar and, more important, was effective in treatment of infected wounds when used topically *(81, 83)*. A significant increase in purity was finally achieved by Chain and his co-workers *(45)*. They demonstrated that the semipurified agent protected mice challenged with lethal doses of several gram-positive bacterial species. Therapeutic effectiveness and a low level of toxicity was subsequently demonstrated in humans *(2, 85)*. As a result of these reports, an intensive effort was made by both British and American workers to improve yields and to further purify the antibiotic. Advances were rapid, principally because of the contributions of a group of investigators headed by Moyer and Coghill of the National Regional Laboratories, Peoria, Illinois. Productivity increased as high-yielding strains of *Penicillium chrysogenum* were selected and grown in improved media under submerged conditions *(52, 53, 144)*. Higher-yielding filtrates permitted a sufficient degree of purification to recognize that a variety of closely related penicillins were being produced. The structures of these are shown in Table I.

This was the first evidence that minor variations in the structure of penicillins could markedly influence their antibacterial potency. Moyer and Coghill *(145)* found that the relative quantities of various penicillins produced during fermentations depended upon the type of medium employed. Penicillins being produced in Great Britain were primarily F and K, whereas, the principal penicillin obtained from United States fermentations was G. The latter was selected as the penicillin of choice, since high yields could be obtained by supplementing the growth medium with phenylacetic acid *(145)*, and since it possessed a high degree of intrinsic activity, particularly against pyogenic cocci. Evidence that structural configuration affected characteristics other than antibacterial potency was the finding that, relative to penicillin G (benzylpenicillin), penicillin K was markedly serum bound *(204)*.

Within a matter of several years, a tremendous amount of information about the therapeutic effectiveness of penicillin G accumulated. As pointed out by Stewart (193), this was accomplished without any serious consequences by virtue of its restriction to use in the highly controlled environment of the military services of the United States and Great Britain. Although the antibiotic was absorbed orally (2, 85), it soon became apparent that a significant amount was being destroyed by gastric juice acidity (156).

TABLE I
PENICILLINS PRODUCED NATURALLY IN *Penicillium* FERMENTATIONS

$$R-\overset{O}{\underset{\|}{C}}-\overset{H}{\underset{|}{N}}\begin{array}{c}S\\ \diagdown\\ -N-\end{array}\begin{array}{c}CH_3\\ CH_3\\ COOH\end{array}$$

Penicillin	R	Potency of Na salt vs. *S. aureus* (units/mg.)
Benzyl (G)	C₆H₅—CH₂—	1667[a]
p-Hydroxybenzyl (X)	HO—C₆H₄—CH₂—	900
2-Pentenyl (F)	$CH_3CH_2CH=CHCH_2-$	1500
n-Amyl (dihydro F)	$CH_3CH_2CH_2CH_2CH_2-$	1670
n-Heptyl (K)	$CH_3CH_2CH_2CH_2CH_2CH_2CH_2-$	2300

[a] Values from U. S. Dispensatory (207).

Despite reports (87, 141) that only a small portion of an orally administered dose of penicillin could be recovered in the urine, significant serum levels (up to 1 unit/ml.) could be obtained after a single dose of 600,000 units (57). Clinical successes with doses of this magnitude had been obtained by various investigators during the late 1940's (61, 79). During the same period, efforts to increase penicillin G yields through supplementation of the medium with phenylacetic acid and related compounds continued. Behrens and his colleagues (25, 26), after establishing with labeled precursors that phenylacetamide derivatives were directly incorporated into the benzylpenicillin molecule, recognized the possibility of preparing new penicillins by inducing the mold to utilize other types of precursors. This possibility was found to be the case, since penicillins with side chains having a substituted benzene ring or certain ring structures other than the phenyl group could be produced. It was also found that the alkyl chain could be "interrupted" by an oxygen or sulfur atom (24). Among compounds produced by this method and studied clinically were phenoxymethylpenicillin (penicillin V), allylthiomethylpenicillin (penicillin O), and butylthiomethylpenicillin (penicillin S) (27, 31, 34). Several years later, Ballio and his co-workers (12) reported that α,ω-dicarboxylic acids could also serve as precursors.

Unfortunately, it soon became apparent that only monosubstituted acetic acid derivatives were acceptable as precursors, and, since none of the resulting penicillins differed appreciably from the natural penicillins in regard to their antibacterial properties, the breakthrough into the realm of more powerful penicillins that had been expected to develop through use of this method failed to materialize.

When the structure of penicillins was finally established as a monocarboxylic acid in amide linkage with a fused β-lactam-thiazolidine ring (51), attempts were made to change biological properties through chemical modification. However, such efforts, because they were largely limited to introduction of functional groups in the side-chain phenyl group, to esterification of the thiazolidine free acid, or to formation of amides, failed to yield significantly improved penicillins. Thus, 20 years after Fleming's original observation, the penicillin of choice, penicillin G, was one of those that had been produced naturally in early fermentations. Neither fermentation studies nor chemical modification had given clinicians an antibiotic that had better acid stability or a broader spectrum than that of penicillin G. Furthermore, the presence of a new, even more serious, problem involving *in vivo* inactivation of penicillin was beginning to be recognized. Production of a bacterial enzyme that could convert penicillins to the corresponding penicilloic acids was first described by Abraham and Chain (1). Hydrolysis of the β-lactam ring by penicillinase eliminates antibacterial activity and thus protects the gram-positive or gram-negative microorganism producing it. Capability to synthesize this enzyme, which may be found either extracellularly or intracellularly or both, is transmitted genetically and cannot be acquired through exposure to penicillins (140). The severity of the problem posed by penicillinase-producing staphylococci was first made apparent by Barber (13). She reported that between 1946 and 1948 in England the incidence of such organisms had increased from 14.1 to 59% (15). Subsequent reports from the United States showed that the occurrence of these strains, particularly in hospitals, ranged as high as 60 to 75% and created such a severe clinical problem that the need for improved penicillins became more acute than ever *(23, 211)*.

During the early 1950's, organic chemists prepared many repository salts of penicillin that were more stable and only slightly soluble in aqueous media. Such salts reduced pain at the injection site, gave more uniform and prolonged serum levels after oral and parenteral administration, and probably lowered the incidence of serious allergenic reactions *(157)*. The most widely utilized compounds were the procaine salt *(177, 197)* and N,N'-dibenzylethylenediamine dipenicillin G or benzathine penicillin G *(56, 201, 212)*.

The first clear-cut evidence that a change in the penicillin side chain could appreciably increase activity against gram-negative bacteria was obtained when the broad-spectrum antibiotic, cephalosporin N *(38)*, was shown in 1953 to be a penicillin *(3)*. Subsequent reports revealed that the side chain of this penicillin was derived from D-α-aminoadipic acid *(147)* and that the compound was identical with synnematin B *(4)*. Compared to benzylpenicillin, cephalosporin N (now called penicillin N) was only 0.01 as active against *Staphylococcus aureus*, but appreciably more inhibitory for gram-negative bacteria *(105)*. Acylation of the free amino group restored some of the activity against gram-positive organisms, but reduced the effect against gram-negative ones *(147)*. Additional evidence that increased gram-negative activity is associated with the free amino group was obtained by the subsequent observation that p-aminobenzylpenicillin had improved gram-negative activity *(205)*. Effects of N-acylation were similar to those obtained with penicillin N. Although the latter (under the designation synnematin B) was found to have therapeutic effectiveness in human infections *(180, 202)*, it has not been extensively utilized because of its poor oral absorbability and weak antibacterial activity.

At about the same time that the structure of penicillin N was elucidated, a discovery of considerable significance was made by the Austrian workers Brandl *et al.* *(35)*. They found that phenoxymethylpenicillin or penicillin V, which Behrens and his co-workers *(27)* had produced by biosynthetic methods some 5 years previously, was quite stable at low pH. The increase in acid stability conferred by the presence of a polar group in the side chain permitted a greater amount of compound to be absorbed after oral administration than was the case with penicillin G. This antibiotic, which was found to be therapeutically effective in a wide variety of infections, has been extensively used since then. It did not entirely replace benzylpenicillin, however, since it is less active against strains of *Neisseria*, *Haemophilus*, and other gram-negative bacteria and since its oral absorbability is somewhat inconsistent *(193)*.

B. DEVELOPMENT OF SYNTHETIC AND SEMISYNTHETIC PENICILLINS

Stimulated by the realization that side-chain modifications could influence characteristics as fundamental as the antibacterial spectrum and acid stability of penicillins, chemists and microbiologists intensified penicillin research efforts, and, as a result, new and profound developments occurred swiftly.

In 1957, Sheehan and Henery-Logan *(181)* achieved total chemical synthesis of penicillin V (phenoxymethylpenicillin). Success in obtaining significant yields of a biologically active penicillin, a goal long sought by others *(72, 198)*, resulted when they were able to cyclize a penicilloic acid salt through use of N,N'-dicyclohexylcarbodiimide. Although chemical synthesis of 6-aminopenicillanic acid was reported by the same authors 2 years later *(182)*, the complexity of the methods and low yields made industrial application unfeasible.

In that same year, Batchelor and his colleagues *(18)* isolated 6-APA from precursor-starved *Penicillium chrysogenum* fermentations. The investigation that resulted in compound isolation stemmed from their observation of a discrepancy between chemical and biological assays of penicillin. Six years previously, Kato *(120, 121)* had made almost precisely the same observation. He noted that values obtained by iodometric assay of penicillinase-treated broth filtrates were considerably higher than assay values for chemically extracted penicillin. Since the discrepancy did not occur when phenylacetic acid was used to supplement the basal medium, Kato reasoned that a portion of the penicillinase substrate present in the original filtrates was the "penicillin nucleus." The possibility was raised that this compound might be identical with "penicin," a compound whose structure had been postulated by Sakaguchi and Murao *(176)*. Penicin was believed by these authors to be present, along with phenylacetic acid, after exposure of benzylpenicillin to penicillin amidases found in the mycelia of *Penicillium chrysogenum* and *Aspergillus oryzae*. Batchelor and the team of scientists from Beecham Research Laboratories in England verified the structure of the "penicillin nucleus" and named it 6-aminopenicillanic acid (6-APA) *(18)*. Reaction of the compound with phenylacetyl chloride and phenoxyacetyl chloride yielded benzylpenicillin and phenoxymethylpenicillin, respectively. Although there was a low yield of 6-APA in filtrates and its extraction was difficult, many new penicillins were synthesized in their laboratories.

Bristol Laboratories then joined with the Beecham Research Laboratories to help develop this area of investigation.

At the same time that significantly improved yields of 6-APA were being achieved at Bristol, a new development occurred. This development stemmed from the recognition that treatment of pure penicillins with an amidase such as that described by Sakaguchi and Murao *(176)* might be a feasible way to produce 6-APA. In 1960, Bristol Laboratories *(50)*, Beecham Research Laboratories *(168)*, Chas. Pfizer & Co., Inc. *(110)*, and Farbenfabriken Bayer A. G. *(122)* independently reported that benzylpenicillin or phenoxymethylpenicillin could be hydrolyzed by microbial amidases to give phenylacetic or phenoxyacetic acid and a good yield of 6-APA. Two distinct classes of amidase were encountered. Those produced by many strains of actinomycetes, yeasts, and filamentous fungi readily split pentyl-, heptyl-, and phenoxymethylpenicillins, but hydrolyzed benzylpenicillin very slowly. The amidases of bacterial origin, particularly those from *Escherichia* and *Alcaligenes* strains, readily attacked benzylpenicillin, but deacylated phenoxymethylpenicillin at a much slower rate.

Semisynthetic penicillins can now be prepared by direct addition of the appropriate acid chloride or acid anhydride *(153)* to 6-APA in aqueous or nonaqueous media. Side-chain acids can also be coupled to 6-APA in the presence of N,N'-dicyclohexylcarbodiimide *(107)*. Finally, acylation of 6-APA has been achieved by enzymatic methods by use of an amidase from *Escherichia coli* *(19, 168)*. The optimal pH for this reaction was about 5.0, as compared to an optimum of 8.0 for deacylation.

III. Structure–Activity Relationships

A. EFFECT OF NUCLEAR MODIFICATIONS ON INTRINSIC ACTIVITY

Since gram-positive bacteria show the highest degree of susceptibility to almost all penicillins, inhibitory effects against these bacteria are considered to stem from "intrinsic" antibacterial activity. [The word "intrinsic" is used in the same sense as used by Stewart *(193)*.] In this review, degree of inhibition of penicillin G-sensitive staphylococci will be utilized to quantitate intrinsic activity.

Doubts as to whether penicillanic acid is the almost ideal moiety for preparation of antibiotics with maximal intrinsic activity have been fairly well eliminated by findings accumulated over the past years. Results of these studies are summarized in Table II. Values shown indicate the relative antistaphylococcal activity of benzylpenicillin derivatives whose fused β-lactam-thiazolidine nucleus has been modified in various ways.

Although removal of the sulfur atom (compound 2) completely eliminates the activity of benzylpenicillin, Guddal et al. (100) re-

TABLE II
EFFECT ON ANTISTAPHYLOCOCCAL ACTIVITY OF MODIFICATIONS IN THE PENICILLANIC ACID MOIETY OF PENICILLIN G

$$R = \langle \rangle - CH_2C(=O)-NH-$$

Compound			Antistaphylococcal activity (% of pen G)	References
Number	Name	Structure		
1	Benzylpenicillin (penicillin G)	[structure]	100	
2	Desthiobenzylpenicillin	[structure]	NIL	117
3	Benzylpenicillin methyl ester	[structure]	NIL	17, 114, 115
4	Benzylpenicilloic acid	[structure]	NIL	
5	Anhydrobenzylpenicillin	[structure]	0.07	214
6	Benzylpenicillin alcohol	[structure]	2	154
7	Benzylpenicillin amide	[structure]	25	55, 111
8	Benzylthiopenicillin	[structure]	90	89
9	2-Ethyl-2-desdimethyl benzylpenicillin	[structure]	Active	40

ported that penicillins in which the sulfur is oxidized to the sulfone or sulfoxide retained some antibacterial activity. Sulfoxide penicillins, for which several methods of preparation have been described (47, 75, 100), may be more stable to acid than unoxidized compounds (41). A complete loss of penicillin activity also occurs when substitutions are made at the C-5 locus (183), when simple esters are formed at the 3-position (compound 3), or when the β-lactam ring is hydrolyzed to penicilloic acid (compound 4). Rearrangement of the thiazolidine ring such as occurs in anhydropenicillins (compound 5) permits only a negligible amount of activity to be retained. Other modifications in the carboxyl group at position 3 give compounds with varied potencies, ranging from 2% for the alcohol (compound 6) to 25% for the amide (compound 7), and up to 90% of that of benzylpenicillin for thio acids (compound 8). The requirement that 3-position esters be hydrolyzed before penicillins containing such substituents demonstrate biopotency has been exploited by some of the pharmaceutical houses. Although simple esters are not suitable substrates for human esterases, some of the more complex esters such as β-diethylaminoethyl (17) and the acetoxymethyl (114) are slowly hydrolyzed *in vivo* and thus give a depot effect. An excellent review of this subject has been prepared by Hamilton-Miller (101). Finally, penicillins (compound 9) in which groups other than methyl are present at the 2-position, reportedly have activity against gram-positive bacteria but are ineffective against gram-negative organisms. Thus all of the above described molecular modifications resulted in compounds whose intrinsic antibacterial activity was inferior to that of penicillins possessing the intact penicillanic acid component. Cephalosporins, of course, which differ from penicillins only by virtue of having a dihydrothiazine rather than thiazolidine ring in the nucleus, also possess marked antibacterial properties. However, because of the large number of semisynthetic derivatives now available, they cannot be considered in the present report.

B. Intrinsic Activity of 6-Substituted Penicillins

The next point to be determined is the extent to which intrinsic activity is imparted by substitutions at the 6-position of penicillanic acid. Table III shows the relative antistaphylococcal activity of a group of compounds representing various classes of substituents.

Compound 1, 6-APA, the nucleus of all natural penicillins, was shown by Rolinson and Stevens (164) to have antibacterial activity, although its inhibitory effect against sensitive staphylococci was less than 1% that of benzylpenicillin. Bromine (compound 2) or chlorine-substituted compounds which have been shown to have an inverted configuration (76) are devoid of activity (48), as are N-substituted aralkyl (compound 3) and aryl derivatives. On the other hand, acylation of 6-APA with either sulfonic or carboxylic acids results in compounds with intrinsic activity greater than that of 6-APA. However, no activity could be shown for those acylated compounds with substitu-

TABLE III. ANTISTAPHYLOCOCCAL ACTIVITY OF 6-SUBSTITUTED PENICILLANIC ACID DERIVATIVES

Number	Name	Structure (R)	Antistaphylococcal activity (% of 6-APA)	References
1	Amino	H₂N—	100	164
2	Bromo	Br— (epimeric)	NIL	48
3	Benzylamino	C₆H₅—CH₂—NH—	NIL	BL-P[a]
4	2,5-Dimethylphenyl sulfonamido	(2,5-(CH₃)₂C₆H₃)—SO₂—NH—	4,000	BL-P
5	Phenylacetamido	C₆H₅—CH₂—C(O)—NH—	20,000	164
6	N-Benzylphenyl acetamido	C₆H₅—CH(C(O)—)—CH₂—N(CH₂C₆H₅)—	NIL	134
7	Benzoylcarboxamido	C₆H₅—C(O)—C(O)—NH—	4,000	BL-P
8	3-(2,6-Dichlorophenyl)-5-methyl-4-isoxazole-carboxamido	(2,6-Cl₂C₆H₃)(5-CH₃-isoxazol-4-yl)—C(O)—NH—	4,000	146
9	n-Propylureido	CH₃(CH₂)₂NH—C(O)—NH—	2,000	153
10	Ethoxycarboxamido	CH₃CH₂O—C(O)—NH—	400	BL-P

[a] Bristol Laboratories, unpublished data.

ents other than hydrogen present on the amide nitrogen (compound 6). As illustrated by compound 4, penicillins prepared from sulfonic acids are, in general, less active than those derived from carboxylic acids. However, the presence of an amide bond in no way assures a high degree of potency since the grouping adjacent to the amide carbonyl greatly influences the level of intrinsic activity. Maximal antistaphylococcal activity is obtained when the neighboring group is methylene as found in compound 5. When, however, the adjacent group is another carbonyl group (compound 7) or a carbon of a resonating ring system (compound 8), the antibacterial effect, although some 40 to 50 times that of 6-APA, is only about one-fourth to one-fifth that of benzylpenicillin. Alkyl- (compound 9) and arylureidopenicillins, which have a substituted amine adjacent to the amide carbonyl, have even less activity, whereas compounds in which an oxygen (compound 10) or sulfur atom is linked to the carbonyl have potencies only slightly greater than that of 6-APA. Thus, for maximal intrinsic activity, the substituent at the 6-position of penicillanic acid must be an α-substituted acetamido group.

1. Monosubstituted Acetic Acid Derivatives

The first series of true penicillins to be examined for antistaphylococcal activity can all be considered monosubstituted acetic acid derivatives. The potencies of representative compounds, given as a percentage of benzylpenicillin, are shown in Table IV. Overall, it is apparent that relative to benzylpenicillin these compounds have excellent antibacterial properties. However, there are significant differences among them since potencies range from 25 to 100% that of benzylpenicillin. In the case of aliphatic R groups, the length of the chain has a definite effect on activity. This is illustrated by the fact that ethylpenicillin (compound 1) with its two-carbon side chain has only one-fourth the activity of n-amylpenicillin (compound 2), which has a five-carbon chain. The potency of the latter is essentially the same as that of benzylpenicillin (compound 3) and many other penicillins that have aromatic structures located in the same side-chain position as the benzene ring of benzylpenicillin. Although the effect on antibacterial activity of substitutions in the benzene ring of benzylpenicillin is usually not striking, significant reductions in activity have been observed. In the present example, m- and p-nitrobenzylpenicillin (compound 4) are both comparable to benzylpenicillin in their antibacterial potency, whereas the o-nitrobenzylpenicillin (compound 5) is about one-half as active.

Penicillins with aralkyl side chains tend to become somewhat less active as alkyl chain length increases. Phenethylpenicillin (compound 6), for example, has somewhat less antistaphylococcal activity than benzylpenicillin. An even further reduction in activity occurs

TABLE IV
ANTISTAPHYLOCOCCAL ACTIVITY OF PENICILLINS
HAVING SIDE CHAINS PREPARED FROM MONOSUBSTITUTED ACETIC ACIDS

Number	Name	R Structure	Antistaphylococcal activity (% of pen G)	References
1	Ethyl	CH_3CH_2-	25	64
2	n-Amyl (dihydro F)	$CH_3CH_2CH_2CH_2CH_2-$	100	207
3	Benzyl (pen G)	$C_6H_5-CH_2-$	100	—
4	3- or 4-Nitrobenzyl	$O_2N-C_6H_4-CH_2-$	100	BL-P[a], 28
5	2-Nitrobenzyl	$2-O_2N-C_6H_4-CH_2-$	50	BL-P
6	Phenethyl	$C_6H_5-CH_2-CH_2-$	75	BL-P
7	β,β-Diphenylethyl	$(C_6H_5)_2CH-CH_2-$	25	BL-P
8	Phenoxymethyl (pen V)	$C_6H_5-O-CH_2-$	100	93

[a] Bristol Laboratories, unpublished data.

when the β-carbon of the side chain of phenethylpenicillin is substituted with another bulky group, for example, a benzene ring (compound 7). Introduction of a polar group such as an oxygen or sulfur atom into the side chain does not appreciably affect the antistaphylococcal potency of penicillins. This is exemplified by phenoxymethylpenicillin or penicillin V (compound 8) which is equal in activity to benzylpenicillin.

2. Disubstituted Acetic Acid Derivatives

Penicillins prepared by acylation of 6-APA with disubstituted acetic acids have, on the average, lower intrinsic activity than monosubstituted derivatives. This observation is supported by the potency values of the penicillins shown in Tables V and VI. Here, potencies relative to that of benzylpenicillin range from 5 to 75%, with the majority in the 20 to 50% range.

Unless both substituents are identical, the α-carbon of the side chain will be asymmetrical and give rise to a mixture of diasterioisomeric penicillins. α-Aminobenzylpenicillin (compound 1), for example, occurs as a mixture of D-(−) and L-(+) epimers and has about 40% of the activity of benzylpenicillin. Against staphylococci, the D-(−) isomer (compound 2) or ampicillin is approximately four times as active as the L-(+) epimer (compound 3).

The next three compounds are penicillins that were derived from D-(−)-α-amino acids. Compound 4, α-amino-2-thenylpenicillin, is equal in activity to ampicillin, whereas the N-substituted methyl derivative of ampicillin (compound 5) is slightly less active. The chain length of α-amino-γ-methylmercaptopropylpenicillin (compound 6) apparently is sufficiently long to give good antibacterial activity.

The remaining compounds in Table V are mixtures of diasterioisomers. These penicillins, which are assumed to contain approximately equal portions of both epimers, demonstrate the influence of chlorine substitution in the benzene ring of DL-α-aminobenzylpenicillin. The presence of the halogen in the p-position or m-position (compound 7) gives compounds with about the same activity as the unsubstituted penicillin (compound 1). Substitution in the o-position (compound 8) however, causes a drastic reduction in intrinsic antibacterial activity. This observation is similar to that of Claesen and his co-workers (49) who found the following antistaphyloccocal activities (as a percentage of penicillin G) for chlorinated derivatives of α-methoxybenzylpenicillin: 3-chloro, 35%; 4-chloro, 65%; and 2-chloro, 5%. Although the obvious implication is that o-substituents in some way hinder attachment of the penicillins to susceptible sites in the bacterial cell, it is difficult to explain why substitution of chlorine in both the 2- and 4-position of α-amino-(compound 9) and α-methoxybenzylpenicillin (49) yields activity comparable to that of the unsubstituted compound. The 3,4-dichloro derivatives (compound

TABLE V. ANTISTAPHYLOCOCCAL ACITIVITY OF PENICILLINS HAVING SIDE CHAINS PREPARED FROM DISUBSTITUTED ACETIC ACIDS

$$R-\overset{O}{\overset{\|}{C}}-\overset{H}{\overset{|}{N}}\cdots\begin{array}{c}S\\ \\N\end{array}\begin{array}{c}CH_3\\CH_3\\COOH\end{array}$$

Number	Name	R Structure	Antistaphylococcal activity (% of pen G)	References
1	D,L-α-Aminobenzyl	C$_6$H$_5$-CH(NH$_2$)-	40	65
2	D-(-)-α-Aminobenzyl	C$_6$H$_5$-CH(NH$_2$)-	75	BL-P[a]
3	L-(+)-α-Aminobenzyl	C$_6$H$_5$-CH(NH$_2$)-	20	BL-P
4	D-(-)-α-Amino-2-thenyl	(2-thienyl)-CH(NH$_2$)-	75	BL-P
5	D-(-)-α-Methylaminobenzyl	C$_6$H$_5$-CH(NHCH$_3$)-	50	BL-P
6	D-(-)-α-Amino-γ-methyl-mercaptopropyl	CH$_3$-S-(CH$_2$)$_2$CH(NH$_2$)-	50	BL-P
7	α-Amino-3- or-4-chlorobenzyl	Cl-C$_6$H$_4$-CH(NH$_2$)-	40-50	BL-P
8	α-Amino-2-chlorobenzyl	2-Cl-C$_6$H$_4$-CH(NH$_2$)-	5	BL-P
9	α-Amino-2,4-dichlorobenzyl	2,4-Cl$_2$-C$_6$H$_3$-CH(NH$_2$)-	40	BL-P
10	α-Amino-3,4-dichlorobenzyl	3,4-Cl$_2$-C$_6$H$_3$-CH(NH$_2$)-	60	BL-P

[a] Bristol Laboratories, unpublished data.

10) were the most active members in both the α-amino- and α-methoxybenzyl series (49, 209).

Additional disubstituted acetic acid derivatives are shown in Table VI. Several of the compounds in this group of α-substituted phenoxy-

TABLE VI. ANTISTAPHYLOCOCCAL ACTIVITY OF PENICILLINS HAVING SIDE CHAINS PREPARED FROM DISUBSTITUTED ACETIC ACIDS

$$R-\underset{O}{\overset{\|}{C}}-\underset{H}{\overset{|}{N}}-\text{[penicillin nucleus]}$$

Number	Name (R)	Structure	Antistaphylococcal activity (% of pen G)	References
1	Phenoxyethyl (phenethicillin)	C₆H₅—O—CH(CH₃)—	50	93
2	Phenoxypropyl (propicillin)	C₆H₅—O—CH(CH₂CH₃)—	50	93
3	Phenoxybutyl	C₆H₅—O—CH(CH₂CH₂CH₃)—	10	93
4	Phenoxybenzyl (phenbenicillin)	C₆H₅—O—CH(C₆H₅)—	25	169
5	Diphenoxymethyl	C₆H₅—O—CH(O—C₆H₅)—	20	BL-P[a]

[a] Bristol Laboratories, unpublished data.

methyl penicillins have received extensive clinical application. Compounds 1–4 have an asymmetric carbon and have generally been studied as diasterioisomeric mixtures. As with ampicillin, one of the epimers usually possesses greater activity than the other. In the case of compound 1, phenethicillin, the L isomer appears to be more active than the D isomer both *in vitro (96, 139)* and *in vivo (74)*. Although this would appear to be an inconsistency in view of the fact that the D epimer of α-aminobenzylpenicillin is the most potent, Bristol Laboratories workers, through application of Fischer-Rosanoff projections and molecular models, were able to furnish an explanation *(96)*. They observed that when the side-chain acids of the most active forms of the two compounds were similarly oriented with respect to their bulky (phenyl or phenoxy) groups, the remaining substituents, the carboxyls as well as methyl and amino groups, had the same spatial relationships and occupied comparable positions.

It is readily apparent from the values in Table VI that antistaphylococcal activity of the phenoxyalkyl series is reduced as the size of the α-substituent is increased, either through lengthening the alkyl chain (compound 3) or through introduction of an aromatic group (compounds 4 and 5).

3. Trisubstituted Acetic Acid Derivatives

Table VII shows antistaphylococcal activity of a series of penicillins prepared from trisubstituted acetic acids. It is obvious that

TABLE VII. ANTISTAPHYLOCOCCAL ACTIVITY OF PENICILLINS HAVING SIDE CHAINS PREPARED FROM TRISUBSTITUTED ACETIC ACIDS

$$R-\overset{O}{\underset{\|}{C}}-\overset{H}{\underset{|}{N}}\diagdown\overset{S}{\diagup}\diagdown\overset{CH_3}{\underset{CH_3}{\diagup}}$$

Number	Name (R)	Structure (R)	Antistaphylococcal activity (% of pen G)	References
1	Dimethylmethoxymethyl	$CH_3-O-\underset{\underset{CH_3}{\mid}}{\overset{\overset{CH_3}{\mid}}{C}}-$	2	BL-P[a]
2	Dimethylbutoxymethyl	$CH_3CH_2CH_2CH_2-O-\underset{\underset{CH_3}{\mid}}{\overset{\overset{CH_3}{\mid}}{C}}-$	10	BL-P
3	Dimethylphenoxymethyl	Ph$-O-\underset{\underset{CH_3}{\mid}}{\overset{\overset{CH_3}{\mid}}{C}}-$	10	BL-P
4	Methyldiphenylmethyl	Ph$_2$$\underset{\underset{}{}}{\overset{\overset{CH_3}{\mid}}{C}}-$	10	62
5	Triphenylmethyl	Ph$_3$C$-$	5	62

[a] Bristol Laboratories, unpublished data.

acylation with these acids does not yield penicillins with high intrinsic activity since none of the compounds has more than 10% the potency of benzylpenicillin. However, as demonstrated by the first three compounds, in which the third substituent on the dimethyl-substituted α-carbon is varied, the nature of the group does influence the degree of antibacterial activity. As is the case with monosubstituted acetic acid derivatives, the presence of a long alkyl chain or an aryl group is required for maximal activity. Introduction of a second aromatic group (compound 4) does not appreciably affect activity; a third, however, as in triphenylmethylpenicillin (compound 5), does bring about a reduction in potency.

4. Carboxylic Acid Derivatives

The final structural class examined for degree of antibacterial potency is comprised of penicillins derived from aromatic and heterocyclic carboxylic acids. Most of the compounds (Table VIII) have intrinsic activities that are equal to or greater than those of penicillins prepared from trisubstituted, but less than those derived from disubstituted acetic acids. The first four penicillins are related compounds which demonstrate the effect on antistaphylococcal activity of methoxy substitution in the benzene ring of phenylpenicillin (compound 1). It can be observed that, although a single methoxy substituent (compound 2) does not influence potency, introduction of additional groups (compounds 3 and 4) causes a progressive loss in antistaphylococcal activity.

The next series of compounds (5 through 9) are 5-methyl-4-isoxazolylpenicillins that have various substituents in the 3-position. The size of this group has an important effect on potency since both the 3-butyl (compound 6) and the 3-phenyl (compound 7) are more active than the 3-methyl (compound 5). Further increases in the activity of 5-methyl-3-phenyl-4-isoxazolylpenicillin can be induced by introduction of chlorine at the 2- (compound 8) or at the 2- and 6- (compound 9) positions of the 3-phenyl group.

Despite the wide range of intrinsic activity demonstrated for the penicillins described in the preceding sections, recent reports suggest that all exert their antibacterial effects through interference with enzymes required for the cross-linking of cell wall glycopeptides (113, 196). Thus, variations in intrinsic activity could stem from the relative degree of difficulty the antibiotic has in reaching the site of enzyme action or from its differing affinity for such enzymes.

C. Antibacterial Spectrum

In addition to its excellent activity against streptococci, pneumococci, and nonpenicillinase-producing staphylococci, penicillin G is highly inhibitory for gonococci and meningococci. Although it is

TABLE VIII. ANTISTAPHYLOCOCCAL ACTIVITY OF PENICILLINS HAVING SIDE CHAINS PREPARED FROM AROMATIC AND HETEROCYCLIC CARBOXYLIC ACIDS.

$$\text{R—C(=O)—NH—[\beta\text{-lactam}]—S—C(CH_3)_2—CH—COOH}$$

Number	Name	R Structure	Antistaphylococcal activity (% of pen G)	References
1	Phenyl		10	BL-P[a]
2	2-Methoxyphenyl		10	BL-P
3	2,6-Dimethoxyphenyl		2	92, 167
4	2,4,6-Trimethoxyphenyl		1	66
5	3,5-Dimethyl-4-isoxazolyl		5	BL-P
6	3-Butyl-5-methyl-4-isoxazolyl		15	BL-P
7	5-Methyl-3-phenyl-4-isoxazolyl (oxacillin)		10	94
8	3-(2-Chlorophenyl)-5-methyl-4-isoxazolyl (cloxacillin)		15	16, 129
9	3-(2,6-Dichlorophenyl)-5-methyl-4-isoxazolyl (dicloxacillin)		20	146

[a] Bristol Laboratories, unpublished data

moderately inhibitory for *Haemophilis* species, most other gram-negative bacilli generally fail to respond to concentrations below 10 μg./ml. and thus are considered resistant. Group D streptococci (enterococci) are also quite refractory to inhibition. The first indication that the spectrum of penicillin G could be extended to include additional gram-negative organisms was given by the finding that penicillin N (D-4-amino-4-carboxybutylpenicillin) had somewhat more activity against *Salmonella* and *Klebsiella* strains than did benzylpenicillin, despite the fact that its gram-positive activity was vastly inferior (3). The relative activity of these two compounds is illustrated graphically in Fig. 1 (compounds 1 and 2). When the reactive free amino group of penicillin N was acylated (compound 3), gram-positive inhibitory effects were enhanced and gram-negative activity was substantially reduced. p-Aminobenzylpenicillin (compound 4), though only slightly less active than penicillin G against gram-positive bacteria, was about 2 to 4 times more active against gram-negative bacilli. The position of the free amino substituent is of considerable importance since neither o- nor m-aminobenzylpenicillin (compound 5) shows an extended spectrum over penicillin G.

Ampicillin or D-(−)-α-aminobenzylpenicillin (compound 6) has been found to have the greatest amount of activity against gram-negative bacteria. However, it is clear from the data reported by Ekstrom *et al.* (73) that the mere presence of a free α-amino group is not sufficient to assure that a penicillin will have maximal gram-negative activity (see Table IX). Although most derivatives were inferior to ampicillin, a particularly poor gram-negative inhibitory effect was obtained with the benzylpenicillin derivative having both methyl- and amino-substituents on the α-carbon (compound 7).

The marked gram-negative activity displayed by α-hydroxybenzylpenicillin (compound 7, Fig. 1) indicates that broad-spectrum antibacterial activity among the penicillins is not restricted to those whose side chains contain an amino group.

p-Aminobenzylpenicillin and ampicillin are much like penicillin N in that acylation or some other N-substitution increases gram-positive activity and reduces gram-negative activity (193, 205). This does not indicate, however, that a free amino or hydroxyl group is essential for enhanced gram-negative activity since some penicillins with a ring nitrogen (compound 9) are considerably more active against gram-negative bacteria than benzylpenicillin.

Most of the structural modifications that result in changes in the spectrum of benzylpenicillin tend to narrow its breadth. Figure 2 shows the spectrum of a variety of structural types of penicillins, including many that are being utilized commercially. Compounds 2–5 are all phenoxyalkylpenicillins. The most active member of the series is phenoxymethylpenicillin or penicillin V (compound 2). Although its inhibitory effect against staphylococci is comparable to that of

		R	Structure	Relative antibacterial effect against			References
Number	Name			Penase- neg. staph.	Strept. and pneumo.	Gram- neg. bacilli	
1	Benzyl		CH$_2$— (phenyl)	▨	▨	▨	—
2	D-4-Amino-4-carboxybutyl (penicillin N)		HOOC—CH(CH$_2$)$_3$— NH$_2$	▨	▨	▨	3, 105
3	D-4-N-Benzoylamino-4-carboxy-n-butyl		HOOC—CH(CH$_2$)$_3$— NH—CO—C$_6$H$_5$	▨	No data	▨	147
4	p-Aminobenzyl		H$_2$N—C$_6$H$_4$—CH$_2$—	▨	▨	▨	205

General structure:

$$\text{R—C(=O)—NH—}\begin{array}{c}\text{penam nucleus with S, CH}_3, \text{CH}_3, \text{COOH}\end{array}$$

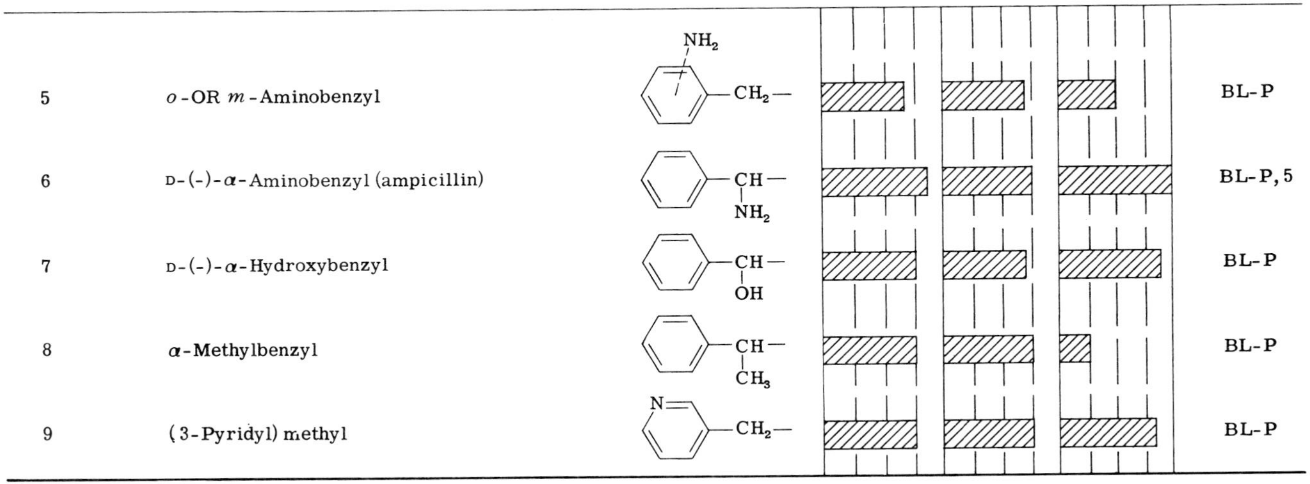

5	*o*-OR *m*-Aminobenzyl			BL-P
6	D-(−)-α-Aminobenzyl (ampicillin)			BL-P, 5
7	D-(−)-α-Hydroxybenzyl			BL-P
8	α-Methylbenzyl			BL-P
9	(3-Pyridyl) methyl			BL-P

FIG. 1. Relative antimicrobial potency of penicillins that possess a gram-negative spectrum. The length of the bar indicates the degree of activity against a particular group of organisms as compared to the most active compound. Each division (| ↔ |) represents a 2-fold difference in activity. BL-P in reference column indicates compound was prepared and tested at Bristol Laboratories.

TABLE IX. ANTIBACTERIAL ACTIVITY OF AMINOPENICILLINS

$$R-\overset{O}{\overset{\|}{C}}-\overset{H}{\overset{|}{N}} \begin{array}{c} S \\ \diagup \\ \diagdown N \end{array} \begin{array}{c} CH_3 \\ CH_3 \\ COOH \end{array}$$

Number	Name	R Structure	Minimum inhibitory concentration (μg/ml.) for *Escherichia coli*
1	α-Aminobenzyl	C₆H₅–CH(NH₂)–	6.25
2	α-Amino-2-chlorobenzyl	2-Cl-C₆H₄–CH(NH₂)–	25
3	α-Amino-3-chlorobenzyl	3-Cl-C₆H₄–CH(NH₂)–	6.25
4	α-Amino-4-chlorobenzyl	4-Cl-C₆H₄–CH(NH₂)–	25
5	α-Amino-2-methoxybenzyl	2-OCH₃-C₆H₄–CH(NH₂)–	25
6	α-Amino-2,4-dichlorobenzyl	2,4-Cl₂-C₆H₃–CH(NH₂)–	62.5
7	α-Amino-α-methylbenzyl	C₆H₅–C(CH₃)(NH₂)–	250
8	α-Amino-3-thenyl	3-thienyl–CH(NH₂)–	2.5

[a] Taken from Ekstrom *et al.* (73).

penicillin G (compound 1), activity against other cocci, both gram-positive and gram-negative, is somewhat less. Its antimicrobial action against gram-negative bacilli is also considerably diminished as compared to penicillin G. The remaining phenoxypenicillins, though possessing basically the same spectrum as penicillin V, show a progressive decrease in antibacterial potency that correlates with the increased size of their α-substituent. Three of these derivatives, phenethicillin (compound 3), propicillin (compound 4), and phenbenacillin (compound 5), although probably not effective against most penicillinase-producing staphylococci, do show increased resistance to β-lactamases as compared to penicillins G and V. Stewart *(193)* considers it possible that the higher level of resistance of this group may lessen the tendency toward staphylococcal superinfection during therapy.

In comparison to the spectrum of benzylpenicillin, the spectra of the remaining compounds in Fig. 2, although ostensibly broadened by addition of inhibitory capability against β-lactamase-producing staphylococci, are actually very much reduced in breadth owing to their relatively poor activity against all gram-negative organisms. Possible exceptions to this generalization are nafcillin (compound 7), which has high resistance to staphylococcal penicillinase but retains fairly good activity against gram-negative cocci, and compound 10, a 3,5-disubstituted isoxazolylpenicillin, which, in addition to possessing a fair degree of β-lactamase resistance, is almost as active (about one-half) against gram-negative bacilli as is benzylpenicillin. Of the penicillins shown, methicillin (compound 6) and quinacillin (compound 8) have the narrowest spectrum. In the case of the former, this is merely a reflection of the overall weakness of its antimicrobial action, whereas the latter demonstrates surprising selectivity since its antibacterial effects are limited almost exclusively to staphylococci. Compound 9, a 3,5-disubstituted isozazolylpenicillin (oxacillin), shows fairly good antibacterial activity against almost all gram-positive organisms, but is quite limited in its effects against gram-negative species. o-Halogenation of the phenyl group of oxacillin, which is located at the 3-position of the five-membered heterocyclic ring, has been found to produce increased antibacterial activity. Both the 2-chlorophenyl (cloxacillin) and 2,6-dichlorophenyl (dicloxacillin) derivatives have greater potency than oxacillin against penicillinase-positive and penicillinase-negative staphylococci as well as streptococci. Antibacterial effects of dicloxacillin may be slightly greater than those of cloxacillin.

Penicillins as diverse as methicillin, ampicillin, and penicillin G all appear to induce abnormal swelling and lysis in bacterial cells by disrupting cell wall synthesis (88, 162, 193). What then is responsible for the fact that ampicillin has excellent gram-negative inhibitory activity, whereas penicillin G has only moderate effectiveness and methicillin virtually no activity against these organisms? In the case of some strains of gram-negative bacilli, the superior activity of ampicillin probably can be attributed to its greater resistance to β-lactamases (10, 199). This cannot account, however, for the greater activity of ampicillin against bacterial strains which do not produce β-lactamase. Examination of the possible role played by bacterial amidases has shown that inactivation by this method probably does not contrib-

Number	Name	Structure (R)	Relative antibacterial effect against					References
			Penase-neg. staph.	Penase-pos. staph.	Strept. and pneumo.	Gram-neg. cocci	Gram-neg. bacilli	
1	Benzyl	–CH$_2$–C$_6$H$_5$						
2	Phenoxymethyl (penicillin V)	C$_6$H$_5$–O–CH$_2$–						90, 125, 210
3	Phenoxyethyl (phenethicillin)	C$_6$H$_5$–O–CH(CH$_3$)–						90, 125, 210
4	Phenoxypropyl (propicillin)	C$_6$H$_5$–O–CH(CH$_2$CH$_3$)–						125, 210
5	Phenoxybenzyl (phenbenicillin)	C$_6$H$_5$–O–CH(C$_6$H$_5$)–						16, 125

General structure:

$$R-\underset{\underset{O}{\|}}{C}-\underset{\underset{H}{|}}{N}-\text{[penam nucleus with S, CH}_3\text{, CH}_3\text{, COOH]}$$

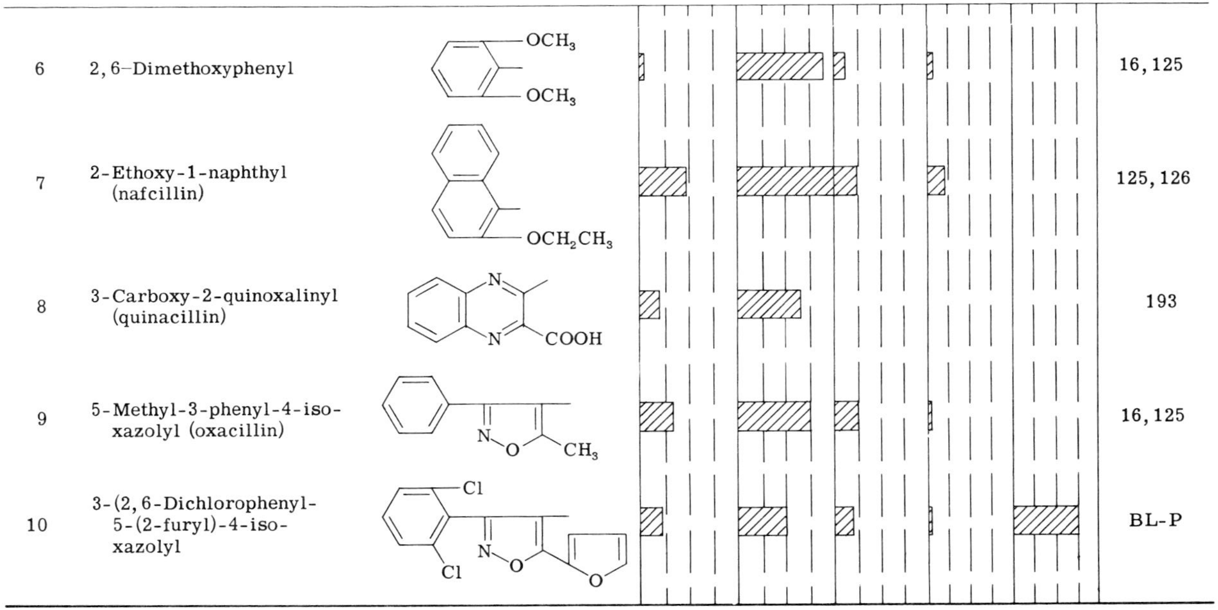

6	2,6-Dimethoxyphenyl				16, 125
7	2-Ethoxy-1-naphthyl (nafcillin)				125, 126
8	3-Carboxy-2-quinoxalinyl (quinacillin)				193
9	5-Methyl-3-phenyl-4-isoxazolyl (oxacillin)				16, 125
10	3-(2,6-Dichlorophenyl)-5-(2-furyl)-4-isoxazolyl				BL-P

FIG. 2. Relative activity of penicillins having an antimicrobial spectrum differing from that of benzylpenicillin. The length of the bar indicates the degree of activity against a particular group of organisms as compared to the most active compound. Each division (| ↔ |) represents a 2-fold difference in activity. BL-P in reference column indicates compound was prepared and tested at Bristol Laboratories.

ute significantly to the relative differences in activity displayed by various penicillins against gram-negative bacteria (54). Undoubtedly of considerable significance, however, is the recent finding that inhibition of cell free *Escherichia coli* transpeptidase activity is achieved at roughly minimal inhibitory concentration levels for ampicillin, methicillin, and penicillin G. The observed differences in activity of these penicillins against transpeptidase could possibly be explained by the hypothesis of Knox (127), who has proposed that receptor sites in gram-negative bacteria are susceptible to penicillins only when there is a side chain on the carbon next to the aryl group and when it comes in close proximity to the sites. Although this fails to account for the moderate effectiveness of several compounds which are not substituted at this position, it would help explain why methicillin and other penicillins with hindered side chains fail to act effectively against cell wall-synthesizing enzymes of gram-negative bacteria.

Although it seems that the very changes that confer staphylococcal β-lactamase resistance tend to eliminate the capability of the penicillin to act on gram-negative bacteria, there are a few compounds that act as exceptions. Compound 10 in Fig. 2, the 3-(2,6-dichlorophenyl)-5-(2-furyl)-4-isoxazolylpenicillin, despite being resistant to staphylococcal penicillinase, retains a fair amount of activity against gram-negative bacilli. Even more striking is the broad-spectrum activity displayed by several α-guanidino-substituted compounds, including the benzyl and 2- and 3-thenylpenicillins. These derivatives, in addition to displaying excellent activity against penicillinase-producing staphylococci and ampicillin-sensitive gram-negative bacteria, are also active against many ampicillin-resistant β-lactamase-producing gram-negative organisms. Because of this last property, they are reviewed in Section G, which is concerned with penicillins that show unusual resistance to gram-negative β-lactamases. Findings such as those described above make it apparent that there is no simple explanation for the spectrum differences possessed by various penicillins.

D. ACID STABILITY AND ORAL ABSORBABILITY

Since Brandl and Margreiter (34) first demonstrated that penicillin V, a structural relative of penicillin G, possesses good acid stability which improves oral absorability, investigators have sought other biologically potent penicillins that have significant resistance to highly acidic gastric juices. Resistance to acid degradation is usually determined by incubating the test compound at 25° to 37°C. and low pH (1.3 to 2.0) for various time periods. Bioactivity is then assayed and the half-life of the compound is calculated. The relative acid stability of a variety of structural types of penicillins is shown in Table X.

TABLE X
THE RELATIVE ACID STABILITY OF A VARIETY OF SEMISYNTHETIC PENICILLINS

	R		Acid stability (half-life	
Number	Name	Structure	in minutes)	References
1	Benzyl (penicillin G)	C₆H₅–CH₂–	< 30	67
2	Phenoxymethyl (penicillin V)	C₆H₅–O–CH₂–	180-210	63
3	β-Phenoxyethyl	C₆H₅–O–CH₂CH₂–	< 30	BL-P[a]
4	α-Methoxybenzyl	C₆H₅–CH(OCH₃)–	60-90	67
5	D-(-)-α-Aminobenzyl (ampicillin)	C₆H₅–CH(NH₂)–	>300	67
6	2,6-Dimethoxyphenyl (methicillin)	2,6-(OCH₃)₂C₆H₃–	< 30	70
7	3-Chloro-2,6-dimethoxyphenyl	3-Cl-2,6-(OCH₃)₂C₆H₂–	30-60	70
8	2,6-Dichlorophenyl	2,6-Cl₂C₆H₃–	>300	70
9	5-Methyl-3-phenyl-4-isoxazolyl (oxacillin)		120-150	68, 94

[a] Bristol Laboratories, unpublished data.

The enhanced stability of penicillin V (compound 2) and other phenoxyalkylpenicillins can be attributed to the presence of the highly polar oxygen atom in the side chain. Comparable increases in stability relative to that of benzylpenicillin (compound 1) can also be achieved by incorporation of sulfur into the side chain. Of interest is the fact that resistance to acidity is lost when the aryloxy group is shifted along the side chain as in β-phenoxyethylpenicillin (compound 3). Certain α-substituted benzylpenicillin derivatives also possess considerable acid stability, for example, α-methoxybenzyl- (compound 4) and α-aminobenzyl- (compound 5) penicillins. Doyle and his co-workers (68) have speculated that the presence of electron-attracting groups in the α-position of benzylpenicillin may interfere with the electron displacement required to cause arrangement to the corresponding biologically inactive penillic acid. The relative effectiveness of the α-substituent appears to correlate with the strength of the side-chain acid from which the penicillin was derived.

Although methicillin (compound 6) and many other phenylpenicillins are susceptible to the action of low pH (70), appropriate substitution may markedly enhance stability. This is illustrated by compounds 7 and 8 which show greater stability than methicillin owing to the introduction of chlorine groups. As was the case with α-substituted benzylpenicillins, a direct correlation between side-chain acid strength and acid stability was shown (70). The final compound in Table X is oxacillin (compound 9), which is only slightly less stable to acid than is penicillin V. Its acid stability is undoubtedly conferred by the presence of nitrogen or oxygen in the isoxazole ring. It is probable that these electronegative atoms are protonated more readily than the β-lactam carbonyl group and thus produce a sparing effect.

Another compound worthy of mention is nafcillin. The variable and modest absorbability possessed by this penicillin may result from its instability to acid (213). However, possession of stability to acid in no way assures that penicillins will be absorbed to the same extent. This is illustrated by the poor absorbability of the acid-stable compound, quinacillin (158), as well as by the variation in peak serum (or blood) levels obtained with the penicillins shown in Table XI following their oral administration to animals and/or humans.

Although penicillin V is absorbed in the gastrointestinal tract more efficiently than penicillin G, it nevertheless gives variable blood and serum levels. A general rule for the phenoxypenicillins is that, the higher the molecular weight, the greater the oral absorbability (93). However, the potentially greater efficacy of the higher molecular weight derivatives may be offset by increased serum binding (32).

Ampicillin and the other α-amino-substituted penicillins are all absorbed considerably better than penicillin G. Of interest is the fact that appropriate disubstitution in the benzene ring of ampicillin (compound 4) significantly increases oral absorbability as does replacement of the benzene ring with thiophene (compound 5). Somewhat prolonged serum concentrations have been demonstrated for another ampicillin derivative, hetacillin *(37)*, which is prepared by condensation of acetone with ampicillin.

Introduction of chlorine atoms into the benzene ring of oxacillin

TABLE XI
RELATIVE ORAL ABSORBABILITY OF A VARIETY OF SEMISYNTHETIC PENICILLINS

Number	Name	Structure	Acid stability	Peak blood or serum level (% of pen G)	References
1	Benzyl (penicillin G)	C₆H₅–CH₂–		100	BL-P[a]
2	Phenoxymethyl (penicillin V)	C₆H₅–O–CH₂–	+	250	BL-P, 193
3	D-(-)-α-Aminobenzyl (ampicillin)	C₆H₅–CH(NH₂)–	+	250	BL-P, 60, 128
4	D-(-)-α-Amino-3-chloro-4-hydroxybenzyl	HO–C₆H₃(Cl)–CH(NH₂)–	+	400	BL-P
5	D-(-)-α-Amino-3-thenyl	thienyl–CH(NH₂)–	+	500	BL-P
6	5-Methyl-3-phenyl-4-isoxazolyl (oxacillin)		+	200	BL-P, 132, 186
7	5-Methyl-3(2,6-dichlorophenyl)-4-isoxazolyl (dicloxacillin)		+	700	BL-P, 193
8	1-Aminocyclohexyl		+	550	BL-P, 109, 172
9	Triphenylmethyl	(C₆H₅)₃C–	+	<100	BL-P, 193

[a] Bristol Laboratories, unpublished data

(compound 6) tends to improve absorption in humans since cloxacillin *(99, 192)* is better absorbed than oxacillin, while dicloxacillin (compound 7) gives even higher serum levels. Rosenblatt *et al. (171)* have indicated that the increase in serum levels attained as one introduces chlorine atoms may result in part from the resulting compounds' progressively lower renal clearance and increased resistance to degradation by the liver.

Results obtained with the final two compounds (8 and 9) illustrate the extent of the differences in oral absorbability that can be found with two acid-stable penicillins. Triphenylmethylpenicillin, possibly because of its bulky nature, is very poorly absorbed, while 1-aminocyclohexylpenicillin possesses a greater degree of absorbability than ampicillin. The higher and more prolonged blood levels obtained with the latter compound can possibly be attributed to the fact that its side-chain acid behaves like that of a closely related one (1-aminocyclopentane carboxylic acid) which has been found to be extremely stable. It is transported like an L-amino acid, is not metabolized, and is excreted at a very slow rate. The stable nature of this side-chain acid *(8)* may endow the entire penicillin molecule with increased stability.

E. Serum Binding

All penicillins appear to be serum bound to some extent. Binding occurs primarily with the serum albumen *(204)* and, although reversible, is thought to interfere with *in vivo* antibacterial activity *(6, 132)*. Tompsett *et al. (204)* were the first to suggest this possibility, after noting that the extent of binding of penicillins G, X, and K correlated extremely well with their *in vivo* efficacy. Commonly used methods for determining the degree to which a given penicillin is bound include equilibrium dialysis *(2)*, ultrafiltration *(1)*, and direct plate assay of solutions of penicillins in buffer and 100% serum in which the bound penicillin can be determined by comparison of the two dose-response curves *(179)*.

Table XII shows the extent to which a variety of penicillins are bound by human serum as measured by ultrafiltration or serum versus buffer agar-diffusion techniques. Alkyl chain length would appear to be a factor since a lower binding value is obtained with *n*-amyl- (compound 2) than with *n*-heptyl- (compound 3) penicillin. Ampicillin (compound 4), as with most hydrophilic penicillins of comparable molecular weight, is only slightly serum bound. In marked contrast is triphenylmethylpenicillin (compound 5) whose bulky substituents result in extensive binding. Molecular weight also appears to be an important factor in binding of compounds 6–9, the phenoxyalkylpenicillins *(91)*, since the degree of serum binding becomes greater as the size of the α-substituent increases.

TABLE XII. PERCENTAGE BINDING TO HUMAN SERUM OF PENICILLINS WITH ACETIC ACID SIDE CHAINS

Number	Name	R Structure	Percentage bound	References
1	Benzyl (penicillin G)	C₆H₅—CH₂—	59	165
2	n-Amyl (dihydro F penicillin)	$CH_3-(CH_2)_3-CH_2-$	63	204
3	n-Heptyl (penicillin K)	$CH_3-(CH_2)_5-CH_2-$	91	204
4	D-(-)-α-Aminobenzyl (ampicillin)	C₆H₅—CH(NH₂)—	18–29	132, 165
5	Triphenylmethyl	(C₆H₅)₃C—	98[a]	—
6	Phenoxymethyl (penicillin V)	C₆H₅—O—CH₂—	80	32, 165
7	Phenoxyethyl (phenethicillin)	C₆H₅—O—CH(CH₃)—	75–83	32, 165
8	Phenoxypropyl (propicillin)	C₆H₅—O—CH(CH₂CH₃)—	89	32, 165
9	Phenoxybenzyl (phenbenicillin)	C₆H₅—O—CH(C₆H₅)—	97	32, 165

[a] Value obtained by serum vs. buffer agar diffusion techniques at Bristol Laboratories.

Results of studies conducted by Gourevitch et al. *(91)* with a series of phenoxyethylpenicillins clearly illustrate the profound effects that serum protein binding has on the efficacy of the compounds in an experimental *S. aureus* infection of mice. Butler *(39)* prepared a plot of these data (see Fig. 3) which he interpreted as follows: (a)

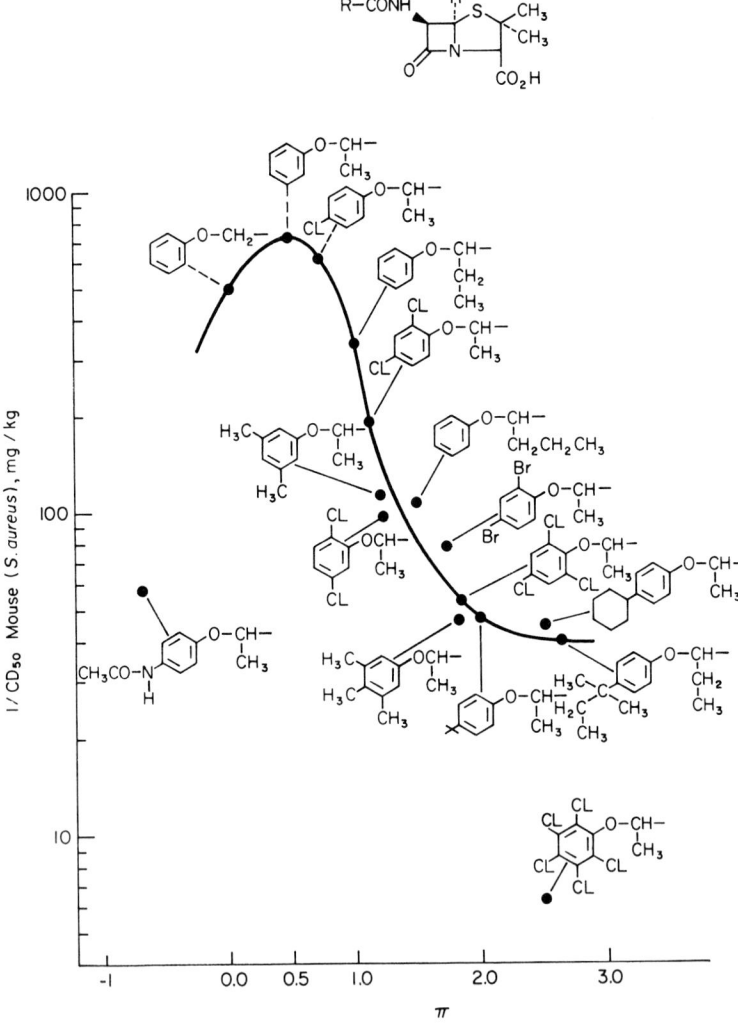

FIG. 3. Taken from Butler *(39)*. Effects of halogenated and nonhalogenated systems. CD, curative dose; π, degree of polarity *(104, 112)*.

there is a simple correlation between lipophilicity, which can be directly equated with serum binding (30, 104), and *in vivo* antibacterial activity; (b) compounds with a low degree of polarity generally have a low degree of activity, although Hansch (104), in reviewing the same data, pointed out that the penicillins with the greatest polarity are not necessarily the most potent; (c) since halogenated and nonhalogenated compounds follow the same curve, electronic properties of the substituents in this series could be considered to be without effect on biological activity.

A listing of serum-binding values found by ultrafiltration or serum versus buffer agar-diffusion techniques for penicillins with various carboxylic acid side chains is given in Table XIII. The first compound in this series, methicillin, shows a lower degree of binding than its monosubstituted analog (compound 2). Also highly bound are phenylpenicillins that have high molecular weight aryloxy, aralkoxy, or alkoxy groups as substituents in the 2- and 6-positions. This is exemplified by the fact that binding in excess of 90% has been found for 2,6-diphenoxyphenyl- (compound 3), 2,6-dibenzyloxyphenyl- and 2,6-dibutoxyphenylpenicillins (Bristol Laboratories, unpublished data). Binding percentages of 90 or greater have also been reported for some of the other carboxylic acid penicillins shown in Table XIII. Of particular interest is the isoxazolylpenicillin series (compounds 5–9), to which addition of one (compound 6) or two (compound 7) chlorine atoms at the *o*-positions of the 3-phenyl substituent of oxacillin (compound 5) causes a progressive increase in the degree of serum binding (132, 193). Compound 8, which differs from dicloxacillin by virtue of a furyl rather than methyl substituent in the 5-position, is bound to an even greater extent.

In any attempt to assess the potential usefulness of a new penicillin, the degree to which it is serum bound must be given a great deal of consideration. This is obvious in view of the poor *in vivo* antibacterial activity demonstrated by such heavily bound compounds as penicillin K, triphenylmethyl-, 2-methoxyphenyl-, 2,6-diphenoxyphenyl-, and 3-(2,6-dichlorophenyl)-5-(2-furyl)-4-isoxazolylpenicillin. However, other factors must be involved since good activity is displayed *in vivo* by propicillin and several of the isoxazolylpenicillins, despite the fact that they are extensively bound. One possible explanation is that set forth by Acred and his colleagues (6). They suggested that the nature of the bond formed with serum proteins by different penicillins may influence binding reversibility. They reported that the percentage of ampicillin and benzylpenicillin retained as part of

TABLE XIII. Percentage Binding to Human Serum of Penicillins Prepared from Aromatic and Heterocyclic Carboxylic Acids

Number	Name	R Structure	Percentage bound	References
1	2,6-Dimethoxyphenyl (methicillin)	2,6-dimethoxyphenyl	49-62	132, 165
2	2-Methoxyphenyl	2-methoxyphenyl	96[a]	–
3	2,6-Diphenoxyphenyl	2,6-diphenoxyphenyl	94[a]	–
4	2-Ethoxy-1-naphthyl (nafcillin)	2-ethoxy-1-naphthyl	89-90	132, 165
5	5-Methyl-3-phenyl-4-isoxazolyl (oxacillin)	isoxazolyl	87-93	132, 165
6	3-(2-Chlorophenyl)-5-methyl-4-isoxazolyl (cloxacillin)	isoxazolyl	94-96	132, 165
7	3-(2,6-Dichlorophenyl)-5-methyl-4-isoxazolyl (dicloxacillin)	isoxazolyl	98	132
8	3-(2,6-Dichlorophenyl)-5-(2-furyl)-4-isoxazolyl	isoxazolyl	99[a]	–

[a] Values obtained by serum vs. buffer agar diffusion techniques at Bristol Laboratories.

an antibiotic-serum protein complex was very much smaller when dialysis was conducted utilizing a flowing rather than a static dialyzate. In the case of triphenylmethylpenicillin, however, the difference was much smaller, leading the authors to suggest that the latter penicillin was involved in "firm" binding, whereas the others were held in a "loose" arrangement.

Other factors which definitely influence the significance of serum binding effects on therapeutic effectiveness are absorbability and intrinsic activity of the penicillin. Despite the higher degree of serum binding by propicillin than by penicillin V, the amount of compound or "free" drug per ml. of serum after oral administration of a comparable dose is equivalent for the two drugs. This is because of the higher absolute blood levels one can obtain with propicillin. For this same reason, serum levels of "free" phenethicillin are much higher than would be expected on the basis of the extent of its binding to serum (32).

This suggests that reporting blood levels in terms of "free" penicillin gives a reliable estimate of *in vivo* efficacy. In practice, however, use of such a system could be misleading; for example, methicillin would seem to be 10 times as effective as isoxazolylpenicillins since the amount of free methicillin in serum is so much higher. In reality, the antibacterial potency of these compounds in serum is fairly comparable (132) owing to the markedly higher intrinsic activity of the isoxazolylpenicillins.

Actually, the desirability of having completely free levels of penicillin in serum has not been established. Binding may be necessary to maintain an equilibrium between tissues and plasma. It also seems probable that completely unbound penicillin would be eliminated at an undesirably rapid rate.

F. Staphylococcal β-Lactamase Resistance

As years of heavy penicillin usage passed, the incidence of infections caused by penicillinase-producing staphylococci in hospital populations rose and virtually reached a point where administration of the available penicillins (G and V) in *any* staphylococcal disease was considered cause for concern. Thus, when the opportunity to synthesize penicillins with new side chains materialized, highest priority was given to the search for those having increased resistance to β-lactamases. The susceptibility of a penicillin to β-lactamases is usually quantitated by measuring its rate of hydrolysis to penicilloic acid. As penicilloic acid is produced, there is a loss in antimicrobial activity, an appearance of a new acidic carboxyl group, a development of capability to decolorize iodine solutions, and a loss in ability to acylate hydroxylamine. Procedures that permit such changes to be assayed are readily available (215).

The relative susceptibility to staphylococcal penicillinase of a group of penicillins derived from acetic acid side chains is shown in Table XIV.

TABLE XIV. RELATIVE SUSCEPTIBILITY OF PENICILLINS WITH
ACETIC ACID SIDE CHAINS TO STAPHYLOCOCCAL β-LACTAMASE

Number	Name	Structure	Response[a]	Hydrolysis rate (% of pen G)	References
1	Benzyl (penicillin G)	Ph–CH$_2$–	S	100	97
2	D-(-)-α-Aminobenzyl (ampicillin)	Ph–CH(NH$_2$)–	S	198	63
3	Phenoxyethyl (phenethicillin)	Ph–O–CH(CH$_3$)–	S	90	98
4	Phenoxybutyl	Ph–O–CH((CH$_2$)$_2$CH$_3$)–	S	85	93
5	Phenoxyisobutyl	Ph–O–CH(CH(CH$_3$)$_2$)–	MOD. R	20	93
6	Triphenylmethyl	Ph$_3$C–	R	<0.5	33
7	Diphenylmethyl	Ph$_2$CH–	S	Approx. 100	33
8	β,β,β-Triphenylethyl	Ph$_3$C–CH$_2$–	S	Approx. 100	33

[a] S, sensitive; R, resistant; and MOD. R, moderately resistant.

Benzylpenicillin (compound 1) and all other monosubstituted acetic acid derivatives are uniformly susceptible to this enzyme as are most penicillins with disubstituted acetic acid side chains. Ampicillin (compound 2), phenethicillin (compound 3), phenoxypropylpenicillin *(14)*, and phenoxybutylpenicillin (compound 4) exemplify the high degree of susceptibility possessed by most penicillins having the latter type of side chain. A striking exception to this was noted by Gourevitch and his co-workers *(93)*, who found that α-phenoxyisobutylpenicillin, in contrast to the n-butyl derivative, had significant resistance to penicillinase. Actually, this was the first hint that the presence of bulky substituents in the vicinity of the side chain amide linkage interferes with enzyme attachment. Further elucidation of the structural characteristics resulting in steric hindrance at the site of enzyme attachment was given in a study of penicillins with trisubstituted acetic acid side chains *(33)*. It was observed that, whereas triphenylmethylpenicillin (compound 6) was highly resistant to staphylococcal β-lactamase, the diphenylmethylpenicillin (compound 7) and the β,β,β-triphenylethyl derivative (compound 8) were not. This indicated that a pair of bulky groups had to be attached to the α-carbon of the aralkyl side chain to prevent hydrolysis by β-lactamases.

These results prompted Doyle *et al.* *(69)* to study penicillins derived from sterically hindered aromatic and heterocyclic carboxylic acids. The relative penicillinase resistance of several such compounds is considered in Table XV. Although phenylpenicillin itself was not effective, 2,6-dimethoxyphenylpenicillin or methicillin (compound 1) was quite resistant. Additional evidence that steric hindrance around the amide linkage is indeed the basis of penicillinase resistance was given by the fact that both the 2,4-dimethoxyphenyl- (compound 2) and 2,6-dimethoxybenzylpenicillin (compound 3) were enzyme sensitive. The hypothesis is also supported by findings with the biphenylpenicillins since the 2-derivative (compound 4) is resistant while the 4-derivative (compound 5) is not. However, the penicillinase resistance produced by introduction of a single bulky group into the o-position of six-membered rings does not occur with a five-membered heterocyclic ring such as that found in the isoxazolylpenicillins. This is amply demonstrated by β-lactamase-sensitive compound 9 (3-phenyl-4-isoxazolylpenicillin). However, introduction of a methyl group at the 5-position in penicillins having a phenyl group in the 3-position (compounds 6 and 7) confers resistance. It is essential that at least one of the groups be bulky, as indicated by the failure of the 3,5-dimethyl-4-isoxazolylpenicillin (compound 8) to show enzyme resistance.

It should be noted that somewhat higher β-lactamase resistance is

TABLE XV
Relative Susceptibility of Penicillins Derived from Aromatic or Heterocyclic Carboxylic Acids to Staphylococcal β-Lactamase

$$R-\overset{O}{\underset{\|}{C}}-\overset{H}{\underset{|}{N}}\diagdown\text{penicillin nucleus}-COOH$$

			Staphylococcal β-lactamase		
Number	Name	Structure	Response[a]	Hydrolysis rate (% of pen G)	References
1	2,6-Dimethoxyphenyl (methicillin)	2,6-(OCH₃)₂-C₆H₃–	R	0.6–0.8	93
2	2,4-Dimethoxyphenyl	2,4-(OCH₃)₂-C₆H₃–	S	–	69
3	2,6-Dimethoxybenzyl	2,6-(OCH₃)₂-C₆H₃–CH₂–	S	–	69
4	2-Biphenylyl	2-C₆H₅-C₆H₄–	R	2.7	97
5	4-Biphenylyl	4-C₆H₅-C₆H₄–	S	–	63
6	5-Methyl-3-phenyl-4-isoxazolyl		R	1.3–1.4	97
7	3-(2-Chlorophenyl)-5-methyl-4-isoxazolyl (cloxacillin)		R	1.1	191
8	3,5-Dimethyl-4-isoxazolyl		S	–	63
9	3-Phenyl-4-isoxazolyl		S	–	71

[a] S, sensitive; and R, resistant.

conferred upon oxacillin (compound 6) by introduction of a chlorine atom into the 2- (compound 7) or 2- and 6-positions of the phenyl ring.

As in any enzyme-substrate interaction, two parameters must be considered when examining the kinetics of the system. The first is a measure of the affinity of the enzyme for the substrate and in terms of the Michaelis-Menten treatment of enzyme kinetics is designated

"K_m." A low K_m value indicates a high degree of affinity of enzyme for substrate. The second parameter is V, the rate of conversion of substrate under saturation conditions to the corresponding penicilloic acid or product. This reaction is usually rate limiting in the case of penicillin-penicillinase interactions.

Penicillins can be roughly classified into three groups on the basis of their susceptibility to staphylococcal penicillinase (98). The first group, which is represented by such highly susceptible substrates as benzylpenicillin and ampicillin, is characterized by low K_m and high V values. Penicillins of intermediate susceptibility, such as α-phenoxyisobutylpenicillin and, to a certain extent, other phenoxypenicillins with large alkyl substituents on the α-carbon, have higher K_m values, but are rapidly hydrolyzed once the enzyme-substrate complex is formed. In the case of resistant penicillins such as methicillin and oxacillin, there is not only a low affinity of enzyme for substrate (high K_m), but also low V values. Some members of the latter group have been shown to act as competitive inhibitors of β-lactamases since they can influence the hydrolysis rate of sensitive penicillins present in the same medium (102, 167).

G. Gram-Negative β-Lactamase Resistance

The incidence of gram-negative bacillary infections has increased 4–8-fold over the past 20 years (58, 80, 138). Although the precise reason for this change is not known, it has been suggested by Rogers et al. (163) that the following may be contributing factors: (a) the advancing age of many patients; (b) the greater use of X-irradiation, cytotoxic drugs, and resistance-lowering steroids; (c) the more frequent use of indwelling catheters and intravenous infusion cannulas; and (d) modifications in the native bacterial flora due to widespread use of antimicrobial agents. Turck (206) implies that the situation is similar to that seen in the 1950's with *Staphylococcus aureus* strains. As an example, he cites the higher percentage of hospital-acquired enteric infections that are now caused by antibiotic-resistant gram-negative bacteria. Also noted was the fact that infections caused by antibiotic-refractory *Pseudomonas* and *Klebsiella* species are almost completely restricted to hospitals. Furthermore, the use of indwelling catheters and intravenous cannulas for infusion offer new routes of entry for opportunistic gram-negative organisms such as drug-insensitive *Herellea* and *Mima* species. Indications that the problem of drug resistance among enterobacteria may assume even more importance in the future is given from worldwide reports (9, 42, 143) which describe isolation of clinical strains carrying episomes that endow the organisms with multiple antibiotic resistance. Since many microbes with such resistance factors are apparently capable of transferring them under *in vivo* conditions to not only members of their own, but to other genera, the implications are obvious. The fact that one of the transferable episomal resistance factors manifests itself by β-lactamase or penicillinase production *(43)* makes the above described phenomenon particularly pertinent to the present discussion. Smith *(188)*, for example, has shown that 3 of 8 *Salmonella* strains carrying such resistance factors could be classified as insensitive to ampicillin. Other genera known to contain strains that produce β-lactamases are *Pseudomonas* (probably the large majority of *Ps. aeruginosa* strains are producers), *Klebsiella*, *Enterobacter*, *Shigella*, *Proteus*, *Escherichia*, and *Herellea*. β-Lactamases of these gram-negative bacilli frequently differ from those of gram-positive bacteria in that they are associated with the soluble cytoplasm rather than the cell wall fraction and as a rule do not have to be induced *(189, 190)*. Furthermore, the β-lactamase of each strain, with the possible exception of *Ps. aeruginosa*, has its own specificity pattern as regards susceptible substrates *(175, 190)*. Since it is now known that some of the enzyme-producing strains have considerable clinical significance, a great deal of emphasis is being given to detection of semisynthetic penicillins that are both resistant to such β-lactamases and inherently active against the organisms producing them. It is recognized, of course, that some clinically important isolates may be resistant to penicillins by virtue of properties other than penicillinase-producing ability *(108)*.

Introduction of ionized groups into the α-position of the benzylpenicillin side chain, as exemplified by the basic amino group of ampicillin and the weakly acidic hydroxyl group of α-hydroxybenzylpenicillin, results in antibiotics with increased inhibitory activity for gram-negative bacteria. Thus, it seemed desirable to investigate the effects of introducing even more highly ionized groups into this position. As a result it has recently been found that several penicillins with highly alkaline or acidic α-substituents, while generally having a somewhat lower degree of potency than ampicillin, are effective against a broader spectrum of gram-negative bacteria. Although it has not been clearly established for each case, the bulk of evidence suggests that the spectrum extension over ampicillin is due in part to the new penicillins' higher level of resistance to the β-lactamases produced by ampicillin-resistant gram-negative organisms.

Figure 4 compares the activity of ampicillin and a series of new α-substituted penicillins against *Ps. aeruginosa*, as well as selected strains of *Proteus* (indol-positive) and *E. coli* that are resistant to ampicillin. Also shown for each of the penicillins is the rate, relative to ampicillin, at which it was hydrolyzed by *Ps. aeruginosa* (A9843) β-lactamase.

Compound 2, which has the highly basic guanidino group in the α-position, has outstanding activity against a high percentage of *Proteus* strains, regardless of their degree of susceptibility to ampicillin. Although it has demonstrated a similar inhibitory effect against some ampicillin-resistant *E. coli* strains, the compound is essentially without effect on *Ps. aeruginosa*. This result is somewhat surprising in view of the fact that it possesses significant and quantitatively similar resistance to all of the *Proteus*, *Escherichia*, and *Pseudomonas* β-lactamase preparations examined. Obviously, factors other than β-lactamase production prevent this penicillin from interfering with the growth of *Ps. aeruginosa* cells.

Results obtained with *Pseudomonas* also contrast sharply with those found for *Staphylococcus aureus*. The penicillin (α-guanidino-2-thenylpenicillin) was hydrolyzed at only a fraction of the rate of ampicillin by a crude enzyme preparation obtained from *Staphylococcus aureus* M-2, and when subjected to *in vitro* and *in vivo* evaluation, proved to be as active as oxacillin against penicillinase-producing staphylococcal strains. Similar findings in experimental staphylococcal infections have been obtained with several closely related compounds, α-guanidinobenzylpenicillin *(133, 203)* and α-guanidino-3-thenylpenicillin (Bristol Laboratories, unpublished data).

The remaining compounds shown in Fig. 4 have, in most cases, highly acidic α-substitutents. Compound 3 (carbenicillin), which is commercially available in Great Britain, has been found to be clinically efficacious in some *Proteus*, *E. coli*, and *Ps. aeruginosa* infections *(36, 161)*. Although most of the successfully treated patients had urinary tract problems, carbenicillin has also been found effective in certain systemic infections. Since it is poorly absorbed by the oral route, possibly because of its lack of acid stability *(7)*, it must be administered by intramuscular or intravenous routes. Bactericidal concentrations for *Ps. aeruginosa* strains are for the most part in the high range of 32 to >125 µg./ml. requiring a dosage level of 8 to 20 gm./day to achieve therapeutic blood and urine levels. The compound does not appear to be highly bound to serum proteins *(166)* and as indicated in Fig. 4, is quite resistant to *Ps. aeruginosa* β-lactamase.

Compound 4, α-(5-tetrazolyl)benzylpenicillin, has a spectrum of antibacterial activity quite comparable to that of carbenicillin. This

Core structure:

$$R-C(=O)-NH-\text{[penam: S, C(CH}_3\text{)(CH}_3\text{), CH-COOH, N, C=O]}$$

Number	Name	R Structure	Relative antibacterial effect against ampicillin-resistant strains of			Pseudomonas aeruginosa β-lactamase Hydrolysis rate (% of ampicillin)	References
			Ps. aeruginosa	Proteus	Escherichia		
1	D-(−)-α-Aminobenzyl (ampicillin)	C$_6$H$_5$−CH(NH$_2$)−	(small)			100	BL-P
2	D-(−)-α-Guanidino-2-thenyl	2-thienyl−CH(NH−C(=NH)−NH$_2$)−		▨	▨	7.1	BL-P
3	α-Carboxybenzyl (carbenicillin)	C$_6$H$_5$−CH(COOH)−	▨	▨		4.0	BL-P, 7
4	α-(5-Tetrazolyl)benzyl	C$_6$H$_5$−CH(5-tetrazolyl)−		▨		<1.0	BL-P

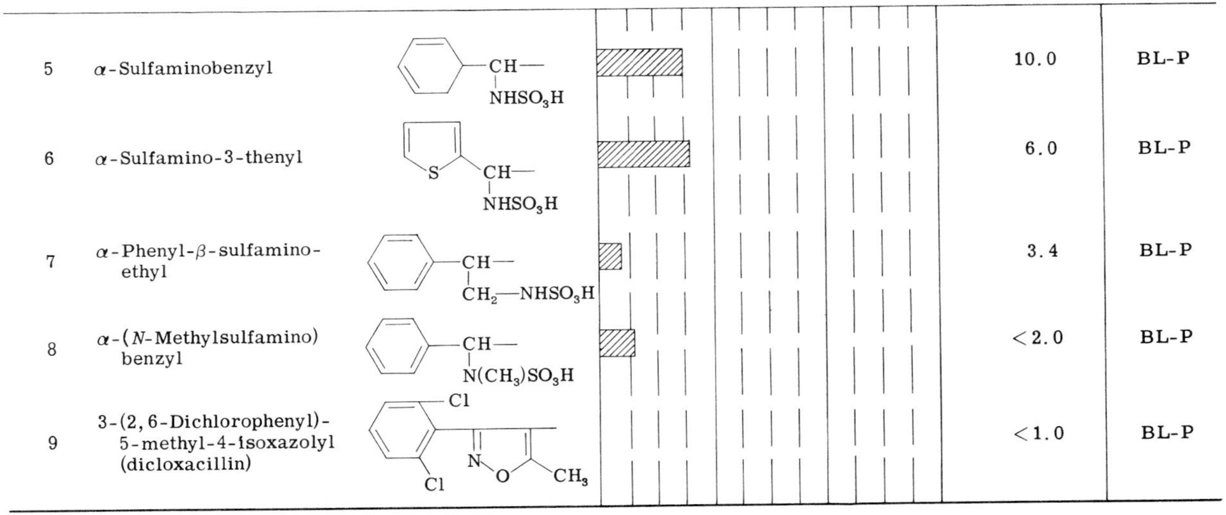
5
α-Sulfaminobenzyl
CH—
NHSO3H
10.0
BL-P
6
α-Sulfamino-3-thenyl
S
CH—
NHSO3H
6.0
BL-P
7
α-Phenyl-β-sulfamino-ethyl
CH—
CH2—NHSO3H
3.4
BL-P
8
α-(N-Methylsulfamino) benzyl
CH—
N(CH3)SO3H
<2.0
BL-P
9
3-(2,6-Dichlorophenyl)-5-methyl-4-isoxazolyl (dicloxacillin)
Cl
Cl
N
O
CH3
<1.0
BL-P

FIG. 4. Relative inhibitory activity of various penicillins for ampicillin-resistant strains of gram-negative bacteria. The length of the bar indicates the degree of activity against a particular group of organisms as compared to ampicillin. Each division (| ↔ |) represents a 2-fold difference in activity. BL-P in reference column indicates compound was prepared and tested at Bristol Laboratories.

is not a surprising finding in view of the fact that the acidic properties of the 5-tetrazolyl group are very similar to those of the carboxyl group *(106)*. It is of interest that the antipseudomonal activity of the compound is no greater than that of carbenicillin despite the fact that it is considerably more resistant to hydrolysis by *Ps. aeruginosa* β-lactamase. Actually this same situation prevails for a whole series of sulfamino-substituted derivatives (compounds 5-8). Although all display considerable resistance to enzymatic hydrolysis, they vary widely in their inhibitory effects against *Ps. aeruginosa* strains. The final compound in Fig. 4, dicloxacillin, which is hydrolyzed by the *Pseudomonas* β-lactamase at a rate less than 1% that of ampicillin, has no measurable antipseudomonal activity.

These results give rise to two important but at present unanswerable questions; first, what is responsible for the wide range in antibacterial activity possessed by the series of enzyme-resistant compounds listed in Fig. 4; and second, how can simple α-substitutents protect the β-lactam ring from enzymatic hydrolysis. No data regarding the first question are available at this time. However, it seems probable that variations in activity can be attributed to factors such as the relative ease with which the compounds reach the site of action and/or the compounds' degree of affinity for the enzymes involved in the cross-linking of cell wall peptidoglycan strands. In regard to the second question, it may be that α-substituents interfere with enzyme attachment to the β-lactam moiety in essentially the same manner that the bulky methoxy groups of methacillin hinder staphylococcal enzymes. Such blocking could occur with these α-substituents because, although small, they are extremely polar and thus are surrounded by a relatively stable and quite large hydration sphere. Since most members of the series shown are not resistant to staphylococcal β-lactamases, it must be presumed that the conformation of the active sites of these enzymes differs from that of *Ps. aeruginosa* enzymes.

H. INHIBITION OF β-LACTAMASES

The earliest report concerning specific antagonists of β-lactamases was that of Saz *et al. (178)*. These workers reported that various dipeptides, especially D-valyl-D-valine, when present in the reaction mixture, produced a modest decrease in the rate of hydrolysis of benzylpenicillin by staphylococcal β-lactamase. Although a hypothesis to account for this action of the dipeptides was not offered, attention was called to the fact that all penicillins possess a valine moiety.

This, of course, suggested that there is competition between inhibitor and penicillin for the active centers on the enzyme. Subsequently, Gourevitch et al. (95) found that methicillin, which is extremely resistant to staphylococcal penicillinase, was capable of irreversibly inactivating the enzyme. This did not occur when a susceptible substrate such as benzylpenicillin was present in the reaction mixture since the affinity of the enzyme for such substrates was markedly greater than for the inhibitor. The failure of penicillinase-resistant penicillins (methicillin and the isoxazole series) to protect susceptible penicillins from the action of staphylococcal penicillinase was confirmed by Hamilton-Miller and Smith (102). However, they did show that one or more of these resistant penicillins could interfere with the hydrolysis of susceptible substrates by crude β-lactamases from *Bacillus cereus, Escherichia coli, Proteus morganii,* and *Enterobacter (Aerobacter) cloacae.*

Results of most studies involving suppression of gram-negative β-lactamase activity strongly suggest that such inhibition is of the competitive type. In the two cases where the kinetics of gram-negative enzyme inhibition were studied (11, 174), it was found that enzyme affinities were in the order of 1000 to 10,000 times greater for the inhibitor than for the β-lactamase-susceptible penicillins. It appears that the β-lactamase-resistant penicillins, which have little or no activity against gram-negative bacteria themselves, permit the hydrolysis-susceptible penicillins to maintain their growth inhibitory effects over a longer period of time. The net result can therefore be considered a synergistic one. Many of the susceptible substrates have significant intrinsic activity against β-lactamase-producing gram-negative bacteria and actually prevent growth and/or kill cells until their concentration is reduced below some critical level by the enzyme. This is exemplified by the results of Sabath and Abraham (174) which are shown in Fig. 5.

Here it can be observed that benzylpenicillin suppresses the growth of *Pseudomonas aeruginosa (pyocyanea)* until a high percentage of the penicillin has been destroyed by the β-lactamase from the organism. Within several hours after this time, culture growth is resumed at an exponential rate. In this same study the authors showed that methicillin caused a delay in the disappearance of susceptible substrates from the *Pseudomonas* culture fluids. Similar observations were made by Kasik (118) who found that oxacillin inhibited *Mycobacterium tuberculosis* β-lactamase and had a sparing effect on benzylpenicillin's disappearance from cultures, and by McKee and Turck *(142)* who reported that dicloxacillin protected the ampicillin analog, hetacillin, from gram-negative β-lactamases.

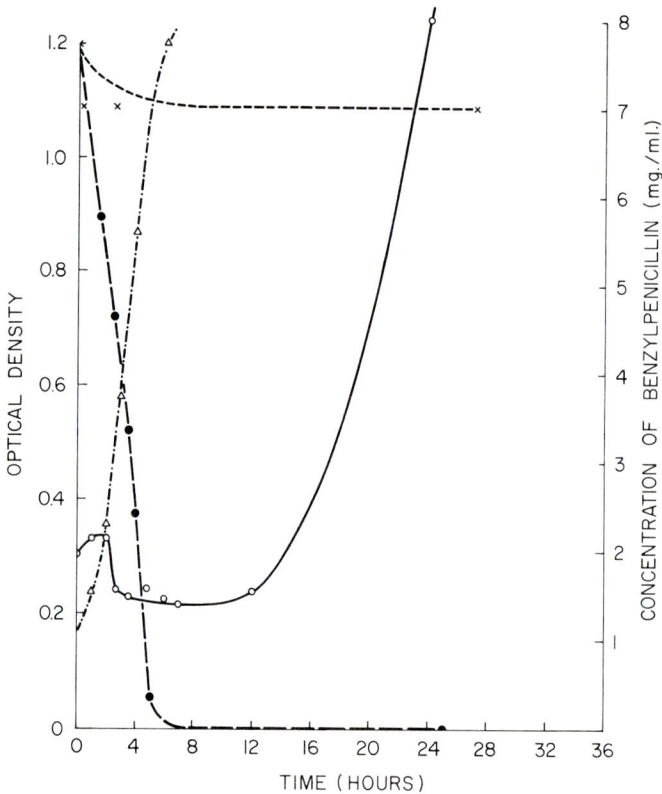

FIG. 5. Taken from Sabath and Abraham *(174)*. Growth of *Ps. aeruginosa* and disappearance of benzylpenicillin from the culture. Δ, Control culture (no antibiotic); O, growth after addition of benzylpenicillin (mg./ml.); ●, concentration of benzylpenicillin in the culture; X, concentration of benzylpenicillin in sterile control.

The precise incidence with which synergism can be demonstrated among penicillin-resistant clinical strains of *Pseudomonas aeruginosa* and various enterobacteria has still not been established. The initial study to investigate this point was conducted by Sutherland and Batchelor *(200)*. They employed combinations of cloxacillin or methicillin with ampicillin or benzylpenicillin. They found 133 ampicillin-resistant strains among 350 clinical isolates. Synergism, which the authors believed to be great enough to have potential clinical usefulness, was observed with ampicillin and benzylpenicillin to the extent of only 4 and 6%, respectively. However, their interest was limited to those cases where the inhibitor (methicillin or cloxacillin) was utilized at 10 µg./ml. and the susceptible penicillin at 5 µg./ml. or less, concentrations readily attainable in human serum. Subsequent investigators *(11, 174)*, however, recognized that the concentrations of penicillins required to give inhibition could, in the majority of cases, be easily achieved in the urine of patients receiving conventional doses of the drugs. On this basis, they suggested that combined penicillin therapy be considered for use in urinary tract infections. Subsequently, several studies have been conducted to determine the incidence of benzylpenicillin- and/or ampicillin-resistant urinary tract isolates that respond in a synergistic fashion to at least one mixture of an inhibitor and susceptible substrate. The percentage responding, although varying from 20 *(ref. 142)* to 87 *(ref. 77)*, is nonetheless remarkably high.

Pathogens that have reportedly responded to penicillin mixtures are: *Enterobacter (Aerobacter) cloacae (11), Enterobacter aerogenes (200), Klebsiella-Enterobacter (77, 175), Escherichia coli (11, 77, 175, 200), Klebsiella pneumoniae (200), Proteus* species *(11, 175, 200)*, and *Pseudomonas aeruginosa* strains *(11, 77, 86, 175, 200)*. In general, it appears that most penicillins possessing resistance to staphylococcal β-lactamase have the capability to protect some enzyme-sensitive penicillins from the β-lactamases of certain gram-negative bacteria. Despite the high degree of specificity demonstrated by the inhibiting penicillins, some do appear to have greater overall effectiveness than others. This is suggested by the data of Table XVI which were obtained when mixtures of ampicillin and various penicillinase-resistant penicillins were evaluated for synergistic effects against an *Enterobacter cloacae* and an *Escherichia coli* strain *(11)*.

It can be seen that the most effective β-lactamase inhibitors were methicillin and the isoxazole penicillins. Some inhibitory action was produced by nafcillin and cephalothin, particularly against the *E. coli* β-lactamase, while propicillin, which has very little resistance to staphylococcal β-lactamase, was without effect on either enzyme. Other penicillinase-resistant penicillins that reportedly act as competitive inhibitors for gram-negative β-lactamases are quinacillin *(103)* and a group of phenoxymethylpenicillin analogs that include the sulfoxide and the dihomo-derivative *(44)*.

Synergism between penicillins has also been demonstrated in experiments with laboratory animals. Infections of mice that responded to synergistic mixtures of penicillins and/or cephalosporins include those caused by *Mycobacterium tuberculosis* H37R$_v$ *(119)*, *Shigella*

TABLE XVI. In Vitro SYNERGISM OF AMPICILLIN WITH
β-LACTAMASE-RESISTANT PENICILLINS (AGAR DILUTION METHOD)[a]

Resistant compound	*Enterobacter cloacae* CDC 2316		*Escherichia coli* M 4888	
	MIC of resistant compound alone	MIC of ampicillin combined with 100 μg/ml of resistant compound	MIC of resistant compound alone	MIC of ampicillin combined with 100 μg/ml of resistant compound
Dicloxacillin	>1,000	8	>1,000	8
Cloxacillin	500	8	500	8
Oxacillin	500	16	500	8
Methicillin	>1,000	2	>1,000	8
Nafcillin	1,000	250	1,000	31
Propicillin	1,000	500	1,000	125
Cephalothin	1,000	1,000	250	62

[a] Taken from Bach *et al.* (11). The minimal inhibitory concentration (MIC) of ampicillin alone was 250 μg/ml for *E. cloacae* and 125 μg/ml for *E. coli*. All MIC values are expressed in μg/ml.

flexneri (11), and *Proteus morganii (148)*. Clinical responses in humans have recently been observed by several investigators. Shirley and Moore *(185)* successfully treated a patient that had a treatment-refractory *Pseudomonas aeruginosa* urinary tract infection with a mixture of methicillin and penicillin G. Sabath and his co-workers *(175)* treated 14 cases of urinary tract infection in humans for 9–17 days with mixtures of penicillins shown to be synergistic in *in vitro* tests. When concentrations exceeding the minimum inhibitory concentration (MIC) level were attained in the urine, organisms were "eliminated or markedly reduced." This occurred with 10 of the 14 patients. As is often the case with such patients, infection recurred in most within a short time after cessation of treatment.

I. HYPERSENSITIVITY

The tendency to induce allergic responses is one of the major problems associated with penicillin usage. A variety of skin reactions, glossitis, stomatitis, drug fever, and anaphylaxis have been found to occur after topical, oral, or parenteral administration of the penicillins *(137)*. Although the precise incidence of hypersensitivity is not known, it has been estimated to be as high as 10% in the United States *(184)*. It is conceivable that this figure is high, but there is little doubt that, among all therapeutic agents, penicillins are the most common cause of allergic reactions *(78)*. Fortunately, less than 1% of penicillin reactors have life-threatening anaphylactic reactions *(208)*.

Penicillins such as penicillin O or allylthiomethyl *(170)*, butylthiomethyl *(159)*, 4-carboxybutyl *(29, 84)*, and 4-carboxy-*n*-butyl *(12)*, as well as several of the newer semisynthetic ones, have at one time or another been considered free from cross-reactivity with benzylpenicillin or to lack antigenicity. All have subsequently demonstrated allergenic potentialities. Evidence that all penicillins should be cross-allergenic was given by Chisholm *et al.* *(46)* who found that 6-APA itself had complete cross-reactivity with penicillins. Subsequent confirmation of the cross-allergenicity among 6-APA derivatives was given by Stewart *(193)*, who administered the antibiotics intradermally into individuals with cutaneous penicillin G hypersensitivity. He found complete cross-sensitivity among benzyl- and *p*-aminobenzylpenicillins, ampicillin, methicillin, cloxacillin, and penicillin N.

It has now been fairly well established that the major antigenic determinant for penicillin hypersensitivity is not penicillin itself, but a penicilloyl derivative. If the penicillin molecule combines with proteins and is not modified, covalent bonding does not occur. As a consequence, the complexes formed are reversible and probably cannot induce hypersensitivity *(149)*. It is thought that an isomer of penicillin, penicillenic acid, which forms *in vivo*, reacts irreversibly with lysine ϵ-amino groups of proteins to form penicilloyl-amino haptenic groups *(59, 135)*. More recently, it has been shown that under certain conditions penicilloyl compounds may also be formed as a direct consequence of a reaction between protein and the unmodified penicillin molecule *(20, 150)*. In either case, the host reacts to the penicilloyl-protein complex by producing antibody.

Detection of individuals that are sensitive to penicillins has generally been a complex and confusing proposition. Use of the complete penicillin molecule alone *(136)* as a skin test antigen fails to predict reliably the hypersensitivity status of the subjects. This has led to development of a penicilloyl-polylysine preparation *(151, 152)* which, when used in conjunction with a complete penicillin, gives a fairly reliable detection method.

The question that has repeatedly risen is whether all penicilloyl derivative formation takes place *in vivo* after penicillin administration or whether some penicilloyl-containing impurities are present in the original penicillin dosage form. An indication that contaminating compounds play a role in cases of penicillin allergy was first suggested by

Silvers *(187)* in 1944. He observed an allergic reaction (cutaneous) following injection of an unpurified, amorphous penicillin sample, but saw no effect when a crystalline preparation was employed. Others *(124, 131)* have also suggested that the degree of purity of the penicillin preparation is a significant factor in penicillin allergies. More recently, it has been suggested that some dosage forms of the natural penicillins, G and V contain a small quantity of highly antigenic penicilloylated protein *(22)*. Stewart *(194)*, Knudsen *et al. (130)*, and Robinson *(160)* found that removal of such contaminants from penicillin G preparations apparently reduced the incidence of allergic responses in sensitive individuals. The fact that such proteinaceous material has not been found in semisynthetic penicillins may help explain why immediate anaphylactoid reactions occur almost exclusively after treatment with natural penicillins *(195)*. One factor complicating the penicillin allergenicity problem is the recent observation that both natural and semisynthetic penicillins may polymerize by internal reactions to produce high-molecular weight substances *(195)*. Thus far, there is no direct evidence that such macromolecules are involved in penicillin allergic reactions.

It has been reported that cephalosporins, which have a 7-aminocephalosporanic acid nucleus, lack cross-allergenicity with penicillins *(123, 193)*. However, Batchelor and his colleagues *(21)* now find in tests involving laboratory animal sera that cephalosporin-protein conjugates show a high degree of cross-reactivity with the benzylpenicilloyl determinant group. Their recommendation that caution be employed in giving these drugs to patients with known penicillin hypersensitivity seems to be justified in view of recent reports *(116, 173)* that describe severe anaphylactic reactions after cephalothin administration to such patients.

IV. Future Objectives

To what extent have we met the goals that were established when routine preparation of semisynthetic penicillins became reality? This question can probably be answered best by examining Fig. 6 which lists the goals and shows the relative degree to which each of the currently available penicillins satisfy them.

Despite the fact that none of the penicillins meets all the goals, there is little question, when one uses penicillin G as a basis for comparison, that significant strides forward have been made. Penicillins with activity against a broad spectrum of gram-negative bacteria, good

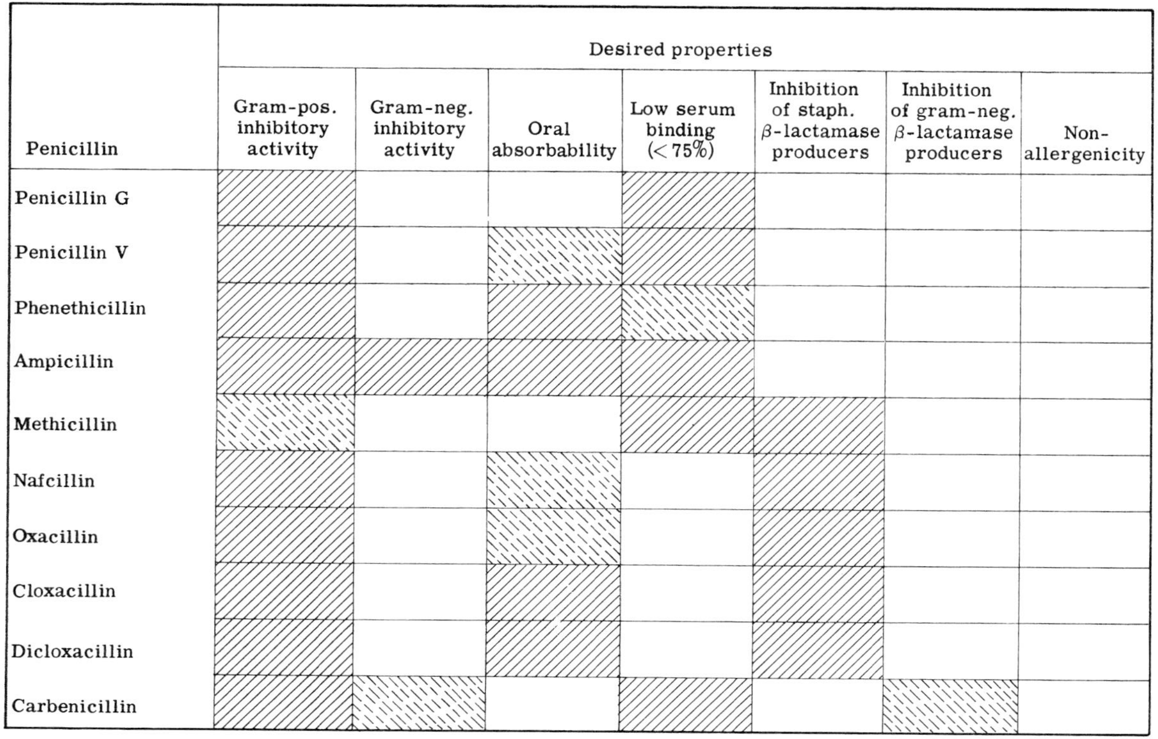

Desired properties

Penicillin	Gram-pos. inhibitory activity	Gram-neg. inhibitory activity	Oral absorbability	Low serum binding (<75%)	Inhibition of staph. β-lactamase producers	Inhibition of gram-neg. β-lactamase producers	Non-allergenicity
Penicillin G	▨			▨			
Penicillin V	▨		▧	▨			
Phenethicillin	▧		▧	▧			
Ampicillin	▨	▨	▧	▨			
Methicillin	▨				▨		
Nafcillin	▨		▧		▨		
Oxacillin	▨		▧		▨		
Cloxacillin	▨		▨		▨		
Dicloxacillin	▨		▨	▨	▨		
Carbenicillin	▨	▧		▨		▧	

FIG. 6. Relative extent to which various commercial penicillins meet requirements for an ideal antimicrobial agent. Shading of bars depict whether the indicated property is adequate, straited; less than adequate, dashed striated; or inadequate, unfilled.

oral absorbability, excellent inhibitory action against penicillinase-producing staphylococci or moderate antipseudomonas activity, are now commercially available. The inability, despite many years of effort, to find one antibiotic that possesses all of these characteristics has fairly well fashioned the direction of current penicillin research. Less effort is now being expended toward finding an all-purpose "super" penicillin derivative. Instead, screening emphasis is being directed toward identifying new compounds that may be effective against a particularly troublesome group of organisms, e.g., carbenicillin against *Pseudomonas aeruginosa*. Unfortunately, the possibility of finding a penicillin that is incapable of inducing hypersensitivity or that fails to cross-react with other penicillins appears very remote.

REFERENCES

1. E. P. Abraham and E. Chain. *Nature* **146**, 837 (1940).
2. E. P. Abraham, E. B. Chain, C. M. Fletcher, H. W. Florey, A. D. Gardner, N. G. Heatley and M. A. Jennings. *Lancet* **ii**, 177–188 (1941).
3. E. P. Abraham, G. G. F. Newton, and C. W. Hale. *Biochem. J.* **58**, 94–102 (1954).
4. E. P. Abraham, G. G. F. Newton, W. Dunn, J. R. Schenck, M. P. Hargie, B. H. Olson, D. M. Schuurmans, M. W. Fisher, and S. A. Fusari. *Nature* **176**, 551 (1955).
5. P. Acred, D. M. Brown, D. H. Turner and M. J. Wilson. *Brit. J. Pharmacol.* **18**, 356–369 (1962).
6. P. Acred, D. M. Brown, T. L. Hardy and K. L. R. Mansford. *Nature* **199**, 758–759 (1963).
7. P. Acred, D. M. Brown, E. F. Knudsen, G. N. Rolinson and R. Sutherland. *Nature* **215**, 25–30 (1967).
8. H. E. Alburn, D. E. Clark, H. Fletcher and N. H. Grant. *In* "Antimicrobial Agents and Chemotherapy – 1967" (G. L. Hobby, ed.), pp. 586–589. Am. Soc. Microbiol., Ann Arbor, Michigan, 1968.
9. F. S. Anderson and N. Datta. *Lancet* **i**, 407–409 (1965).
10. G. A. J. Ayliffe. *J. Gen. Microbiol.* **30**, 339–348 (1963).
11. J. A. Bach, N. Buono, D. Chisholm, K. E. Price, T. A. Pursiano and A. Gourevitch. *In* "Antimicrobial Agents and Chemotherapy – 1966" (G. L. Hobby, ed.), pp. 328–336. Am. Soc. Microbiol., Ann Arbor, Michigan, 1967.
12. A. Ballio, E. B. Chain, F. Di Accadia, M. F. Mastropietro-Cancellieri, G. Morpurgo, G. Serlupi-Crescenzi and G. Sermonti. *Nature* **185**, 97–99 (1960).
13. M. J. Barber. *J. Pathol. Bacteriol.* **59**, 373–384 (1947).
14. M. Barber and R. Novick. *Lancet* **i**, 1059–1060 (1961).
15. M. Barber and M. Rozwadowska-Dowzanko. *Lancet* **ii**, 641–644 (1948).
16. M. Barber and P. M. Waterworth. *Brit. Med. J.* **1**, 1159–1164 (1962).
17. R. L. Barnden, R. M. Evans, J. C. Hamlet, B. A. Hems, A. B. A. Jansen, M. E. Trevett and G. B. Webb. *J. Chem. Soc.* pp. 3733–3739 (1953).
18. F. R. Batchelor, F. P. Doyle, J. H. C. Nayler and G. N. Rolinson. *Nature* **183**, 257–258 (1959).

19. F. R. Batchelor, E. B. Chain, M. Richards and G. N. Rolinson. *Proc. Roy. Soc. (London)* **B154**, 522–531 (1961).
20. F. R. Batchelor, J. M. Dewdney and D. Gazzard. *Nature* **206**, 362–364 (1965).
21. F. R. Batchelor, J. M. Dewdney, R. D. Weston and A. W. Wheeler. *Immunology* **10**, 21–33 (1966).
22. F. R. Batchelor, J. M. Dewdney, J. G. Feinberg and R. D. Weston. *Lancet* **i**, 1175–1177 (1967).
23. A. W. Bauer, D. M. Perry and W. M. M. Kirby. *J. Am. Med. Assoc.* **173**, 475–480 (1960).
24. O. K. Behrens. *In* "The Chemistry of Penicillin" (H. T. Clarke, J. R. Johnson, and R. Robinson, eds.), pp. 657–679. Princeton Univ. Press, Princeton, New Jersey, 1949.
25. O. K. Behrens, J. Corse, R. G. Jones, M. J. Mann, Q. F. Soper, F. R. Van Abeele and Ming-Chien Chiang. *J. Biol. Chem.* **175**, 751–764 (1948).
26. O. K. Behrens, J. Corse, R. G. Jones, E. C. Kleiderer, Q. F. Soper, F. R. Van Abeele, L. M. Larson, J. C. Sylvester, W. J. Haines and H. E. Carter. *J. Biol. Chem.* **175**, 765–769 (1948).
27. O. K. Behrens, J. Corse, J. P. Edwards, L. Garrison, R. Jones, Q. F. Soper, F. R. Van Abeele and C. W. Whitehead. *J. Biol. Chem.* **175**, 793–809 (1948).
28. O. K. Behrens, J. W. Corse, R. G. Jones and Q. F. Soper. U. S. Patent Spec. 2,479,295 (1949).
29. G. H. Berryman and J. C. Sylvester. *Antibiot. Ann.* pp. 521–525 (1960).
30. A. E. Bird and A. C. Marshall. *Biochem. Pharmacol.* **16**, 2275–2290 (1967).
31. W. P. Boger and W. W. Wilson. *Proc. Soc. Exptl. Biol. Med.* **69**, 458–460 (1948).
32. J. M. Bond, J. W. Lightbown, M. Barber and P. Waterworth. *Brit. Med. J.* **2**, 956–961 (1963).
33. E. G. Brain, F. P. Doyle, K. Hardy, A. A. W. Long, M. D. Mehta, D. Miller, H. H. C. Nayler, M. I. Soulal, E. R. Stove and G. R. Thomas. *J. Chem. Soc.* pp. 1445–1452 (1962).
34. E. Brandl and H. Margreiter. *Oestrr. Chemiker-zt.* **55**, 11–21 (1954).
35. E. Brandl, M. Giovannini and H. Margreiter. *Wien. Med. Wschr.* **103**, 602–607 (1953).
36. W. Brumfitt, A. Percival and D. A. Leigh. *Lancet* **i**, 1289–1293 (1967).
37. P. A. Bunn, S. Milicich and J. S. Lunn. *In* "Antimicrobial Agents and Chemotherapy–1965" (G. L. Hobby, ed.), pp. 947–950. Am. Soc. Microbiol., Ann Arbor, Michigan, 1966.
38. H. S. Burton and E. P. Abraham. *Biochem. J.* **50**, 168–174 (1951).
39. K. Butler. *In* "Encyclopedia of Chemical Technology" (A. Standen, ed.), pp. 652–707. Wiley, New York, 1967.
40. F. H. Carpenter, G. W. Stacy, D. S. Genghof, A. H. Livermore and V. du Vigneaud. *J. Biol. Chem.* **176**, 915–927 (1948).
41. C. J. Cavallito and J. H. Harley. *J. Org. Chem.* **15**, 815–819 (1950).
42. Y. A. Chabbert and L. LeMinor. *Presse Med.* **74**, 2407–2410 (1966).
43. Y. A. Chabbert and L. LeMinor. *Presse Med.* **74**, 2479–2484 (1966).
44. S. M. Chaikovskaya, R. A. Makarova, A. E. Tebyakina, E. M. Kleiner, A. S. Guseva and A. S. Khokhlov. *Antibiotiki* **13**, 155–158 (1968).
45. E. Chain, H. W. Florey, A. D. Gardner, N. G. Heatley, M. A. Jennings, J. Orr-Ewing and A. G. Saunders. *Lancet* **ii**, 226–228 (1940).

46. D. R. Chisholm, A. R. English and N. A. MacLean. *J. Allergy* **32**, 333-342 (1961).
47. A. W. Chow, N. M. Hall and J. R. E. Hoover. *J. Org. Chem.* **27**, 1381-1383 (1962).
48. G. Cignarella, G. Pifferi and E. Testa. *J. Org. Chem.* **27**, 2662-2669 (1962).
49. M. Claesen, P. J. Van Dijck, H. Vanderhaeghe and P. De Somer. *Antibiot. Chemotherapy* **12**, 187-191 (1962).
50. C. A. Claridge, A. Gourevitch and J. Lein. *Nature* **187**, 237-238 (1960).
51. H. T. Clarke, J. R. Johnson and R. Robinson (eds.). "The Chemistry of Penicillin." Princeton Univ. Press, Princeton, New Jersey, 1949.
52. R. D. Coghill. *Chem. Eng. News* **22**, 588-593 (1944).
53. R. D. Coghill and R. S. Kock. *Chem. Eng. News* **23**, 2310-2316 (1954).
54. M. Cole and R. Sutherland. *J. Gen. Microbiol.* **42**, 345-356 (1966).
55. D. E. Cooper and S. B. Binkley. *J. Am. Chem. Soc.* **70**, 3966-3967 (1948).
56. L. L. Coriell, R. M. McAllister, E. Preston III and A. D. Hunt. *Antibiot. Chemotherapy* **3**, 357-367 (1953).
57. G. A. Cronk and D. E. Naumann. *Antibiot. Ann.* pp. 111-117 (1955).
58. H. P. Dalton and M. J. Allison. *Appl. Microbiol.* **15**, 804-814 (1967).
59. A. L. De Weck. *Intern. Arch. Allergy* **21**, 20-37 (1962).
60. M. A. Dolan, A. Bondi, J. R. E. Hoover, R. Tumilowicz, R. C. Stewart and R. J. Ferlauto. *In* "Antimicrobial Agents and Chemotherapy – 1961" (M. Finland and G. M. Savage, eds.), pp. 648-654. Am. Soc. Microbiol., Ann Arbor, Michigan, 1962.
61. H. F. Dowling, G. Rotman-Kavka, H. H. Hussey and H. L. Hirsch. *Am. J. Med. Sci.* **213**, 413-417 (1947).
62. F. P. Doyle and J. H. C. Nayler. British Patent Spec. 878,233 (1961).
63. F. P. Doyle and J. H. C. Nayler. *In* "Advances in Drug Research" (N. J. Harper and A. B. Simmonds, eds.), Vol. I, pp. 1-69. Academic Press, New York, 1964.
64. F. P. Doyle, J. H. C. Nayler and G. N. Rolinson. British Patent Spec. 870,395 (1961).
65. F. P. Doyle, J. H. C. Nayler and G. N. Rolinson. British Patent Spec. 873,049 (1961).
66. F. P. Doyle, J. H. C. Nayler and G. N. Rolinson. British Patent Spec. 880,400 (1961).
67. F. P. Doyle, J. H. C. Nayler, H. Smith and E. R. Stove. *Nature* **191**, 1091-1092 (1961).
68. F. P. Doyle, A. A. W. Long, J. H. C. Nayler and R. R. Stove. *Nature* **192**, 1183-1184 (1961).
69. F. P. Doyle, D. Hardy, J. H. C. Nayler, M. J. Soulal, E. R. Stove and H. R. J. Waddington. *J. Chem. Soc.* pp. 1453-1459 (1962).
70. F. P. Doyle, J. H. C. Nayler, H. R. J. Waddington, J. C. Hanson and G. R. Thomas. *J. Chem. Soc.* pp. 497-506 (1963).
71. F. P. Doyle, J. C. Hanson, A. A. W. Long, J. H. C. Nayler and E. R. Stove. *J. Chem. Soc.* pp. 5838-5845 (1963).
72. V. Du Vigneaud, J. L. Wood and M. E. Wright. *In* "The Chemistry of Penicillin" (H. T. Clarke, J. R. Johnson and R. Robinson, eds.), pp. 892-908. Princeton Univ. Press, Princeton, New Jersey, 1949.
73. B. Ekstrom, A. Gomez-Revilla, R. Mollberg, H. Thelin and B. Sjoberg. *Acta Chem. Scand.* **19**, 281-299 (1965).
74. A. R. English and T. J. McBride. *In* "Antimicrobial Agents and Chemotherapy – 1961" (M. Finland and G. M. Savage, eds.), pp. 636-641. Am. Soc. Microbiol., Ann Arbor, Michigan, 1962.

75. J. M. Essery, K. Dadabo, W. J. Gottstein, A. Hallstrand and L. C. Cheney. *J. Org. Chem.* **30,** 4388–4389 (1965).
76. E. Evrard, M. Claesen and M. Vanderhaeghe. *Nature* **201,** 1124–1125 (1964).
77. W. E. Farrar, Jr., N. M. O'Dell and J. M. Krause. In "Antimicrobial Agents and Chemotherapy – 1966" (G. L. Hobby, ed.), pp. 316–320. Am. Soc. Microbiol., Ann Arbor, Michigan, 1967.
78. S. M. Feinberg. *J. Am. Med. Assoc.* **178,** 815–818 (1961).
79. M. Finland, M. Meads and E. M. Ory. *J. Am. Med. Assoc.* **129,** 315–320 (1945).
80. M. Finland, W. F. Jones, Jr., and M. W. Barnes. *J. Am. Med. Assoc.* **170,** 2188–2197 (1959).
81. A. Fleming. *Brit. J. Exptl. Pathol.* **10,** 226–236 (1929).
82. A. Fleming, *J. Pathol. Bacteriol.* **35,** 831–841 (1932).
83. A. Fleming and I. H. MacLean. *Brit. J. Exptl. Pathol.* **11,** 127–134 (1930).
84. H. W. Florey. *Ann. Internal Med.* **43,** 480–490 (1955).
85. M. E. Florey and H. W. Florey. *Lancet* **i,** 387–397 (1943).
86. M. A. Fraher and E. Jawetz. In "Antimicrobial Agents and Chemotherapy – 1967" (G. L. Hobby, ed.), pp. 711–715. Am. Soc. Microbiol., Ann Arbor, Michigan, 1968.
87. A. H. Free, J. R. Leonards, D. R. McCullagh and B. E. Biro. *Science* **100,** 431–432 (1944).
88. H. Gooder and W. R. Maxted. *Brit. Med. J.* **1,** 205 (1961).
89. W. J. Gottstein, R. B. Babel, L. B. Crast, J. M. Essery, R. R. Fraser, J. C. Godfrey, C. T. Holdrege, W. F. Minor, M. E. Neubert, C. A. Panetta and L. C. Cheney. *J. Med. Chem.* **8,** 794–796 (1965).
90. A. Gourevitch, G. A. Hunt and J. Lein. *Antibiot. Ann.* pp. 111–118 (1960).
91. A. Gourevitch, G. A. Hunt and J. Lein. *Antibiot. Chemotherapy* **10,** 121–127 (1960).
92. A. Gourevitch, J. A. Luttinger and J. Lein. In "Antimicrobial Agents Annual 1960" (P. Gray, B. Tabenkin, and S. G. Bradley, eds.), pp. 6–9. Plenum Press, New York, 1961.
93. A. Gourevitch, G. A. Hunt, J. R. Luttinger, C. C. Carmack and J. Lein. *Proc. Soc. Exptl. Biol. Med.* **107,** 455–458 (1961).
94. A. Gourevitch, G. A. Hunt, T. A. Pursiano, C. C. Carmack, A. J. Moses and J. Lein. *Antibiot. Chemotherapy* **11,** 780–789 (1961).
95. A. Gourevitch, T. A. Pursiano and J. Lein. *Nature* **195,** 496–497 (1962).
96. A. Gourevitch, S. Wolfe and J. Lein. In "Antimicrobial Agents and Chemotherapy – 1961" (M. Finland and G. M. Savage, eds.), pp. 576–580. Am. Soc. Microbiol., Ann Arbor, Michigan, 1962.
97. A. Gourevitch, C. T. Holdrege, G. A. Hunt, W. F. Minor, C. C. Flanigan, L. C. Cheney and J. Lein. *Antibiot. Chemotherapy* **12,** 318–324 (1962).
98. A. Gourevitch, T. A. Pursiano and J. Lein. In "Antimicrobial Agents and Chemotherapy – 1962" (J. C. Sylvester, ed.), pp. 318–322. Am. Soc. Microbiol., Ann Arbor, Michigan, 1963.
99. C. F. Gravenkemper, D. R. Sweedler, J. L. Brodie, S. Sidell and W. M. M. Kirby. In "Antimicrobial Agents and Chemotherapy – 1963" (J. C. Sylvester, ed.), pp. 231–236. Am. Soc. Microbiol., Ann Arbor, Michigan, 1964.
100. E. Guddal, P. Morch and L. Tybring. *Tetrahedron Letters* **9,** 381–385 (1962).
101. J. M. T. Hamilton-Miller. *Chemotherapia* **12,** 73–88 (1967).
102. J. M. T. Hamilton-Miller and J. T. Smith. *Nature* **201,** 999–1001 (1964).
103. J. M. T. Hamilton-Miller, J. T. Smith and R. Knox. *Nature* **208,** 235–237 (1965).
104. C. Hansch and A. R. Steward. *J. Med. Chem.* **7,** 691–694 (1964).

105. N. G. Heatley and H. W. Florey. *Brit. J. Pharmacol.* **8**, 252–258 (1953).
106. R. N. Herbst. *In* "Essays in Biochemistry" (S. Graff, ed.), pp. 141–155. Wiley, New York, 1956.
107. D. C. Hobbs and A. R. English. *J. Med. Pharm. Chem.* **4**, 207–210 (1961).
108. R. J. Holt and G. T. Stewart. *J. Gen. Microbiol.* **36**, 203–213 (1964).
109. M. W. Hopper, J. A. Yurchenko and G. H. Warren. *In* "Antimicrobial Agents and Chemotherapy – 1967" (G. L. Hobby, ed.), pp. 597–601. Am. Soc. Microbiol., Ann Arbor, Michigan, 1968.
110. H. T. Huang, A. R. English, T. A. Seto, G. M. Shull and B. A. Sobin. *J. Am. Chem. Soc.* **82**, 3790–3791 (1960).
111. H. T. Huang, T. A. Seto, J. M. Weaver, A. R. English, T. J. McBride and G. M. Shull. *In* "Antimicrobial Agents and Chemotherapy – 1963" (J. C. Sylvester, ed.), pp. 493–499. Am. Soc. Microbiol., Ann Arbor, Michigan, 1964.
112. J. Iwasha, T. Fujita and C. Hansch. *J. Med. Chem.* **8**, 150–153 (1965).
113. K. Izaki, M. Matsuhashi and J. L. Strominger. *Proc. Natl. Acad. Sci. U. S.* **55**, 656–663 (1966).
114. A. B. A. Jansen and T. J. Russell. *J. Chem. Soc.* pp. 2127–2132 (1965).
115. D. A. Johnson. *J. Am. Chem. Soc.* **75**, 3636–3637 (1953).
116. S. A. Kabins, B. Eisenstein and S. Cohen. *J. Am. Med. Assoc.* **193**, 165–166 (1965).
117. E. Kaczka and K. Folkers. *In* "The Chemistry of Penicillin" (H. T. Clarke, J. R. Johnson, and R. Robinson, eds.), pp. 243–268. Princeton Univ. Press, Princeton, New Jersey, 1949.
118. J. E. Kasik. *In* "Antimicrobial Agents and Chemotherapy – 1964" (J. C. Sylvester, ed.), pp. 315–320. Am. Soc. Microbiol., Ann Arbor, Michigan, 1965.
119. J. E. Kasik, M. Weber, E. Winberg and W. R. Barclay. *Am. Rev. Respiratory Diseases* **94**, 260–261 (1966).
120. K. J. Kato. *J. Antibiotics (Tokyo)* **A6**, 130–136 (1953).
121. K. J. Kato. *J. Antibiotics (Tokyo)* **A6**, 184–185 (1953).
122. K. Kaufmann and K. Bauer. *Naturwissenschaften* **47**, 474–475 (1960).
123. J. W. Kislak, B. W. Steinhauer and M. Finland. *Am. J. Med. Sci.* **251**, 433–488 (1966).
124. D. K. Kitchen, C. R. Rein, E. W. Thomas and H. J. Spoor. *Am. J. Syph.* **35**, 578–582 (1951).
125. J. O. Klein and M. Finland. *New Engl. J. Med.* **269**, 1–19 (1963).
126. J. O. Klein and M. Finland. *Am. J. Med. Sci.* **246**, 10–26 (1963).
127. R. Knox. *Nature* **192**, 492–496 (1961).
128. E. T. Knudsen and G. N. Rolinson. *Lancet* **ii**, 1105–1109 (1959).
129. E. T. Knudsen, D. M. Brown and G. N. Rolinson. *Lancet* **ii**, 632–634 (1962).
130. E. T. Knudsen, O. P. W. Robinson, E. A. P. Croydon and E. C. Tees. *Lancet* **i**, 1184–1188 (1967).
131. K. H. Koster, C. G. Lund and K. Pederson-Bjergaard. *Nord. Med.* **42**, 1540–1543 (1949).
132. C. M. Kunin. *Clin. Pharmacol. Therap.* **7**, 166–179 (1966).
133. W. J. Leanza, B. G. Christensen, E. F. Rogers and A. A. Patchett. *Nature* **207**, 1395–1396 (1965).
134. T. Leigh. *J. Chem. Soc.* pp. 3616–3619 (1965).
135. B. B. Levine and Z. Ovary. *J. Exptl. Med.* **114**, 875–904 (1961).
136. J. Love and R. Weir. *Antibiot. Ann.* pp. 521–528 (1956).

137. W. J. Martin. *Lancet* **86,** 159–168 (1966).
138. W. R. McCabe and G. G. Jackson. *Arch. Internal Med.* **110,** 847–864 (1962).
139. C. G. McCarthy and M. Finland. *New Engl. J. Med.* **263,** 315–326 (1960).
140. W. McDermott. *Yale J. Biol. Med.* **30,** 257–291 (1958).
141. W. McDermott, P. A. Bunn, M. Benoit, R. DuBois and M. E. Reynolds. *J. Clin. Invest.* **25,** 190–210 (1946).
142. W. M. McKee and M. Turck. *In* "Antimicrobial Agents and Chemotherapy – 1967" (G. L. Hobby, ed.), pp. 705–710. Am. Soc. Microbiol., Ann Arbor, Michigan, 1968.
143. S. Mitsuhashi. *Gunma J. Med. Sci.* **14,** 169–209 (1965).
144. A. J. Moyer and R. D. Coghill. *J. Bacteriol.* **51,** 57–78 (1946).
145. A. J. Moyer and R. D. Coghill. *J. Bacteriol.* **53,** 329–341 (1947).
146. P. Naumann and B. Kempf. *Arzneimittel-Forsch.* **15,** 139–144 (1965).
147. G. G. F. Newton and E. P. Abraham. *Biochem. J.* **58,** 103–111 (1954).
148. C. H. O'Callaghan and P. W. Muggleton. *J. Gen. Microbiol.* **48,** 449–460 (1967).
149. C. W. Parker. *Am. J. Med.* **34,** 747–752 (1963).
150. C. W. Parker and J. A. Thiel. *J. Lab. Clin. Med.* **62,** 482–491 (1963).
151. C. W. Parker, A. L. DeWeck, M. Kern and H. N. Eisen. *J. Exptl. Med.* **115,** 803–819 (1962).
152. C. W. Parker, J. Shapiro, M. Kern and H. N. Eisen. *J. Exptl. Med.* **115,** 821–838 (1962).
153. Y. G. Perron, W. F. Minor, L. B. Crast and L. C. Cheney. *J. Org. Chem.* **26,** 3365–3367 (1961).
154. Y. G. Perron, L. B. Crast, J. M. Essery, R. R. Fraser, J. C. Godfrey, C. T. Holdrege, W. F. Minor, M. E. Neubert, R. A. Partyka and L. C. Cheney, *J. Med. Chem.* **7,** 483–487 (1964).
155. K. E. Price, A. Gourevitch and L. C. Cheney. *In* "Antimicrobial Agents and Chemotherapy – 1966" (G. L. Hobby, ed.), pp. 670–698. Am. Soc. Microbiol., Ann Arbor, Michigan, 1967.
156. C. H. Rammelkamp and C. S. Keefer. *J. Clin. Invest.* **22,** 425–437 (1943).
157. P. P. Regna. *In* "Antibiotics, Their Chemistry and Non-Medical Uses" (H. S. Goldberg, ed.), pp. 58–173. Van Nostrand, Princeton, New Jersey, 1959.
158. H. C. Richards, J. R. Houseley and D. F. Spooner. *Nature* **199,** 354–356 (1963).
159. G. Risman and W. P. Boger. *J. Allergy* **21,** 425–431 (1950).
160. O. P. W. Robinson. *In* "Antimicrobial Agents and Chemotherapy – 1967" (G. L. Hobby, ed.), pp. 550–552. Am. Soc. Microbiol., Ann Arbor, Michigan, 1968.
161. O. P. W. Robinson. *In* "Antimicrobial Agents and Chemotherapy – 1967" (G. L. Hobby, ed.), pp. 614–618. Am. Soc. Microbiol., Ann Arbor, Michigan, 1968.
162. H. J. Rogers and J. Mandelstam. *Biochem. J.* **84,** 299–303 (1962).
163. D. E. Rogers, M. G. Koenig and K. K. Holmes. *Southern Med. J.* **58,** 1391–1396 (1965).
164. G. N. Rolinson and S. Stevens. *Brit. Med. J.* **2,** 191–196 (1961).
165. G. N. Rolinson and R. Sutherland. *Brit. J. Pharmacol.* **25,** 638–650 (1965).
166. G. N. Rolinson and R. Sutherland. *In* "Antimicrobial Agents and Chemotherapy – 1967" (G. L. Hobby, ed.), pp. 609–613. Am. Soc. Microbiol., Ann Arbor, Michigan, 1968.
167. G. N. Rolinson, S. Stevens, F. R. Batchelor, J. Cameron-Wood and E. B. Chain. *Lancet* **ii,** 564–567 (1960).

168. G. N. Rolinson, F. R. Batchelor, D. Butterworth, J. Cameron-Wood, M. Cole, G. C. Eustace, M. V. Hart, M. Richards and E. B. Chain. *Nature* **187**, 236–237 (1960).
169. I. M. Rollo, G. F. Somers and D. M. Burley. *Brit. Med. J.* **1**, 76–80 (1962).
170. C. L. Rose, P. N. Harris, O. K. Behrens and K. K. Cheu. *J. Lab. Clin. Med.* **34**, 126–131 (1949).
171. J. E. Rosenblatt, A. C. Kind, J. L. Brodie and W. M. M. Kirby. *Arch. Internal Med.* **121**, 345–348 (1968).
172. S. B. Rosenman, L. S. Weber, G. Owen and G. H. Warren. *In* "Antimicrobial Agents and Chemotherapy—1967" (G. L. Hobby, ed.), pp. 590–596. Am. Soc. Microbiol., Ann Arbor, Michigan, 1968.
173. P. D. Rothschild and D. B. Doty. *J. Am. Med. Assoc.* **196**, 372–373 (1966).
174. L. D. Sabath and E. P. Abraham. *Nature* **204**, 1066–1069 (1964).
175. L. D. Sabath, C. E. McCall, N. H. Steigbigel and M. Finland. *In* "Antimicrobial Agents and Chemotherapy—1966" (G. L. Hobby, ed.), pp. 149–155. Am. Soc. Microbiol., Ann Arbor, Michigan, 1967.
176. K. Sakaguchi and S. Murao. *J. Agr. Chem. Soc. Japan* **23**, 411 (1950).
177. C. J. Salivar, F. H. Hedger and E. V. Brown. *J. Am. Chem. Soc.* **70**, 1287–1288 (1948).
178. A. K. Saz, D. L. Lowery and L. J. Jackson. *J. Bacteriol.* **82**, 298–304 (1961).
179. W. Scholtan and J. Schmid. *Arzneimittel-Forsch.* **12**, 741–750 (1962).
180. B. Schwimmer and N. D. Henderson. *Brit. J. Venereal Diseases* **35**, 258–259 (1959).
181. J. C. Sheehan and K. R. Henery-Logan. *J. Am. Chem. Soc.* **79**, 1262–1263 (1957).
182. J. C. Sheehan and K. R. Henery-Logan. *J. Am. Chem. Soc.* **81**, 3089–3094 (1959).
183. J. C. Sheehan and G. D. Laubach. *J. Am. Chem. Soc.* **73**, 4376–4380 (1951).
184. W. B. Shelley. *J. Am. Med. Assoc.* **184**, 171–178 (1963).
185. R. L. Shirley and J. W. Moore. *New Eng. J. Med.* **273**, 283 (1965).
186. S. Sidell, R. J. Bulger, J. L. Brodie and W. M. M. Kirby. *Clin. Pharmacol. Therap.* **5**, 26–34 (1964).
187. S. H. Silvers. *Arch. Dermatol. Syphilis* **50**, 328–329 (1944).
188. D. H. Smith. *New Eng. J. Med.* **275**, 625–630 (1966).
189. J. T. Smith. *J. Gen. Microbiol.* **30**, 299–306 (1963).
190. J. T. Smith and J. M. T. Hamilton-Miller. *Nature* **197**, 976–978 (1963).
191. J. T. Smith, J. M. T. Hamilton-Miller and R. Knox. *Nature* **195**, 1300–1301 (1962).
192. G. T. Stewart. *Lancet* **ii**, 634–640 (1962).
193. G. T. Stewart. "The Penicillin Group of Drugs." Elsevier, Amsterdam, 1965.
194. G. T. Stewart. *Lancet* **i**, 1177–1183 (1967).
195. G. T. Stewart. *In* "The Pediatric Clinics of North America—Antimicrobial Therapy" (B. M. Kagan, ed.), pp. 13–29. Saunders, Philadelphia, Pennsylvania, 1968.
196. J. L. Strominger and D. J. Tipper. *Am. J. Med.* **39**, 708–721 (1965).
197. N. P. Sullivan, A. T. Symmes, H. C. Miller and H. J. Rhodehamel, Jr. *Science* **107**, 169–170 (1948).
198. O. Sus. *Ann. Chem.* **571**, 201–225 (1951).
199. R. Sutherland. *J. Gen. Microbiol.* **34**, 85–98 (1944).
200. R. Sutherland and F. R. Batchelor. *Nature* **201**, 868–869 (1964).
201. J. L. Szabo, C. D. Edwards and W. F. Bruce. *Antibiot. Chemotherapy* **1**, 449–503 (1951).

202. J. D. Thayer, F. W. Field, M. I. Perry, J. E. Martin, Jr., and W. Garson. *In* "Antimicrobial Agents Annual 1960" (P. Gray, B. Tabenkin, and S. G. Bradley, eds.), pp. 352–356. Plenum Press, New York, 1961.
203. E. H. Thiele and H. J. Robinson. *Appl. Microbiol.* **16,** 228–231 (1968).
204. R. Tompsett, S. Schultz and W. J. McDermott. *J. Bacteriol.* **53,** 581–595 (1947).
205. A. L. Tosoni, D. G. Glass and L. Goldsmith. *Biochem. J.* **69,** 476–480 (1958).
206. M. Turck. *In* "Antimicrobial Agents and Chemotherapy – 1966" (G. L. Hobby, ed.), pp. 265–273. Am. Soc. Microbiol., Ann Arbor, Michigan, 1967.
207. U. S. Dispensatory, 25th ed., p. 982. Lippincott, Philadelphia, Pennsylvania, 1955.
208. P. P. VanArsdel, Jr. *J. Am. Med. Assoc.* **191,** 238–239 (1965).
209. H. Vanderhaeghe, P. Van Dijck, M. Claesen and P. De Somer. *In* "Antimicrobial Agents and Chemotherapy – 1961" (M. Finland and G. M. Savage, eds.), pp. 581–587. Am. Soc. Microbiol., Ann Arbor, Michigan, 1962.
210. G. Wallmark. *Arch. Internal Med.* **100,** 787–793 (1962).
211. G. Wallmark and M. Finland. *J. Am. Med. Assoc.* **175,** 886–897 (1961).
212. H. Welch, W. A. Randall and F. D. Hendricks. *Antibiot. Chemotherapy* **3,** 1053–1062 (1953).
213. A. C. Whitehouse, J. G. Morgan, J. Schumacher and M. Hamburger. *In* "Antimicrobial Agents and Chemotherapy – 1962" (J. C. Sylvester, ed.), pp. 384–392. Am. Soc. Microbiol., Ann Arbor, Michigan, 1963.
214. S. Wolfe, J. C. Godfrey, C. T. Holdrege and Y. G. Perron. *J. Am. Chem. Soc.* **85,** 643–644 (1963).
215. D. A. Wolff and M. Hamburger. *J. Lab. Clin. Med.* **59,** 469–480 (1962).

Resistance to Antimicrobial Agents

J. S. Kiser, G. O. Gale, and G. A. Kemp

Agricultural Division, American Cyanamid Company, Princeton, New Jersey

I.	Introduction	77
II.	Evolution of Drug Resistance	78
III.	Biochemical Basis of Drug Resistance	80
IV.	Genetic Basis of Drug Resistance	83
	A. Chromosomal Resistance	83
	B. Extrachromosomal Resistance	86
	C. Episomal Transfer of Drug Resistance	87
V.	The Hospital Environment and Drug Resistance	91
VI.	The Agricultural Environment and Drug Resistance	93
VII.	Summary	98
	Note Added in Proof	98
	References	99

I. Introduction

The survival of a parasite depends, in part, on its ability to adapt to changing environmental conditions. Species which cannot adapt are often replaced by others, better suited to the existing environment. A great variety of forces may act to modify the environment in which microorganisms exist. Depletion of essential nutrients, changes in temperature, pH, light, etc., are all examples of such forces. The presence of antimicrobial agents in the environment is an example of a force which is usually introduced by man in an effort to eliminate the pathogen.

Any substance, regardless of its origin, which acts to inhibit or kill microorganisms, may be termed an antimicrobial agent. An antibiotic is an antimicrobial agent which is produced by a living organism. Of the hundreds of natural and synthetic compounds which have been shown to have antimicrobial activity, less than fifty have found widespread use in the prevention and treatment of bacterial disease in animals and man. To be useful for the treatment of microbial disease in men or animals, antimicrobials must demonstrate selectivity in toxic action and a mechanism of action which the bacteria cannot easily circumvent. Lack of the first property excludes literally thousands of antimicrobial agents from consideration in the chemotherapy of infectious disease. Some aspects of the second will be discussed later in this review.

It is not difficult to find materials which have *in vitro* antibacterial activity. The difficult task is to find a substance which has sufficient selective toxicity for bacterial pathogens so that it will inhibit or kill the pathogens at a dose below that which would be toxic to the host directly, or which would harm the host indirectly by unfavorably affecting the indigenous nonpathogenic microbial flora of the host. Selectivity in toxic action may be manifested in several ways. Drugs which interfere with processes in bacterial pathogens which either do not exist in the host or which can be interrupted temporarily without irreversible damage to the host would be expected to be less toxic than drugs which inhibit processes equally vital to pathogen and host. Penicillin, for example, is a drug with very low toxicity for humans; it interferes specifically with the formation of cell walls in sensitive bacteria; mammalian cells do not possess a structure analogous to the bacterial cell wall. This drug may also be cited as an example of a different manifestation of toxicity. It has been known for many years that guinea pigs are killed by very small doses of penicillin. While this was at first dismissed as a peculiar host sensitivity to the drug, recently the real basis of this toxicity has been determined. Farrar and Kent (1965) and Farrar *et al* (1966) have shown that the normal, overwhelmingly gram-positive intestinal microflora of guinea pigs is rapidly destroyed by penicillin or bacitracin. Superinfection with gram-negative organisms follows, resulting in death of the host usually within 3 to 6 days after the introduction of penicillin or bacitracin. It is indeed fortunate that the intestinal microflora of most animals, including man, differs significantly from the guinea pig in this respect. It is also fortunate that the guinea pig was not chosen as the laboratory animal for use in initial tests of penicillin against experimental infections.

In addition to possessing selective toxicity, an antimicrobial agent should also be capable of absorption and distribution throughout the tissues of the host.

II. Evolution of Drug Resistance

The bacterial pathogens, which are important causes of disease in man and animals today, have evolved over thousands of years. They have emerged because they were strains and species best suited to certain very specialized environments. This not only includes the ability to invade, grow, and multiply in the host without destroying the host, but also the ability to maintain themselves in the presence of other microorganisms either in the tissues of or apart from the host.

Pathogens which have emerged as best fitted to cope with their environment are often highly adapted organisms; so highly adapted have some become that their host range is limited to one or, at best, a few species and may further be restricted to specific tissues within these species. Another reason they compete successfully is because they have evolved ways to conserve energy. They have, in many cases, eliminated the "excess baggage" of unneeded mechanisms of synthesis of cell components which can be drawn from host tissues. It is likely that organisms replacing those inhibited by drugs will be less specifically adapted to the ecological niche formerly occupied by the pathogen. When these highly specialized virulent strains are eliminated from the host by therapy with antibacterials, they are replaced by others, less highly specialized, which have retained means of circumventing the toxic action of the antibacterial agent, but which must expend energy to maintain those means. Thus, they may grow more slowly or be less virulent and more susceptible to host nonspecific defense mechanisms.

A number of widely used antibacterial drugs primarily affect one major class of bacterial pathogens, e.g., the macrolides have activity against the gram-positive group, but have little or no effect against many other pathogens and nonpathogens. Thus, these drugs would act as a selective pressure, favoring the disappearance of the gram-positive cocci whose adaptation had formerly permitted them to flourish. The case of the broader spectrum drugs, such as the tetracyclines and sulfonamides, is fundamentally the same. They suppress the most highly adapted disease flora and allow its elimination by the host and replacement by a flora not quite as well adapted.

In mentioning the replacement of the highly adapted disease flora by a less well adapted flora we have implied that the less well adapted flora was insensitive to, i.e., resistant to, the antimicrobial drug. This selective pressure exerted by antimicrobial drugs and the emergence of organisms resistant to this pressure was observed at the very beginning of chemotherapy by Ehrlich working with trypanocidal dyes and is reviewed elsewhere (Schnitzer and Grunberg, 1959). It seems likely, however, that this selective pressure antedates man's deliberate use of antimicrobial agents in the prevention and treatment of disease.

Antibiotics are classed as secondary metabolites (Lechevalier and Lechevalier, 1967). Their roles, if any, in the life processes of the organisms which produce them are not well understood. Many antibiotics are produced by microorganisms whose natural habitat is soil. Numerous unsuccessful attempts have been made to demonstrate formation of antibiotics in soil under natural conditions. These difficulties have led to scepticism as to whether antibiotics are ever produced under natural conditions. In their review of the biology of the actinomycetes, Lechevalier and Lechevalier (1967) suggest it is unlikely that natural production of antibiotics plays a major role in the ecological relationships of microorganisms. On the other hand, an article by Pollock (1967) presents evidence indicating that this view may be incorrect. Pollock reported that penicillin has been isolated from a strain of *Penicillium chrysogenum* growing on the surface of sterilized soil. He also listed 9 genera, comprising 36 species of fungi which are capable of producing penicillins and/or cephalosporins. Pollock argues that organisms producing antibiotics and those which develop means to inactivate antibiotics coexist in soil and water. Thus, natural selective pressure by antibiotics may exist on a wide scale.

Pollock also reported an experiment which indicates that drug resistance is associated not only with man's use of antimicrobial agents, but that this phenomenon antedates by some time the era of chemotherapy which began with Ehrlich and his co-workers. Spores of *Bacillus licheniformis* were obtained from within soil samples attached to the roots of plant specimens which had remained untouched in the British Museum since 1689. Upon revival of these spores, it was found that the penicillinases produced were physiologically and immunologically similar to those existing at present.

Regardless of its origin the selective pressure exerted by antimicrobial drugs and the development of resistance to this pressure is best understood in the context of the mechanisms of action of these drugs on bacteria and the counterreactions of bacteria to these drugs.

III. Biochemical Basis of Drug Resistance

A number of reviews have been devoted to the subject of the mechanisms of action of antimicrobial drugs (Goldberg, 1965; Gottlieb and Shaw, 1967; Strominger and Tipper, 1965). Essentially these compounds act by interfering with the structure and/or function of the outer surfaces of bacteria, the cell wall and cell membrane, or by inhibiting an internal metabolic process. Penicillin, bacitracin, vancomycin, cephalosporins, and D-cycloserine are all inhibitors of cell wall synthesis. Bacitracin, colistin, polymyxin, and streptomycin are among those antibiotics which have been shown to interfere with cell membrane function.

Among the internal mechanisms which may be affected are deoxyribonucleic acid metabolism (actinomycin, novobiocin, griseofulvin), ribonucleic acid metabolism (neomycin, kanamycin, streptomycin), protein synthesis (tetracyclines, chloramphenicol, erythromycin, lincomycin), purine synthesis (azaserine, psicofuranine), co-factor synthesis (sulfonamides), and respiration (nigericin, flavensomycin, and pyocyanine). It is obvious that some antibiotics, e.g., streptomycin and bacitracin, may have more than one major site of action. On the other hand, some evidence which appears to point to multiple sites of action may merely be observations of secondary effects which follow sequentially from the primary action of the drug.

In order to cause disease, bacteria must multiply, at least initially. Multiplication involves the performance of a complex series of reactions beginning at the nuclear or chromosomal level and ending with the division of the parent cell into two daughter cells. The general order of these steps from translation and transcription of the genetic message in DNA, through messenger and transfer RNA, to protein formation on the ribosome are now well known, although the details are still not clear (Falkow, 1965). Inhibition of any of these steps will cause inhibition of bacterial growth.

Bacteria can adapt in several ways to avoid the action of an antibiotic. They may produce an enzyme which destroys the drug. Penicillinase is the most familiar example (Pollock, 1967). In addition, enzymatic drug inactivation has been reported as a basis for resistance to chloramphenicol (Shaw, 1967; Suzuki and Okamoto, 1967), colistin (Sebek, 1967), and dihydrostreptomycin and kanamycin (Okamoto and Suzuki, 1965). Bacterial drug resistance may be manifested by changes in internal structures. Erythromycin normally binds to the 50 S ribosomal subunits of sensitive *Bacillus subtilis* strains but fails to bind to the same ribosomal subunits of an erythromycin-resistant mutant of *B. subtilis.* Cell wall permeability may be altered to prevent entry of the antibiotic into the cell. This has been shown to be a basis for resistance to tetracycline, streptomycin, and chloramphenicol (Unowsky and Rachmeler, 1966). In some strains of *Escherichia coli*, the system for the synthesis of ribosomal protein in cells with altered permeability is still sensitive to the action of tetracyclines (Franklin, 1963). Several workers (Franklin, 1967; Izaki *et al.*, 1966; Unowsky and Rachmeler, 1966) have observed an apparent adaptive phenomenon associated with the phenotypic expression of tetracycline resistance in *E. coli* strains where resistance was based on decreased permeability of the drug. Genotypically resistant cells cultured in the absence of tetracyclines exhibited what Franklin termed "low level resistance," i.e., a level of 10 μg. tetracycline/ml.

had little effect on protein synthesis, while a level of 50 µg./ml. produced about 50% inhibition. After incubation with subinhibitory concentrations of tetracycline for periods as short as 15–30 minutes these cells became very much more resistant. The amount of antibiotic absorbed by cells preincubated with tetracycline was six- to tenfold less than that absorbed by cells which had been preincubated without tetracycline, when both types were subsequently exposed to 200 µg./ml. of the drug.

β-Apooxytetracycline, which has little antibacterial activity per se, acts as an inducer for the system which regulates tetracycline uptake by these cells.

An extremely short preincubation time (1 minute) with 10 µg./ml. of oxytetracycline followed by growth for 15–30 minutes in drug-free medium produced a marked fall in the absorption of the drug by the resistant cells. Preincubation with very low concentrations of oxytetracycline (0.05 µg./ml.) for 30 minutes produced a similar effect. This ability to exclude oxytetracycline is retained by preincubated resistant cells after growth for 2 hours in drug-free medium. However, after growth for 16 hours in drug-free medium the cells absorb oxytetracycline freely. Chloramphenicol and proflavin inhibit the adaptive decrease in tetracycline absorption. Conversion of resistant cells to spheroplasts reduced their ability to respond adaptively to preincubation with tetracyclines.

The enzymatic basis for this adaptive phenomenon has not yet been elucidated. However, the speed with which this phenomenon operates argues for enzyme induction as its basis rather than a selection of cells of differing genotypes from within the population. Similar findings were reported by Gale and Hall (1959) some years ago in tetracycline-resistant staphylococci exposed to submaximal concentrations of drug. As information on the precise mechanism of decreased tetracycline absorption is obtained it may offer clues as to means of overcoming this particular type of bacterial resistance. These experiments also point up one aspect of drug resistance that is often overlooked.

Drug resistance in bacteria is often measured in relation to the smallest amount of drug which completely prevents growth or inhibits visible growth in culture tubes. These end points are designated as minimum lethal (or bactericidal) and minimum inhibitory (or bacteriostatic) concentrations. These are abbreviated to MLC and MIC, respectively. For example, if the MIC for a given drug against a given strain of bacteria is determined to be 20 µg./ml. that strain is said to be resistant to levels less than 20 µg./ml. of the drug. It is tacitly assumed by some that drug levels below the MIC will be without real effect on the resistant organism.

The experiments with the adaptive phenomenon in some tetracycline-resistant cells indicate that this is not necessarily true. Drug concentrations below the MIC may have a profound effect on a resistant culture, e.g., as reflected by decreased rate of protein synthesis (Franklin, 1967) and a pronounced increase in the lag phase of growth (Gale and Hall, 1959).

This partial effectiveness of a drug at levels below the MIC may be of utility in an *in vivo* situation, where the natural defenses of the host play an important role in inhibiting bacterial growth in addition to the action of a drug. We are not implying that *in vitro* sensitivity tests have no relevance to the outcome of a therapeutic situation. We are implying that the mere determination of *in vitro* resistance does not always indicate therapeutic failure on the part of the drug in question; such a situation has been duplicated experimentally in *Salmonella*-infected chickens (Garside et al., 1960). We are further implying that one reason for this might be that some "resistant" cells are not completely unaffected by levels of drug below that determined to be the minimum inhibitory or lethal concentration for that given organism in a given *in vitro* test system.

IV. Genetic Basis of Drug Resistance

A. CHROMOSOMAL RESISTANCE

The occurrence of drug resistance in a previously sensitive organism represents a change, which if it is transferable from generation to generation, may be a chromosomal change. Such a change (mutation) does not necessarily improve an organism's ability to cope with its environment. It may in fact be lethal (Oginsky and Umbreit, 1959). In any population of drug-sensitive bacteria a spontaneous chromosomal change may occur in one cell leading to resistance. The mutation rate is a measure of the probability of this spontaneous change occurring per bacterium per division in this population. This probability is usually quite low. Braun (1965) lists mutation rates ranging from 3×10^{-6} for resistance to isoniazid in *Mycobacterium ranae* to 6×10^{-10} for streptomycin resistance in *Haemophilus pertussis*. Let us assume a chromosomal change causing a mutation specific for one antibiotic or group of closely-related antibiotics (or synthetic antimicrobials) as having a probability of 1×10^{-6} to 1×10^{-8}.

A separate mutation at another point on the chromosome would be required to make the cell resistant to an unrelated antibiotic. The probability of a cell mutating to become resistant to two different drugs simultaneously would be the product of the probabilities for each single mutation or 1×10^{-12} to 1×10^{-16}.

These are infinitesimally small probabilities and argue for the use of a combination of antimicrobial drugs. The mutation rate, i.e., the rate of new mutational events, must be distinguished from mutant frequency, i.e., the number of mutants in a given population (Srb *et al.*, 1965). The frequency of mutants in a population at any one time is related to but not identical to the rate at which new mutants are formed. Application of different selective factors can produce different mutant frequencies in cultures of a single strain all with identical mutation rates.

The question of therapy with combinations of antibiotics has been widely discussed, and has been succinctly reviewed by Dowling (1965). No rationale exists for the indiscriminate use of combinations of antibacterial drugs, but combinations are the treatment of choice in enterococcal endocarditis and tuberculosis in humans. In both instances, one of the cardinal reasons for the use of combinations is to prevent or delay the emergence of resistant organisms. Our own work (Gale *et al.*, 1963) indicated that a combination of chlortetracycline and a sulfonamide also prevents or delays emergence of resistance to the sulfonamide by a *Salmonella gallinarum* infection in chickens. One of the limitations to this rationale is that resistance to different antimicrobials need not arise solely as the result of independent mutational events which occur only at rare intervals. This subject will be dealt with in some detail in a subsequent section.

That a mutation to resistance may be unfavorable is demonstrated by the fact that resistant organisms have often been shown to grow more slowly than their sensitive parents. In a careful study of bacterial populations in continuous culture, Zamenhof and Eichhorn (1967) found that an azide-sensitive strain of *Bacillus subtilis* completely overgrew an azide-resistant mutant derived from it in 24 generations when both were grown in the absence of sodium azide. A wild type of *B. subtilis* sensitive to 5-methyltryptophan overgrew a derepressed 5-methyltryptophan resistant mutant in 42 generations when both were grown in continuous culture in unsupplemented medium.

Schnitzer and Grunberg (1959) summarized fifteen studies on comparative growth rates of sensitive and resistant strains of several genera of pathogens resistant to streptomycin, nitrofurans, sulfonamides, penicillin, tetracyclines, or isoniazid. The growth rates of organisms resistant to penicillin, isoniazid, or tetracyclines all decreased. These included strains of *Streptococcus, Staphylococcus, Diplococcus, Escherichia, Aerobacter, Klebsiella,* and *Mycobacterium.* Organisms resistant to sulfonamides showed either decreased or unchanged growth rates compared with the parent sensitive strains. The growth rates of *Staphylococcus* and *Escherichia* strains resistant to nitrofurazone were unchanged. Streptomycin-resistant coliforms and staphylococci, however, showed three patterns of behavior. Their growth rates compared to those of the sensitive parent strains were unchanged, decreased, or increased.

A mutation to resistance can be accompanied by a change in virulence. Schnitzer and Grunberg also summarized a number of studies on the influence of drug resistance on the virulence of pathogenic organisms. The drugs used were streptomycin, sulfapyridine, chlortetracycline, and penicillin. The pathogens were strains of *Staphylococcus, Diplococcus, Streptococcus, Pasteurella,* and *Neisseria.* Twenty-six strains among these genera showed a measurable decrease in virulence on becoming resistant; twelve strains showed no change in virulence when they became resistant. No strain showed an increase in virulence. Shmidov (1965) found that *Salmonella choleraesuis* which had been selected for stable resistance to several antibiotics showed changes in colonial morphology, antigenic structure, and fermentative activity which could be characterized as a change from smooth to rough and that these rough, drug-resistant variants had a significantly decreased virulence for mice. Again, we do not wish to imply that a change to resistance will usually result in a loss of virulence. The evidence does indicate, however, that a mutation to resistance may lead to no change or to a decrease in virulence. We know of no evidence indicating an increase in virulence accompanying a mutation to drug resistance in a pathogen.

Garside, Gordon, and Tucker (1960) studied antibiotic resistance in strains of *Salmonella typhimurium* administered to chickens receiving medicated rations. They reported the development of resistance *in vitro* to chlortetracycline by *S. typhimurium.* In several experiments, mortality due to infection with the resistant strain was about the same as mortality due to the sensitive strain in chickens receiving nonmedicated rations. The values reported were 92/217 (42.4%) mortality for the resistant organism and 90/185 (48.6%) for the sensitive organism. When the same two strains were used to infect chickens fed 100 gm. chlortetracycline per ton of feed, mortality rates were as follows: 26/164 (15.8%) of chlortetracycline-fed chickens receiving the resistant strain died; 44/188 (23.4%) of chicks fed chlortetracycline and receiving the sensitive strain died. Thus, the strain which was shown to be resistant by *in vitro* testing was certainly no more virulent than the parent sensitive strain. Second, infections due to both the resistant and sensitive strains responded equally well to treatment with chlortetracycline. As discussed previously, an *in vitro* demonstration of drug resistance does not necessarily mean that the drug would be without therapeutic effect in a disease situation.

Up to this point only resistance due to changes in genes on bacterial chromosomes has been considered; changes which are passed on only to the daughter cells of a mutant. Genetic material in bacteria may also be extrachromosomal, that is, it may be carried in the cytoplasm replicating independently of the chromosome. Both chromosomal and cytoplasmic genetic information can be transferred among different strains, different species, or even different genera of bacteria. Several mechanisms for this transfer are known.

B. Extrachromosomal Resistance

One method of genetic exchange in bacteria is *transformation* (Falkow, 1965). Deoxyribonucleic acid (DNA) extracted from cells of one strain can induce a heritable change in cells of another strain. Isolated viral DNA has been used to infect bacterial cells, resulting in the production of complete virus. This process, akin to transformation, has been termed *transfection* (Spizizen *et al.,* 1966). A third method of exchange of genetic material is *transduction* (Zinder and Lederberg, 1952), a process mediated by bacterial viruses. In this process virulent bacteriophage multiply in and disrupt host cells, releasing numbers of virus particles which in turn infect and disrupt more host cells. Phage which exist within a cell without disrupting it are called temperate phage. Temperate phage may integrate with the chromosome of the host cell and replicate with it. This is the prophage state. Temperate phage may revert to the virulent state and in so doing may carry along a portion of the chromosome in which they were integrated. If the virulent phage carrying a portion of the chromosome of its former host invades a new host, the new host may respond to the genetic message of the previous host and is then said to have been transduced. For example, the ability to form flagella has been transferred between strains of *Salmonella* by transduction.

A fourth method of genetic transfer in bacteria is *conjugation* (Falkow, 1965; Meynell and Datta, 1966a,b). Here, cell-to-cell contact appears to be required via a cytoplasmic bridge or tube between mating bacteria. The capacity for conjugation is controlled by certain sex factors contained in the donor (male) cells. These sex factors promote their own transfer to the recipient (female) cells which may then themselves become competent donors. The principal sex factor in bacteria is called F. The notation F^+ is used for male cells possessing F and F^- is used for female cells lacking it. Sex factors may, however, exist in any one of a number of alternative states in the bacterial cell. Thus, when integrated into the bacterial chromosome it promotes not only its own transfer but also that of a portion or all of the chromosome in linear order. In this state the cell is called Hfr (high frequency of recombination). The sex factor may revert from the integrated (Hfr) to the autonomous cytoplasmic state (F^+). In the process it may take with it a small portion of the host chromosome. This type of particle which contains both a chromosomal marker and the sex factor is termed F'.

Donor bacteria have different surface properties from recipients. They have an altered surface charge and, in addition, form specialized structures called sex pili or fimbriae. These threadlike structures appear necessary for the mating process to occur. Their removal from a donor results in sterility. Recent studies (Lawn et al., 1967) indicate that most sex pili fall into two main classes termed F and I according to morphology, antigenic structure, and ability to adsorb specific bacteriophage. Other pili are also found on gram-negative organisms. These "common" pili do not appear to play a role in conjugation.

Conjugation experiments can be conducted in the laboratory simply by mixing large numbers of competent donor and recipient bacteria together. The conjugal transfer of genetic material can be interrupted by separating the mating pairs mechanically in a food blender (Falkow, 1965). Conjugation has been demonstrated only in gram-negative bacteria, particularly members of the Enterobacteriaceae. Transduction and transformation occur in both gram-negative and gram-positive bacteria. Extrachromosomal elements such as sex factors and certain transducing phage which may exist either integrated into the bacterial chromosome or autonomously in the cytoplasm are called episomes; those which have not been observed to exist in the integrated state are called plasmids. The distinction is tenuous and some authors appear to use the terms interchangeably.

C. Episomal Transfer of Drug Resistance

The transfer of drug resistance among different species of gram-negative bacteria was first observed in Japan in 1959 during studies on shigellosis in humans (Watanabe, 1963). This transfer took place by a process of conjugation. Conjugation between donor and recipient cells is promoted by an extrachromosomal element called RTF (resistance transfer factor). Attached to the RTF may be one or more fragments of genetic material which confer drug resistance. These have been called resistance determinants and can be carried into a recipient cell along with the RTF. The RTF with resistance determinants is termed an R factor. These episomes have characteristics similar to those of the sex factor F. They exist in an independent state in the cytoplasm, determine the synthesis of specific pili, promote their own transfer, and have recently been demonstrated to be capable of integrating stably with chromosomes as does F, and to be capable of transferring a portion of the bacterial chromosome from a fixed locus (Pearce and Meynell, 1968) similar to Hfr. Like F, they may be lost spontaneously.

An important difference between F and R is the frequency with which they promote conjugation. Almost all of the cells containing F can conjugate and transfer F alone or, in the Hfr state, chromosomal genetic material as well. The actual occurrence of genetic recombinants for chromosomal markers is quite rare, however, with a frequency of only 10^{-5} per conjugating pair (Meynell and Datta, 1967). With the R factors, conjugation itself is a rare event which may occur with a frequency of only 10^{-3} to 10^{-7}. It has been observed that cells newly infected with R factors can transfer these factors to other recipients at a very high rate, for about five generations. Such populations are called HFT (high frequency transfer). Meynell and Datta postulated that the usual lower level of R factor-mediated conjugation resulted from repression of this function by regulator genes. The transient high frequency transfer of R factors in newly infected cells could be explained by assuming that transfer proceeds at a high rate until sufficient repressor has been synthesized to bring the rate back down to the usual 0.1% or less. Two classes of R factors have been described (Meynell and Datta, 1966a). The distinction depends partly on their interaction with F. One category, fi^+ R factors, inhibit formation of F pili in F^+ cells, and the formation of R pili in R cells. They restrict the susceptibility of the cell to "male-specific" phage, but do not inhibit the attachment of other phage. The second category, fi^- R factors, do not inhibit F pilus formation in F^+ cells and do not determine the formation of an F type R pilus but do apparently determine formation of fi^- R pili (which belong to the I group). They do not restrict the susceptibility of the F^+ cell to male specific phage but do inhibit the attachment of other phage. Meynell and Datta isolated both fi^+ and fi^- derepressed mutants in which the frequency of R factor conjugation approached that of F factor conjugation. The repression mechanism in $F^- fi^+$ R cultures is due to suppression of F-like pilus formation. $F^- fi^-$ R cultures do not synthesize F-like pili (Pearce and Meynell, 1968); however, they do synthesize an I-type pilus which is normally repressed, but which is susceptible to filamentous I phages (Meynell and Datta, 1967; Meynell and Lawn, 1968).

The molecular nature of R factors has been examined by Rownd (Rownd et al., 1966; Rownd, 1967) and Falkow (1967). Hybridization and density gradient experiments indicate that the DNA of R factors from *E. coli* and *Proteus mirabilis* differs from the chromosomal DNA of these organisms. In *E. coli* there appears to be one copy of R factor produced for each copy of the chromosome. In *P. mirabilis* one can get several copies of the resistance episome for each copy of the chromosome. The number of episomes per chromosome is directly proportional to the generation time in logarithmically grown cells. In *P. mirabilis*, replication of R factors continues after chromosomal replication ceases and during the stationary phase of growth as many as sixty R factors can exist in a single cell.

Transferable or infectious drug resistance has been shown to occur in many genera of gram-negative bacteria including *Shigella, Salmonella,* coliforms, *Serratia, Vibrio,* and *Pasteurella* (Watanabe, 1963). Gram-positive organisms are not known to conjugate; however, transfer of episomal drug resistance has been observed in staphylococci. The transfer process in gram-positive cells is transduction, mediated by bacteriophage.

Another characteristic of the infectious transfer of drug resistance is that, unlike chromosomal mutants, resistance to several antibacterial drugs is often transferred simultaneously from donor to recipient. It seems unlikely that exposure to a single antibiotic induces the development of multiple resistance. It is more probable that isolation of multiply resistant cells after exposure to a single antimicrobial agent represents selection of cells which were already multiply resistant.

The discovery of episomal transfer of resistance is relatively recent but the phenomenon itself has apparently existed for a long time. It may have been going on for almost as long as the bacteria involved have existed. Smith has reported identification of an fi^+ R factor mediating resistance to streptomycin, tetracycline, and bluensomycin in a strain of *E. coli* which had remained lyophilized since 1946 (D. H.

Smith, 1967a). This represents the earliest reported isolate infected with an R factor. In earlier reports Smith found that many R factors, including those originally isolated in Japan, mediated resistance to such things as mercuric salts and spectinomycin (D. H. Smith, 1967b,c), a very recently introduced antibiotic. These observations indicate that R factors mediate resistance to agents not previously employed in treatment of disease. Thus, one cannot presume that man's use of antibiotics has been responsible for the spread of R factors from a parent genotype which originated in Japan in the 1950's and was subsequently distributed throughout the world. Smith's results strongly suggest that R factors are not a recent phenomenon (D. H. Smith, 1967d). Work in our laboratory has shown that apparent demonstrations of *in vivo* transfers may be misleading (Jarolmen and Kemp, 1969b). Weanling swine which had never received antibiotics were shown to carry *E. coli* with R factors in their intestines. These swine were infected experimentally with a drug-sensitive strain of *S. choleraesuis*. Cultures were then made from fecal matter and selected tissues from these swine. If the cultures were first incubated in broth and then plated on selective media, colonies of the infecting organism resistant to antibiotics could easily be recovered; however, if specimens were plated directly on agar, resistant *Salmonella* were not recovered. Thus, the transfer most likely occurred *in vitro*, not in the gut of the pig.

The episomal transfer of resistance has many variables. All, some, or none of the resistance determinants with or without the transfer factor in a multiply resistant donor may be passed on to a recipient. The level of resistance of the recipient may be the same as or different from that of the donor; in certain cases of streptomycin resistance this has been shown to result from the presence of both an episomal resistance determinant and a chromosomal resistance gene in the same cell with only the episomal determinant being transferred at conjugation (D. H. Smith, 1967a). After becoming resistant, a recipient cell may become a competent donor if it has received the transfer factor as well as resistance determinants. Alternatively, it may lose the transfer factor and/or the resistance determinants spontaneously.

Thermosensitive replication of a specific resistance factor has been reported by Terewaki *et al.* (1967). They obtained a strain of *Proteus vulgaris* from the urinary tract of a patient with postoperative pyelonephritis. The *Proteus* isolate was resistant to sulfonamides, streptomycin, tetracycline, and kanamycin. Only kanamycin resistance was transferred upon conjugation with a sensitive *E. coli* recipient. The rate of R factor-mediated transfer of kanamycin resistance at 25°C. was 100,000 times greater than the rate at 37°C. When *P. vulgaris* was grown at 42°C. the kanamycin R factor was spontaneously eliminated. These experiments emphasize the role which the environment may play in determining the frequency with which a resistance transfer factor may be passed from one cell to another.

Most of the experiments on episomal transfer of resistance have been done *in vitro*. An understanding of the extent to which this phenomenon occurs in nature is most important. Transfer experiments have been reported in animals (Walton, 1966a; Salzman and Klemm, 1968). However, in these experiments either germ-free animals or animals whose intestinal flora had been altered drastically by massive doses of antibiotics were used. Moreover, very large numbers of resistant donor and sensitive recipient types were given. Braude and Siemienski (1968) established experimental *E. coli* infections in the kidneys of rats by injecting a donor organism in one kidney, and a sensitive recipient in the other. They found transfer of colicinogeny and of antibiotic resistance occurred in the bladder but not in the kidneys of these animals. The relationship of these experiments to transfer in nature, as shown in the study of shigellosis in Japan (Watanabe, 1963) and elsewhere (Kabins and Cohen, 1966), or their significance to public health is still unclear. On the other hand, it is clear that mortality due to shigellosis declined sharply in Japan during the 1950's even though episomal transfer of drug resistance was occurring there at the time. One recent observation made in our laboratory (Jarolmen and Kemp, 1969a) bears on this point. A strain of *S. choleraesuis* virulent for swine has been isolated from the intestinal tract of pigs and found to be present in two forms; an antigenically smooth virulent variant and an antigenically rough variant with greatly reduced virulence for mice. There is greater than a hundredfold difference in the ability of these two variants to act as recipients of R factors with the rough form being more competent. This may well be part of the reason why the increased reports of R factors have not been accompanied by a corresponding increase in the occurrence of bacterial diseases or a failure of their therapy.

V. The Hospital Environment and Drug Resistance

Several studies have indicated that gram-positive bacteria, particularly pneumococci and β-hemolytic streptococci, were responsible for a majority of serious pyogenic infections in humans in the preantibiotic and presulfonamide era. During the late 1940's and early 1950's staphylococci appeared to have become the most important bacterial pathogens for humans. Many of the staphylococci isolated in hospitals were resistant to one or more antibiotics, particularly the penicillins. In the past decade, however, it has become apparent that a change has taken place in the pattern of nosocomial infections. Rogers (1959) compared the origin of fatal infections in the New York Hospital in 1938–1940 with those seen in 1957–1958. He reported that 53 of 57 fatal bacterial infections in the preantibiotic era were community acquired whereas 15 of 28 such infections in 1957–1958 arose within the medical wards. These latter individuals were already compromised by serious disease and often receiving therapies which were known to affect host resistance to infection. Gram-negative enteric bacteria have become increasingly important as etiologic agents of hospital-acquired infections. The relative frequency and importance of staphylococcal infection has decreased, although staphylococci still remain a significant cause of nosocomial infections in hospitals.

The principal gram-negative species involved in hospital-acquired infection are *E. coli, Klebsiella, Aerobacter, Herellea, Proteus, Pseudomonas,* and *Serratia*. A considerably higher incidence of *Klebsiella* isolations in hospital- versus community-acquired infections has been reported. In a study at Boston City Hospital, Barrett et al. (1968) found *Klebsiella* responsible for 16.7% (30/180) of hospital-acquired infections versus 7.8% (16/205) of community-acquired infections. In a study at the Research and Educational Hospital of the University of Illinois, Kessner and Lepper (1967) found that *Klebsiella* accounted for 19.4% (19/98) of the hospital-acquired infections whereas only 2% (1/49) of the community-infected controls yielded *Klebsiella* on culture.

The reasons for the increase in importance of gram-negative bacilli and the relative decrease in the importance of staphylococc· in hospital-acquired infections are not well understood. One important factor may be an increase in the number of aged people undergoing hospitalization and, more particularly, undergoing surgery while hospitalized.

The elderly (and children under 1 year of age) are particularly susceptible to infections with gram-negative bacilli. Prior malignant disease predisposes toward infection with these organisms. Han *et al.* (1967) reported that salmonellosis was encountered only once in 9000 patients with nonmalignant disease hospitalized at Roswell Park Memorial Institute during the 7-year period from 1959 to 1965.

In contrast, salmonellosis was observed in 16 of 5384 patients with malignant lymphoma, leukemia, carcinoma of the sigmoid colon or rectum, or other pelvic neoplasms.

Risk of infection also appears to be associated with the hospital service to which the patient is admitted. Kessner and Lepper reported gram-negative infection rates per 1000 patients of 23.1, 20.6, and 15.0 for the neurosurgical, urologic, and gynecologic services, respectively, in their hospital. In contrast, the rates per 1000 patients on the plastic surgical, pediatric, and obstetric services were 4.4, 2.1, and 1.2, respectively.

It is apparent that the hospital environment contributes significantly to the incidence of bacterial infection in humans. Unfortunately, many of these organisms are resistant to antimicrobial agents. David Smith (1967e) reported that resistant bacteria caused 108/249 major bacterial infections in patients surveyed at the Children's Hospital Medical Center in Boston, between January 1 and April 30, 1967. He also emphasized that gonococci, meningococci, and tubercle bacilli were important causes of disease as well as enterobacteria and staphylococci. Among the resistant coliforms and *Klebsiella-Aerobacter* species, 80–90% owed their resistance to R factors. The percentage of R factor-mediated resistance in *Proteus* was lower but still significant. Very few resistant pseudomonads were shown to contain R factors; presumably their resistance was due to chromosomal mutations. More recently, Gardner and Smith (1968) reported on an unusually high incidence of hospital-acquired infections due to antibiotic-resistant *Klebsiella* on a general surgical service at the Massachusetts General Hospital. Here, patients were free of the resistant *Klebsiella* on admission but acquired multiply resistant organisms while patients. The isolates were predominantly of one serotype, however, and contained essentially identical multiple resistances. These authors concluded that it was highly unlikely that the R factor was acquired *in vivo* or that it was transmitted to other bacteria subsequently, but rather that the resistant *Klebsiella* itself was the infective organism and was responsible for the infections.

Thus, the hospital environment not only is a major reservoir of infectious bacteria, it is a major reservoir of drug-resistant bacteria of both the chromosomal and episomal types.

VI. The Agricultural Environment and Drug Resistance

Some conjecture has arisen regarding the possible role of enterobacteria, particularly *E. coli*, from animals fed antibiotics, as disseminators of infectious drug resistance to pathogens, particularly salmonellae (H. Smith, 1967). Certain strains of *E. coli* are pathogenic for man and animals, and while these seem to be largely species specific (Sojka, 1965) they can act as donors, transferring multiple drug resistance to recipients such as *Salmonella* which may be more pathogenic (H. Smith, 1967). Certainly the use of antibacterials in animal feeds exerts a selective pressure on sensitive pathogens and nonpathogens alike. H. Williams Smith has stated that the widespread use of antimicrobial drugs in animals has led to emergence of drug-resistant strains and that this now constitutes a definite complication to the satisfactory treatment of animal disease. In contrast, Kampelmacher has reported (1967) that mortality due to *Salmonella* in humans has remained essentially constant in the Netherlands over the period from 1957 to 1966 and that the use of drugs in animal feeds has not adversely affected therapy of salmonellosis and *E. coli* infections in animals.

The possibility that widespread use of antibiotics may have increased the incidence of or mortality from enteropathogenic *E. coli* (EEC) should be considered. That the reverse is true, i.e., that there was a decrease in infant mortality due to EEC during the period from 1948 to 1957, was shown by Taylor (1966). The mortality rate due to gastroenteritis, which includes EEC, in infants under 1 year in Great Britain dropped from 2.90 per 1000 live births in 1948 to 0.44 in 1956. This does not mean that EEC are no longer of importance, as emphasized by a recently reported outbreak of gastroenteritis in Great Britain in which, among others, *E. coli* serotype 0128 has been implicated. Strains of this serotype are known to be pathogenic for humans. They have been isolated from healthy human carriers, domestic animals, and rather exotic sources such as kangaroos and mutton birds.

Anderson and Lewis (1965) and Anderson (1968) have expressed concern that drug-resistant salmonellae gain access to the intestinal tract of animals in which "the normal fecal commensals" have been eliminated or reduced by the use of drugs to which the salmonellae are resistant. Walton (1966b) has reported on the isolation of drug-resistant *E. coli* from healthy farm animals in Great Britain. Resistant coliforms were found in animals on farms where antibiotics were in use. They were also found in healthy animals from farms which were not using antibiotics. With regard to the influence of these coliforms on salmonellosis in humans, Walton stated that while many human cases of infection with *Salmonella typhimurium* are probably caused by contaminated meat, the high incidence of drug-resistant coliforms in farm animals which might be expected to lead to an increase in the reservoir of drug-fast S. *typhimurium* has apparently caused no increase in the number of incidents of this type of food poisoning in man. As a matter of fact, the number of incidents of S. *typhimurium* in man decreased from about 3200 in 1955 to 1721 in 1965 (Walton, 1966b) in Great Britain. Schroeder *et al.* (1968) found that among 400 strains of *Salmonella* isolated during 1967 from clinical sources throughout the United States, only 22% were resistant to one or more drugs and that only 41 of these possessed the RTF. He concluded that there was no significant change in the incidence of resistant *Salmonella* during the past 5 years.

The major source of salmonellosis in humans appears to be contaminated foods. Convalescent and asymptomatic human carriers are important reservoirs for contamination of foods and also as direct sources of infection for other humans. Direct transmission of salmonellosis from animals to humans has been demonstrated in infections originating in pet turtles, extensively handled at home. Fish *et al.* (1967) reported a case of salmonellosis in a 12-year-old Canadian farm boy which appeared to be traceable to contact with an infected cow. *Salmonella typhimurium* strains of identical phage-type fermentation pattern and similar drug sensitivity patterns were isolated from both the boy and the cow. The boy was hospitalized for 4 days and was given neomycin orally. His course was uneventful and the child was asymptomatic at discharge.

In a study of drug resistance in Holland, Guinee (1967) reported an increase in isolation of resistant strains of *Salmonella panama* in humans. This resistance was always due to a transferable fi$^-$, phage restricting, R factor. Among the tetracycline-resistant *E. coli* strains found in human patients or in healthy animals being fed low levels of antibiotics, relatively few strains were found to harbor the same type of R factor as that which occurred in *S. panama*. The majority of these strains harbored fi$^+$ R factors or fi$^-$ factors which exerted another pattern of phage restriction than that observed with the *S. panama* factor. There was no evidence that the *in vitro* transfer rate of the *E. coli* fi$^+$ R factors was lower than that of the *S. panama* fi$^-$ R factors. The discrepancy between the type of R factor occurring in *S. panama* of human origin and the type found predominantly in *E. coli* indicates that no unlimited natural transfer of R factors from *E. coli* to *S. panama* had occurred.

Salmonellae are often found in cattle at slaughter. Brownlie and Grau (1967) obtained strains of *S. chester*, *S. typhimurium*, *S. oranien-*

berg, *S. muenchen, S. anatum,* and *S. adelaide* from the rumen of cattle at slaughter. These organisms were subcultured and then introduced into the rumen of healthy cattle to determine their ability to produce and multiply in competition with normal rumen microflora. It was consistently found that these salmonellae were rapidly eliminated from the rumen when the cattle were on full feed. Reduction of feed intake to one-third of normal or removal of feed for one or more days favored growth of *Salmonella* (and *E. coli*) in the rumen. A partial explanation for this might be that there is a bactericidal or bacteriostatic action of volatile fatty acids at low pH levels in the rumen of a fully fed animal. Brownlie and Grau suggest that conditions which obtain during the transport and holding of cattle before slaughter would appear to predispose them toward *Salmonella* and *E. coli* infection.

Currently, we are investigating the role which antibiotic-resistant organisms play in the ecology of domestic animals. We have worked with both chickens and swine.

We found that *E. coli* strains resistant to chlortetracycline are present in 1-day-old chickens which have never received antibiotics. Resistant organisms continued to be present in birds at 3, 6, and 9 weeks of age whether or not they received chlortetracycline in the feed for the first 3 weeks of life. Both groups appeared clinically normal. At least some of the resistant strains were capable of transferring this resistance, by conjugation, to a sensitive recipient strain of *Salmonella typhi*. Subsequent to the completion of this experiment we have learned that workers in the U. S. Food and Drug Administration Research Laboratories have also isolated antibiotic-resistant strains of *E. coli* from chickens which have never received antibiotics (Roegner and Pocurull, 1968).

Our earlier chicken experiments with *Salmonella gallinarum* (cf. p. 84) were conceived at a time when we were not aware of the possibility of infectious transfer of multiple resistance. In retrospect, if this phenomenon occurred at all, it did not produce a detectable change in the dynamics of the *S. gallinarum* infection employed. In theory, the continuous feeding of a combination of chlortetracycline and sulfaethoxypyridazine could exert selective pressure to promote growth of organisms with episomal resistance to both agents. The use of the combination should not have been successful in preventing development of resistance to those drugs if transfer of multiple resistance had occurred.

We have found *Salmonella choleraesuis* var. *kunzendorf* to be a useful organism for establishing experimental infections in swine (Culbreth and Messersmith, 1968). These infections are necessary to evaluate the prophylactic and therapeutic efficacy of selected chemotherapeutic agents. They have formed the basis for a model to study R factor transmission *in vivo* (Jarolmen and Kemp, 1969b). In a recent experiment, replicate groups of weanling swine (5/group) were infected with *S. choleraesuis* var. *kunzendorf* administered in the feed at one feeding. This procedure produced an acute progressive infection which closely resembled a very severe field outbreak of salmonellosis. The pigs used in this test were bred at our own facilities. They were not given antibiotics at any time; however, the sows from which they were farrowed were routinely maintained on a ration containing a prophylactic level (50 gm./ton) of chlortetracycline. The *Salmonella* culture employed was sensitive to chlortetracycline but resistant to streptomycin. The streptomycin resistance of this organism is not transferable by conjugation and appears to be of chromosomal origin.

Prior to infection and medication a total of 30 cultures were isolated from fecal samples of the pigs on the experiment. Of these strains, 18 were *E. coli* found to possess transferable R factors.

The *S. choleraesuis* var. *kunzendorf* strain RC221 used as an infecting organism was shown, *in vitro*, to be a competent recipient for R factors of *E. coli* origin. It was also demonstrated that the infecting strain could transfer the resistance factors it had acquired to a recipient *E. coli*.

A total of 217 *Salmonella* cultures was isolated from lung, liver, spleen, lymph nodes, intestine, and feces throughout the course of the experiment (4 weeks). These were screened for the presence of transferable R factors using selective media. Only one of the 217 cultures harbored a transferable R factor. However, this organism, which was isolated from an uninfected, untreated control pig, was serologically different from the infecting organism. It was also shown *in vitro* that the infecting *Salmonella* was a competent recipient for the *E. coli* R factors present in the swine coliforms. At the termination of the test, it was again determined that *E. coli* carrying transmissible R factors were still present in the pigs. Thus, although *S. choleraesuis* var. *kunzendorf* RC221 was a competent recipient of R factors *in vitro*, in no case was the infective organism shown to have picked up a transferable R factor. One obvious omission from this experiment was the lack of selective pressure from a drug to which the indigenous coliform fecal flora of the pigs were resistant.

VII. Summary

The microbial flora of man and animals is adapted to and in dynamic equilibrium with the host. This equilibrium can be altered by the intervention of a bacterial pathogen which has evolved in a way to permit it to invade and multiply in the animals, displacing the so-called "normal" flora. This maximally efficient pathogen may be suppressed by an effective antimicrobial drug, but mutants of the pathogen may emerge which are resistant. These resistant mutants may be less well adapted to survive in the host in the presence of the normal bacterial flora.

The discovery that resistance to antibacterial drugs may be transferred by extrachromosomal genetic material, while initially alarming, became less so when it was shown that the episomes containing the resistance factor had existed in bacteria before the commercialization of antibiotics. Infectious transfer of resistance must, therefore, have been possible since the earliest use of these drugs. This concern was further reduced by the fact that mortality from shigellosis in Japan decreased dramatically from 1949 to 1960, though infectious resistance was known to be occurring at that time. Infant mortality in Great Britain due to enteropathogenic *E. coli* decreased almost to the vanishing point from 1948 to 1957, a period when the use of antibiotics was increasing very rapidly, and the incidents of human illness due to *S. typhimurium* decreased by about half between 1955 and 1965, at a time when antibiotic usage was at a maximum. There was a dramatic increase in the number of tetracycline-resistant *Salmonellae* isolated in the Netherlands between 1959 and 1966 but no corresponding increase in morbidity or mortality in humans due to salmonellosis. It thus seems most improbable that the widespread use of antibiotics, and particularly their use in animal feeds, has created any public health problem.

Finally, antibiotics continue to be highly effective, even after 10 or more years of use in single locations where, if the appearance of resistance caused a loss of activity, they should have become ineffective.

Note Added in Proof

Two additional pieces of work concerning transferable resistance have come to our attention since the preparation of this manuscript. Watanabe (1969) has found that the introduction of R factors into *S. typhimurium* LT-2 results, without exceptions, in a reduction of the mouse virulence of these microorganisms. H. W. Smith (1969) found that R-factor containing *E. coli* from meat animals were incapable of permanently colonizing the alimentary tract of a human—even when administered at a concentration of a billion cells per day for a week; further, that when an R factor from one of these donors was transferred to an *E. coli* resident in the human, it was unable to persist in competition with its non-R-factor-containing counterparts.

REFERENCES

Anderson, E. S. (1968). *Ann. Rev. Microbiol.* **22**, 131–180.
Anderson, E. S., and Lewis, M. J. (1965). *Nature* **206**, 579–583.
Barrett, F. F., Casey, J. I., and Finland, M. (1968). *New Engl. J. Med.* **278**, 5–9.
Braude, A. I., and Siemienski, I. C. (1968). *J. Clin. Invest.* **47**, 1763–1773.
Braun, W. (1965). "Bacterial Genetics," 2nd ed., p. 82. Saunders, Philadelphia, Pennsylvania.
Brownlie, L. E., and Grau, F. H. (1967). *J. Gen. Microbiol.* **46**, 125–134.
Culbreth, W., and Messersmith, R. E. (1968). *Bacteriol. Proc.*, p. 92.
Dowling, H. F. (1965). *Am. J. Med.* **39**, 796–803.
Falkow, S. (1965). *Am. J. Med.* **39**, 753–765.
Falkow, S. (1967). "The Molecular Nature of R Factors," paper presented Intern. Symp. Infectious Multiple Drug Resistance, Georgetown University, Washington, D.C., May, 1967.
Farrar, W. E., Jr., and Kent, T. H. (1965). *Am. J. Pathol.* **47**, 629–642.
Farrar, W. E., Jr., Kent, T. H., and Elliott, V. B. (1966). *J. Bacteriol.* **92**, 496–501.
Fish, N. A., Finlayson, M. C., and Carere, R. P. (1967). *Can. Med. Assoc. J.* **96**, 1163–1165.
Franklin, T. J. (1963). *Biochem. J.* **87**, 449–453.
Franklin, T. J. (1967). *Biochem. J.* **105**, 371–378.
Gale, G. O., and Hall, R. H. (1959). *Antibiot. Ann.*, pp. 1040–1046.
Gale, G. O., Kiser, J. S., and McNamara, T. F. (1963). *Avian Diseases* **7**, 457–466.
Gardner, P., and Smith, D. H. (1968). *Abstr. 8th Intersci. Conf. Antimicrobial Agents and Chemotherapy*, Am. Soc. Microbiol., Ann Arbor, Michigan. p. 76.
Garside, J. S., Gordon, R. F., and Tucker, J. F. (1960). *Res. Vet. Sci.* **1**, 184–199.
Goldberg, I. M. (1965). *Am. J. Med.* **39**, 722–752.
Gottlieb, D., and Shaw, P. D. (eds.) (1967). "Antibiotics I. Mechanisms of Action." Springer-Verlag, New York.
Guinee, P. A. M. (1967). "Veterinary Aspects of Infectious Multiple Drug Resistance," paper presented Intern. Symp. Infectious Multiple Drug Resistance, Georgetown University, Washington, D.C., May, 1967.
Han, T., Sokal, J. E., and Neter, E. (1967). *New Engl. J. Med.* **276**, 1045–1052.
Izaki, K., Kiuchi, K., and Arima, K. (1966). *J. Bacteriol.* **91**, 628–633.
Jarolmen, H., and Kemp, G. A. (1969a). *J. Bacteriol.* **97**, 962–963.
Jarolmen, H., and Kemp, G. (1969b). *J. Bacteriol.* **99**, 487–490.
Kabins, S. A., and Cohen, S. (1966). *New Engl. J. Med.* **275**, 248–252.
Kampelmacher, E. H. (1967). "Foods of Animal Origin as a Vehicle for Transmission of Drug-Resistant Organisms to Animals and Man," paper presented NRC-NAS Symp. on The Use of Drugs in Animal Feeds, Washington, D.C., June, 1967.
Kessner, D. M., and Lepper, M. H. (1967). *Am. J. Epidemiol.* **85**, 45–60.
Lawn, A. M., Meynell, E., Meynell, G. G., and Datta, N. (1967). *Nature* **216**, 343–346.
Lechevalier, H. A., and Lechevalier, M. P. (1967). *Ann. Rev. Microbiol.* **21**, 89–90.

Meynell, E., and Datta, N. (1966a). *Genet. Res. (Cambridge)* **7**, 134–140.
Meynell, E., and Datta, N. (1966b). *Genet. Res. (Cambridge)* **7**, 141–148.
Meynell, E., and Datta, N. (1967). *Nature* **214**, 885–887.
Maynell, G. G., and Lawn, A. M. (1968). *Nature* **217**, 1184–1186.
Oginsky, E. L., and Umbreit, W. (1959). "An Introduction to Bacterial Physiology," 2nd ed., p. 146. Freeman, San Francisco, California.
Okamoto, S., and Suzuki, Y. (1965). *Nature* **208**, 1301–1303.
Pearce, L. E., and Meynell, E. (1968). *J. Gen. Microbiol.* **50**, 173–176.
Pollock, M. R. (1967). *Brit. Med. J.* **4**, 71–77.
Roegner, F. R., and Pocurull, D. (1968). *Bacteriol. Proc.* p. 84.
Rogers, D. E. (1959). *New Engl. J. Med.* **261**, 677–683.
Rownd, R. (1967). "The Molecular Nature and Control of the Replication of R Factors," paper presented Intern. Symp. on Infectious Multiple Drug Resistance, Georgetown University, Washington, D.C., May, 1967.
Rownd, R., Nakaya, R., and Nakamura, A. (1966). *J. Mol. Biol.* **17**, 376–393.
Salzman, T. C., and Klemm, L. (1968). *Proc. Soc. Exptl. Biol. Med.* **128**, 392–394.
Schnitzer, R. J., and Grunberg, E. (1959). "Drug Resistance in Microorganisms," pp. 158–174. Academic Press, New York
Schroeder, S. A., Terry, P. M., and Bennett, J. V. (1968). *J. Am. Med. Assoc.* **205**, 87–90.
Sebek, O. (1967). *In* "Antibiotics I. Mechanism of Action" (C. D. Gottlieb and P. D. Shaw, eds.), p. 750. Springer-Verlag, New York.
Shaw, W. V. (1967). *J. Biol. Chem.* **242**, 687–693.
Shmidov, P. N. (1965). *Veterinarya* **7**, 129–136.
Smith, D. H. (1967a). *J. Bacteriol.* **94**, 2071–2072.
Smith, D. H. (1967b). *Science* **156**, 1114.
Smith, D. H. (1967c). *Lancet* **i**, 252–254.
Smith, D. H. (1967d). *In* "Antimicrobial Agents and Chemotherapy – 1966" (G. Hobby, ed.), 274–280. Am. Soc. Microbiol., Ann Arbor, Michigan.
Smith, D. H. (1967e). "The Influence of Drug-Resistant Organisms on the Health of Man," paper presented NRC-NAS Symp. on The Use of Drugs in Animal Feeds, Washington, D.C., June, 1967.
Smith, H. W. (1967). *Vet. Record* **80**, 464–469.
Smith, H. W. (1969). *Lancet* **i**, 1174–1176.
Sojka, W. J. (1965). "*Escherichia coli* in Domestic Animals and Poultry," pp. 65–183. Commonwealth Agricultural Bureaux, Farnham Royal, Bucks.
Spizizen, J. B., Reilly, E., and Evans, A. H. (1966). *Ann. Rev. Microbiol.* **20**, 391.
Srb, A. M., Owen, R. D., and Edgar, R. S. (1965). "General Genetics," pp. 243–244. Freeman, San Francisco, California.
Strominger, J. L., and Tipper, D. J. (1965). *Am. J. Med.* **39**, 708–721.
Suzuki, Y., and Okamoto, S. (1967). *J. Biol. Chem.* **242**, 4722–4730.
Taylor, J. (1966). *J. Appl. Bacteriol.* **29**, 1–12.
Terewaki, Y., Takayasu, H., and Akiba, T. (1967). *J. Bacteriol.* **94**, 687–690.
Unowsky, J., and Rachmeler, M. (1966). *J. Bacteriol.* **92**, 358–365.
Walton, J. R. (1966a). *Nature* **212**, 312–314.
Walton, J. R. (1966b). *Lancet* **ii**, 1300–1302.
Watanabe, T. (1963). *Bacteriol. Rev.* **27**, 87–115.
Watanabe, T. (1969). Personal communication.
Zamenhof, S., and Eichhorn, H. H. (1967). *Bacteriol. Proc.*, p. 59.
Zinder, N. D., and Lederberg, J. (1952). *J. Bacteriol.* **64**, 679–699.

Micromonospora Taxonomy

GEORGE LUEDEMANN

Department of Microbiology,
Schering Corporation,
Bloomfield, New Jersey

I.	Introduction	101
	A. Some Thoughts on the Problem	101
	B. The Unnatural Isolate	102
	C. Objectives and Limits	103
II.	Review of the Taxonomic Literature	103
	A. The Rare Actinomycetes	103
	B. *Streptothrix chalcea*	104
	C. Jensen's Soil Isolates	104
	D. *Actinomyces gallicus*	106
	E. The Russian Species	107
	F. The Aquatic Habitat	108
	G. *Micromonospora* sp. ATCC 10026 and Micromonosporin	109
	H. *Actinomyces caballi*	110
	I. The Anaerobic *Micromonospora*	111
	J. The Microcins and Actinomycin	112
	K. T'ao Thesis	112
	L. *Micromonospora melanosporea*	113
	M. *Micromonospora echinospora*	114
	N. *Micromonospora narashinoensis*	120
III.	Relationships	122
	A. Relationships within the Genus	122
	B. Relationships to Other Actinomycete Genera	123
IV.	Impressions and Reflections	126
	A. Background	126
	B. Early Scheme for Differentiation	126
	C. Traditional Taxonomic Methods	127
	D. Transfer to Transfer Variation	128
	E. A Method for Studying Morphology	129
	References	131

I. Introduction

A. SOME THOUGHTS ON THE PROBLEM

A biological classification is the systematic arrangement of organisms into groups or categories based upon some definite scheme. Taxonomy is a classification of organisms according to their natural relationships (Simmons, 1966). We know nothing of the natural relationships of *Micromonospora* species, the fact is we are having a great deal of difficulty in trying to determine what a species of *Micromonospora* is.

Is a species concept in microorganisms a macromyth (Cowan, 1962)? Whether a species concept is fact or fiction (it is neither, but a little bit of each) we humans are content only when we can name and classify things. If we define thing literally, as an inanimate object, we find the chemist able to pursue his science with a sense of exactness, he can characterize and measure his elements and compounds. However, the biologist is not dealing with things but with lesser beings capable of growth, adaptability, mutation, and a profound inherent ability to confuse those who would attempt to make them a thing, a definable and static entity.

A species has been defined as a group having certain common and permanent characteristics which clearly distinguish it from other groups. In theory we attempt to label a microorganism with a finite name and a set of attributable characteristics; in practice we assign name and characteristics to a microbial population and hope that the population will behave according to the limits we have assigned.

In characterizing a microorganism we search for constants by which to identify similar organisms found in nature. Morphology appears to be the primary characteristic agreed upon by most actinomycete authorities (Waksman, 1957; Pridham, 1959; Kriss, 1939; Ørskov, 1923; Baldacci *et al.*, 1953; Krasilńikov, 1960; Lechevalier, 1964). Actinomycetes bear a resemblance to filamentous fungi and lend themselves to morphologic characterization. They are, as Lechevalier (1964) has said "... bacteria with fungal morphology." The physiologic processes result under amenable conditions in reproduction by means of characteristic morphologic units. While an actinomycete may temporarily lose the ability to reproduce by identifiable morphologic units, it does not mutate or adopt characteristic reproductive units of another organism. Cultural, physiological, and biochemical properties are used as secondary characteristics, and require a consideration of physiological adaptation and degeneration. On the other hand, morphology may play a secondary role where we are dealing with nonfilamentous organisms similar in size and shape such as budding yeasts or fission bacteria. In such cases of morphologic similarity, physiological and biochemical properties attain prime importance.

B. The Unnatural Isolate

When we isolate a single colony of a micromonospora on a dilution plate and transfer it to the "germfree" environment of an agar slant or petri dish we may find that the culture requires time in which adaptation to the new environment occurs. The organism often requires a number of months before a vigorous, stable adaptant is secured (Kriss, 1939). It appears that with some cultures we not only have to await the appearance of a vigorously growing sport (variant) but one which is resistant to lysis perhaps due to the transmission of phage or other unknown lethal or debilitating factors. This initial period of stabilization or selection, which is necessary in some isolates, is important because often physiological data collected after only a few subcultures may not be representative of fully domesticated strains. Once a culture has adapted to laboratory conditions physiological tests can be performed and reproducible data obtained. In this respect it is comforting to find in our laboratory that cultures maintained under lyophilization for 10 years are constant in their response year after year to the series of tests performed with them.

C. Objectives and Limits

The genus *Micromonospora* has been clearly described by Ørskov (1923) but the species have most often been described with few clear or significant characteristics useful for comparative purposes. The objective of this review is to acquaint the reader with the taxonomic literature or in some instances ecological studies which help establish the place of the micromonosporae in the balance of nature. Much of the literature on micromonosporae has appeared in publications generally difficult to obtain or requiring translation. I am appreciative of the diligence with which the library staff tracked down many of these journals.

The review will be limited to discussing mesophilic species of *Micromonospora* and not concern itself with thermophilic forms which are often placed in other genera (Küster and Locci, 1964). For general discussions of the genus *Micromonospora* the reader is referred to Waksman's texts on the actinomycetes and other reviews (Waksman, 1950, 1959, 1961, 1967; Waksman and Lechevalier, 1962), Erikson (1949), and Lechevalier and Lechevalier (1967).

II. Review of the Taxonomic Literature

A. The Rare Actinomycetes

H. L. Jensen (1930) prophetically entitled a paper "The Genus *Micromonospora* Ørskov, a Little Known Group of Soil Microorganisms." The group is still little known and characterization of many good species probably still await description. It is of interest that specialists familiar with *Streptomyces* species have often overlooked *Micromonospora*. Baldacci *et al.* (1953) mentioned that he had not been able to find in more than 2000 isolations from the soil any strains of micromonosporae. That micromonosporae rarely occur is a myth due to insufficient incubation of isolation plates and unfamiliarity with what to look for (Umbreit and McCoy, 1941); to this might be added inability to maintain the culture once it has been isolated (Kriss, 1939). The cosmopolitan nature of the micromonosporae has been demonstrated by Lechevalier (1964) in a table entitled "Distribution of the 'Rare' Forms of Actinomycetes in 16 Soil Samples." Of 132 isolates 70 were micromonosporae and 62 represented the remaining 6 genera.

B. *Streptothrix chalcea*

The genus *Micromonospora* was proposed by Ørskov (1923) based upon a microscopic study of the morphology of an organism obtained from the Kràl Collection (Vienna). This organism had been isolated from the air by Foulerton (1905) and described as *Streptothrix chalcea*. Foulerton did not describe the microscopic morphology of this organism. Foulerton described his culture of *Streptothrix chalcea* using essentially bacteriological techniques of macroscopic colony description and physiological reactions and Ørskov used the mycological technique of studying the morphologic development of the spore-bearing structure. However, a combination of both descriptions is not adequate to identify the large number of micromonosporae which could be comfortably accommodated by the criteria of these two authors.

C. Jensen's Soil Isolates

H. L. Jensen (1932) was the first to discover a natural habitat for micromonosporae. He isolated 67 strains of *Micromonospora* from Australian soils and of these, 60 were classified as *Micromonospora chalcea*.[1] The remaining 7 isolates were divided into three new species. Jensen characterized *Micromonospora parva* by its scant

---

[1] Technically Jensen (1932) made the new combination *Micromonospora chalcea* (Foulerton) however, Ørskov (1923) made his intent sufficiently clear in proposing the genus *Micromonospora* based upon his observations of *Streptothrix chalcea* to credit him with the new combination.

growth and sporulation, pink to orange colored vegetative mycelium, failure to invert sucrose, and relative scarcity, only 3 strains having been isolated. *Micromonospora fusca* was characterized by heavy, orange vegetative growth rapidly changing in color to deep brown and then nearly black. Illustrations of spore clusters characteristic of this isolate appear in an earlier publication (Jensen, 1930). A significant characteristic of the culture was believed to be the production of a dark brown to nearly black soluble pigment on agar media. This soluble pigment was believed to differ from that seen in streptomycetes because very little of this pigment was produced in gelatin or milk, media in which the chromogenic streptomycetes were found to produce an abundance of dark diffusible pigment. Sucrose was inverted, this being determined by a reduction of Fehling's solution. Only 2 of the 67 strains produced such a diffusible pigment and one concludes that it is a rather rarely encountered organism. I have encountered a small number of dark brown to black diffusible pigment producing micromonosporae which are suggestive of Jensen's *Micromonospora fusca*, they occur with a frequency similar to that noted by Jensen. Cultures designated *Micromonospora fusca* present in the CBS and NRRL culture collections, are considered to fall short of Jensen's original description and more will be said about this further on.

The last of Jensen's proposed species was called *Micromonospora coerulea*. This culture was characterized by slowly developing colonies which produced a dense vegetative mycelium of a deep greenish blue, equivalent to "dusky green blue" of Ridgeway (Jensen, 1932). After 25 days the mycelium became nearly black. Sporulation was very scant and sometimes the colony surface was covered with a thin grayish veil resembling an aerial mycelium but devoid of spores (Jensen, 1930). Sucrose was not inverted and growth did not occur on potato. Two of the 67 strains of *Micromonospora* were assigned to this species. I have seen only two soil isolates of *Micromonospora* fitting the dusky green-blue Ridgeway color. This color is unusual and I suspect these isolates are similar to the *Micromonospora coerulea* isolates of Jensen. These soil isolates grow on potato and utilize sucrose, and should, in my estimation, be described possibly as subspecies of *M. coerulea*. Rarely can color be used as a primary characteristic (Kriss, 1939) but in this particular color case the characteristic seems unusual enough to make an identification. Color variants of some species of *Micromonospora* have been obtained (Ilavsky, 1968), however none of the color variants approach the dark greenish blue color of these coerulea-like isolates. The illustration (Jensen, 1932) of the spore bearing structures and spores of *M. coerulea* is similar to those observed in my blue-green isolates. Krasilńikov (1941) describes the only other green (green and black) colored *Micromonospora, M. bicolor*. This organism grew well on synthetic media utilizing ammonium salts and nitrates but no growth occurred on protein media, potato, gelatin, or milk.

Jensen's criteria for speciation were not considered relevant by Kriss (1939) but I belive that his characterizations of *Micromonospora fusca* and *M. coerulea* have sufficient merit to be able to attempt to identify certain soil isolates with his descriptions. Unfortunately, *Micromonospora parva* does not seem to be founded upon much more than poor growth and poor sporulation, neither character is particularly unusual for such organisms.

D. *Actinomyces gallicus*

Erikson (1935) described an organism which she named *Actinomyces gallicus*. Sporulation in this organism was noted to be similar to *Micromonospora chalcea* as illustrated by Ørskov (1923) but a limited segmentation of the substrate mycelium and a few aerial hyphae were sufficient, she thought, to prevent it being placed in the genus *Micromonospora*. The organism studied by Erikson had been isolated by Gibson from a blood culture in a case of Banti's disease and bore the NCTC strain number 4582 (Lister Institute). The organism was listed in the 6th Edition of Bergey's Manual (1948) as *Streptomyces gallicus*. Erikson refers twice to this culture in later papers, in 1941 stating that *Actinomyces gallicus* should probably be regarded as a micromonospora of slow growth with scant lateral spore formation, resembling *Micromonospora parva*. Erikson (1949) again attempted to correct the nomenclature of this organism in the following statement: "The reviewer cannot let pass this opportunity of remarking that in erecting a new species for a micromonospora-like organism, on the grounds of partial segmentation of the vegetative mycelium and the infrequent production of spores, Erikson erred. This was later admitted, when the writer had occasion to study a number of undoubted micromonosporae. The organism having now reached sanctuary under the name *S. gallicus* among the streptomycetes in Bergey's Manual (1948), it has been thought advisable to mention the matter again" (see Waksman, 1967). It is of interest that Erikson noted segmentation of the mycelium and chose to call the culture *Actinomyces*. Shidara (1955) made a similar mistake on the basis of segmented mycelia and coccoid bodies and proposed the name *Nocardia narashinoensis* for an organism found to produce an antibiotic named nocardorubin (Aiso *et al.*, 1954). This organism was reclassified as *Micromonospora narashinoensis* by Arai and Kuroda (1965).

Whether *Micromonospora gallica* is a distinct species from *M. parva* can not easily be ascertained from the meager and negative data accompanying the description of each of these species.

E. THE RUSSIAN SPECIES

The problems of nomenclature may be appreciated in the case of *Micromonospora globosa*. N. A. Krasilńikov (1938) published a book on the Actinomycetes in which (p. 134) he illustrates by drawings "*Micromonospora globosa* n.sp." He does not describe *M. globosa* but mentions on pp. 132 and 133 that this organism had been studied by A. E. Kriss and the work was (or was to be) published by Kriss. Kriss (1939) published his data on this species stating that 8 strains of micromonosporae that he had isolated from soils of the Saratov region all belonged to one species, the only differences noted were in the rapidity of development or in some physiological reactions. Kriss states: "On the ground of its characteristic property to form spheric conidia, I give this species the name: *Micromonospora globosa* n.sp." The publication of Kriss' revealed an intimate knowledge of the isolation and cultural problems presented by micromonospora isolates and also a knowledge of the problems of attempting to characterize isolates in terms of their pigmentation. In spite of the excellent observations made by Kriss the ball-shaped spores used to characterize this species appear of little diagnostic aid in attempting to separate this organism from most other micromonosporae which also produce spherical conidia.

N. A. Krasilńikov (1941) described *Micromonospora globosa* n.sp. and used the same illustrations which appear in his 1938 work and which also appear in the publication of A. E. Kriss (1939). Krasilńikov, in 1949, again referred to *Micromonospora globosa* Krasilńikov 1941. In a strict interpretation of the rules of nomenclature N. A. Krasilńikov's 1938 publication date is the earliest record of the name *Micromonospora globosa* (which I have found) with illustrations used to characterize the species. No description accompanies this name and the work of Kriss on this organism is mentioned. Obviously *Micromonospora globosa* Krasilńikov 1941 cannot be considered as a publication date. A nomenclatorial decision should be made as to the author and date to be accepted for *Micromonospora globosa*. A possibility is to list it as follows if a strict interpretation of the law of priority is upheld: *Micromonospora globosa* Krasilńikov 1938, 134; emend, Kriss 1939, 179.

Krasilńikov (1941) described three new species of *Micromonospora*. *Micromonospora bicolor* appeared to be similar in its green color to *M. coerulea* Jensen differing from this species in the production of a dark brown to black layer of conidia and poor growth on protein media, while good growth was obtained on synthetic media. *Micromonospora elongata* was characterized by poor growth on common nutrient media and no growth on potato, gelatin, or milk. Conidia were oval and produced on short conidiophores. Colonies on nutrient media were gray brown or straw colored and poorly developed. *Micromonospora globosa* was listed as a new species but since it was previously described by Kriss (1939) and referred to by Krasilńikov (1938) this publication cannot be considered the original reference date for this species.

F. The Aquatic Habitat

The publication of Umbreit and McCoy (1941) did not add any new species of *Micromonospora* for the taxonomist's consideration but the study of the microbial ecology of Wisconsin lake and lake bottom samples is important for its discovery of the high proportion of micromonosporae present in this environment. The relationship of the micromonosporae to other bacterial chromogens and a seasonal periodicity were other interesting factors uncovered by this study. Umbreit and McCoy concluded that the predominance in aquatic environments of micromonosporae (10–50% of total microbial population) together with the lack of other forms of actinomycetes, "makes it seem probable that these organisms are an integral part of the population of lake bottom deposits." These authors believed that micromonosporae were a distinct minority in soils and comprised but the merest fraction of the microbial population there. Micromonosporae were believed to act as mineralizers of lake bottom deposits similar to the role of the streptomycetes on land. Umbreit and McCoy made no taxonomic commitments in regard to the identity of their aquatic isolates stating that the study of this group had been neglected.

Colmer and McCoy (1943) continued this study on the relationship of micromonosporae to the microbial populations of Wisconsin lakes. These authors refer to a study by Harden (1941) entitled "Studies on the Micromonosporae" which I have located through Dr. McCoy but have not as yet seen (the University of Wisconsin Library no longer has this thesis on file, personal communication, 1964).

Erikson (1941) studied ten strains of micromonosporae from Wisconsin lake muds and concluded that although considerable individual variation was found among these strains they all essentially agreed in cultural characteristics with the large species group considered to be represented by *Micromonospora chalcea*. Erikson found that her strains of micromonosporae grew on a number of natural substrates such as cellulose, lignin, and chitin. Such sugars and sugar alcohols as sorbose, melezitose, rhamnose, heptose, erythritol, quercitol, dulcitol, and *d*-arabitol were also utilized. Erikson noted that a soluble red pigment was formed in tyrosine media which indicated that these micromonosporae were able to produce a precursor to melanin. Erikson also found that they would grow under reduced oxygen tension, the growth being poor and very slow but this indicated that the organisms could survive under the low oxygen tensions found at the bottom of a lake. Erikson observed that single terminal spores were most frequently produced but that any strain in the same microscope field might show two, three, or more spores in close juxtaposition or in botrytis-like clusters. Fragmentation of the hyphae did not occur in young cultures. Branching of the hyphae was essentially monopodial but a dichotomous appearance was occasionally observed. Sparse aerial hyphae were only rarely noted under unfavorable culture conditions. On Czapek's sucrose agar colonies developed as dense compact structures with sporulation developing from the center outward. Under cultivation in liquid media colonies were small and poorly sporulating. Erikson sectioned colonies growing on agar and noted that sporulation occurred not only on the surface of the colony but in patches even below the surface of the colony. She concluded from a study of these sections that a definite colonial organization characterized by zonation of sporulation was present.

G. *Micromonospora* SP. ATCC 10026 AND MICROMONOSPORIN

Welsch (1942) reported antibiotic activity in a culture of *Micromonospora* isolated from lake mud. Waksman *et al.* (1947) isolated this antibiotic which was named micromonosporin and briefly characterized its properties. The organism was deposited in the ATCC as *Micromonospora* sp. 10026 (Rutgers 3450). Recent literature (Waksman, 1961) cites *Micromonospora fusca* as the producer of the antibiotic micromonosporin. G. A. de Vries of the Centraalbureau voor Schimmelcultures has traced the authors making this identification (personal communication, 1968) to Pridham and Gottlieb (1948) who in a table of actinomycetes list culture, source, and identity. Two micromonospora cultures IMRU 3450 and 3451 are identified as *Micromonospora fusca*. In 1954 *Micromonospora* sp. (3450) was listed in the CBS catalog as *Micromonospora fusca*. de Vries believes this was probably on the basis of the Pridham and Gottlieb paper. I believe this was an unfortunate identification because *Micromonospora* sp. ATCC 10026 falls short of the characterization of the type species described by H. L. Jensen (1932) for the following reasons. *Micromonospora fusca* was characterized by the production of a dark brown to brownish black diffusible pigment on agar media. *Micromonospora* sp. ATCC 10026 does not produce a dark brown diffusible pigment but on a number of agar media produces a characteristic bright yellow diffusible pigment which is fluorescent under ultraviolet light, becoming a more intense yellow. *Micromonospora* sp. ATCC 10026 was isolated from an aquatic environment. I have frequently isolated a similar culture from aquatic samples and found its significant characteristics of mannitol utilization, chromatographically similar antibiotics and morphology to agree with *Micromonospora* sp. ATCC 10026. Jensen's *Micromonospora fusca* was isolated from Australian soil and comprised only two strains of 67 micromonosporae isolated. Infrequently we have isolated a number of micromonosporae from soil which are characterized by the production of a brown to brownish black soluble pigment and a characteristic glycerol utilization. I believe these isolates might better serve as a neotype for *Micromonospora fusca* Jensen. Shull (1959) in a U. S. patent on steroid transformations identified *Micromonospora* sp. ATCC 10026 as a strain of *Micromonospora chalcea*. I believe Shull could accommodate his soil isolate, *Micromonospora chalcea* ATCC 12452, within the species description given for *Micromonospora chalcea* but I cannot agree that *Micromonospora* sp. ATCC 10026 should also be included in this species. Both of these organisms were compared to each other by Luedemann and Brodsky (1964) and appear to be recognizably different in respect to morphology and physiology.

H. *Actinomyces caballi*

Morquer and Comby (1943) described an organism thought to be a micromonospora which was isolated from a cutaneous actinomycosis of a horse. This organism produced aerial hyphae in which short chains of microarthrospores were produced. Lateral spores were produced which were 0.6–0.8 μm in diameter and were released by a gelatinization (lysis?) of the hyphae. In this paper the photomicrograph is labeled *Actinomyces caballi;* however, the authors believed this organism had affinity with the genus *Micromonospora* due to the production of lateral monospores and described the organism as *Micromonospora caballi.* This publication concerns an organism I believe to be similar to the case of *Actinomyces odontolyticus* mentioned by Bissett (1957) in an article entitled: "Some Observations upon the Mode of Sporulation and Relationship of Monosporous Actinomycetes." Bissett says "The appearance of the immature and maturing spores of *Micromonospora chalcea* and *Actinomyces odontolyticus* suggests that these develop in the same way as enlargements upon an existing filament, and not by outgrowth from the tip." Bissett believed micromonosporae exhibited a specialized case of spore formation while *Actinomyces odontolyticus* represented a degenerate parasitic form of development. Bisset's photomicrographs of monosporous development in *Actinomyces* and *Streptomyces* point out some of the pitfalls of interpreting the production of monospores alone as a generic characteristic. Until *Micromonospora caballi* can be redescribed with a few more micromonosporoid characteristics it might better be considered a doubtful species in respect to generic placement.

I. The Anaerobic *Micromonospora*

Hungate (1946) described *Micromonospora propionici,* an anaerobic micromonospora isolated from the intestinal tract of a worker termite. The colonies appeared white and no colored pigments were formed. The organism was an obligate anaerobe and the fermentation products of glucose or cellulose were CO_2, propionic acid, and acetic acid. Hungate believed glucose was the chief product of cellulose digestion. This is an interesting ecological niche for micromonosporae capable of cellulose digestion. Aerobic micromonosporae have been found in paper mill systems by Wang (1965) and diseased cotton bolls (J. Peterson, personal communication, 1963).

Sebald and Prévot (1962) confirmed Hungate's discovery of the presence of micromonosporae in the intestinal tract of termites. Their isolates were obtained from the termite *Reticulitermes lucifugus* var. *saintonnensis,* a different genus and species from the one Hungate made his isolations *(Amitermes minimus).* Sebald and Prévolt called their species *Micromonospora acetoformici.* This organism was anaerobic, the colonies were white to ochre in color, but failed to digest cellulose and produced formic and lactic acids. An observation made by these authors deserving further investigation was that CaCO$_3$ appeared necessary for growth and might constitute the essential source of carbon. Whether these two anaerobic species producing monospores are taxonomically related to (have natural affinities with) the aerobic species or merely fit into this form genus should be an interesting problem for further investigation.

J. THE MICROCINS AND ACTINOMYCIN

Taira and Fujii (1951) reported two antibiotics, microcin A and B, from a soil isolate believed to be similar to *Micromonospora fusca*. The culture on glucose-asparagine agar first appeared colorless but as the mycelium developed it became a light yellow which later turned gray. The center of the colony was raised and cream colored, its perimeter was slightly raised and tinged with gray; the circumference of the colony was flat and white. At the tips of short mycelial-like branches were observed spores. On glucose-peptone agar at 28°C. there slowly developed a white aerobic (aerial?) mycelium. On the colony reverse a purplish-brown soluble pigment was formed. On potato plug the colony was wrinkled and white developing a grayish white aerobic mycelium. The reverse was at first yellowish brown, later becoming dark brownish-purple. Gelatin was not liquefied. The description of this culture by Taira and Fujii in respect to the white to grayish colored colony having white aerobic mycelium and its failure to liquefy gelatin leads one to suspect that it may not be a micromonospora just as species of *Actinobifida* are not *Micromonospora* even though their spores are produced singly on short (antlerlike) aerial hyphae. The culture, as the authors mention, needs further investigation.

Fisher *et al.* (1951) reported an actinomycin produced from a strain of micromonospora isolated from a Costa Rican soil sample and believed to be similar to *Micromonospora globosa*. The organism produced a yellow to orange, waxy colony composed of slender mycelia 0.7 μm in diameter. Spores were spherical, 1.4 μm in diameter, and were produced on short conidiophores in grapelike clusters.

K. T'AO THESIS

One of the most detailed taxonomic studies made on isolates of micromonosporae is recorded in the masters thesis of T'ao (1958). Many of the problems and conclusions concerning micromonospora isolates that T'ao discusses are familiar to those who have been frustrated in their attempts to identify micromonosporae. I am sympathetic with her conclusions which I believe sum up quite adequately the species identification problem in this genus when she says "Micromonosporae are difficult to identify because of their delicacy of structure, the slight morphological differences between species and the variation found in physiological reactions. In carrying out the identification of an unknown micromonospora it is not possible to use the descriptive literature solely as a basis of identification, as the literature is inconsistent and incomplete. The taxonomic criteria found now in the literature on micromonosporae are confusing and inadequate." T'ao classified 40 isolates of micromonosporae, mostly obtained from the waters of Flathead Lake by Potter and Baker (1956), with five known species of *Micromonospora* on the basis of comparative morphology and physiology. I would be critical of T'ao's acceptance of the CBS *Micromonospora fusca* as Jensen's isolate (it appears it is Waksman's 1948 deposit of Rutgers 3450, ATCC 10026) and her identification of a mesophilic culture occasionally producing a white aerial mycelium but otherwise pink to yellow nonaerial mycelium as *Micromonospora vulgaris*. Erikson (1953) has made a sharp distinction between the mesophilic and thermophilic micromonosporae and has commented upon the occasional appearance of a sparse white aerial mycelium in mesophilic forms. I believe T'ao had a point when she differentiated *Micromonospora fusca* from *Micromonospora chalcea* morphologically by the botryoid (clustered) form of mycelium in the former and the long, slightly curved mycelium in the latter. T'ao also made some progress with a diagnostic carbohydrate utilization method; however, since she used an inorganic nitrogen source in some instances her dextroxe positive control was poor and comparative results are difficult to evaluate.

L. *Micromonospora melanosporea*

Baldacci and Locci (1961) made a new combination and subspecies, *Micromonospora melanosporea* (Krainsky 1914) Baldacci and Locci 1961, 19 and *M. melanosporea* subsp. *corymbica* Baldacci and Locci 1961, 19. This new combination is of interest for several reasons. Baldacci (1958) has used color of aerial and vegetative mycelium extensively in his keys to the *Actinomyces (Streptomyces)*. Baldacci (1938) studied a culture isolated from the blood of a woman affected with a mycetoma of the knee. He identified this culture with one of Krainsky's (1914) species, essentially a colored colony description as many of Krainsky's descriptions are. Krainsky's species was called *Actinomyces (Streptomyces) melanosporeus* and the colony description does resemble the colony coloration of a micromonospora. Baldacci desposited this culture at the CBS as *Actinomyces melanosporeus;* however, in 1956 G. A. de Vries studied this culture and suggested to Baldacci that the isolate should be assigned to the genus *Micromonospora.* Baldacci reinvestigated this culture along with 9 other strains of micromonosporae isolated from soil and water and concluded that all the strains showed a rather remarkable degree of cultural homogeneity and that his *Actinomyces melanosporeus* identification was indeed a micromonospora as de Vries had suggested. Baldacci and Locci differentiated their 10 strains upon the basis of morphology into those producing branched sporophores (6 strains) and those producing clustered sporophores (4 strains). Those producing the clustered sporophores were given the subspecies epithet *corymbica,* from the Latin *corymbus,* a flower cluster. The authors did not believe that either the color of the vegetative mycelium or the minor physiological variations could be used as distinguishing characteristics.

If Baldacci and Locci are correct in their new combination, one wonders perhaps how many other essentially macroscopic colony descriptions of *Actinomyces (Streptomyces)* might on microscopic examination be found to be *Micromonospora?* Waksman (1959) believed *Actinomyces keratinolytica* of Acton and McGuire (1931) appeared from illustrations of these authors to be a micromonospora. I have looked at a culture reported to be the Acton and McGuire isolate of *A. keratinolytica* and considered it to be a typical *Micromonospora parva* type. Two Millard and Burr species that I have studied also appear to be poorly sporulating micromonosporae, these are *Actinomyces salmonicolor* IMRU 3377 and *Nocardia maculata* IMRU 3376.

M. *Micromonospora echinospora*

Luedemann and Brodsky (1964) described two species of *Micromonospora* producing the antibiotic complex gentamicin (Weinstein et al., 1963). The species *Micromonospora echinospora* was characterized by spores with blunt spines 0.1–0.2 μm in length. This organism had a significant carbohydrate utilization pattern which included good growth on l-rhamnose and poor growth on α-melibiose, d-mannitol and d-ribose. Two subspecies *M. echinospora* subsp. *pallida* and *M. echinospora* subsp. *ferruginea,* were described which varied from *M. echinospora* in respect to color and secondary physiological criteria. *Micromonospora purpurea* was described as a second

PLATE I

(1) *Micromonospora chalcea* NRRL-B 944. Well-stained spores, broth culture, oil immersion.

(2) *Micromonospora chalcea* ATCC 12452. Loosely woven, sparsely branched mycelial pattern, broth culture, low magnification.

(3) *Micromonospora carbonacea* NRRL 2972. Sympodial type sporophore, broth culture, oil immersion.

(4) *Micromonospora carbonacea* NRRL 2972. Long laterally branched vegetative mycelium, broth culture, oil immersion.

(5) *Micromonospora halophytica* var. *nigra* NRRL 3097. Single spores and spores in juxtaposition, broth culture, oil immersion.

(6) *Micromonospora halophytica* var. *nigra* NRRL 3097. Loosely woven branched mycelial pattern, broth culture, low magnification.

(7) *Micromonospora halophytica* NRRL 2998. An unusual spore cluster produced about a single mycelial strand composed of sessile spores and spores produced on very short sporophores, broth culture, oil immersion.

(8) *Micromonospora halophytica* NRRL 2998. Loosely woven mycelial pattern, broth culture, low magnification.

(9) *Micromonospora* sp. ATCC 10026. Clusters of spores replacing producing mycelium typical of this organism, broth culture, oil immersion.

(10) *Micromonospora* sp. ATCC 10026. "Barbed wire" appearance of mycelium in water culture, low magnification.

(11) Same picture as in (10) but under oil immersion. Some of the short spikes appear to be sessile spores.

(12) *Micromonospora* sp. ATCC 10026. Tightly woven mycelial colony with "barbed wire" fringe, water culture, low magnification.

(13) *Micromonospora* sp. ATCC 10026. Mycelial pattern of short lateral spikes and spores, water culture, oil immersion.

(14) and (15) *Streptomyces salmonicolor* (Millard and Burr, 1926) Waksman, 1953 IMRU 3377. Short mycelial fragments and monospores, broth culture, oil immersion.

(16) *Micromonospora echinospora* subsp. *pallida* NRRL 2996. Rough walled spores above, spores attached to sporophores below, broth culture, oil immersion.

(17) *Micromonospora echinospora* subsp. *pallida* NRRL 2996. Spores and sporophores, note sympodial development at top of photomicrograph. Tomato paste oatmeal agar plus $CaCO_3$, smear, oil immersion.

(18) *Micromonospora echinospora* subsp. *pallida* NRRL 2996. Vegetative mycelium, mostly straight sparsely branched, broth culture, oil immersion.

(19) *Micromonospora echinospora* NRRL 2985. Electron micrograph of echinulate spores and sporophore, broth culture.

(20) *Micromonospora* isolate no. 12. Intercalary chlamydospores, broth culture, oil immersion.

(21) *Micromonospora parva* NRRL B 2676. Indeterminate chlamydospore, broth culture, oil immersion.

(22) *Micromonospora halophytica* var. *nigra* NRRL 3097. Intercalary chlamydospores, broth culture, oil immersion.

(23) *Micromonospora chalcea* ATCC 12452. Terminal and intercalary chlamydospores, broth culture, oil immersion.

(24) *Micromonospora echinospora* subsp. *ferruginea* NRRL 2995. Intercalary chlamydospores one of which appears to have lysed, broth culture, oil immersion.

(25) *Promicromonospora citrea* (Russian strain). Typical fragmented mycelium and monospores, broth culture, oil immersion.

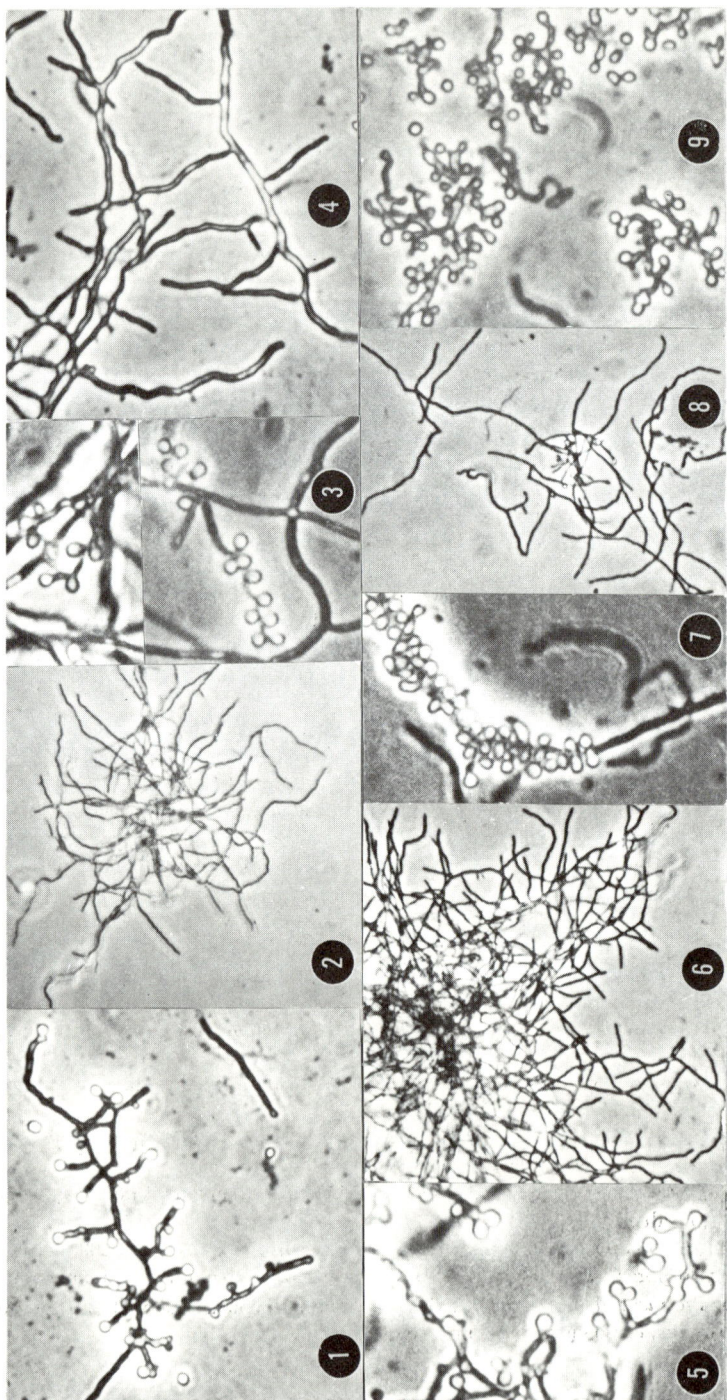

Plate I. (1)–(9). For legend see p. 115.

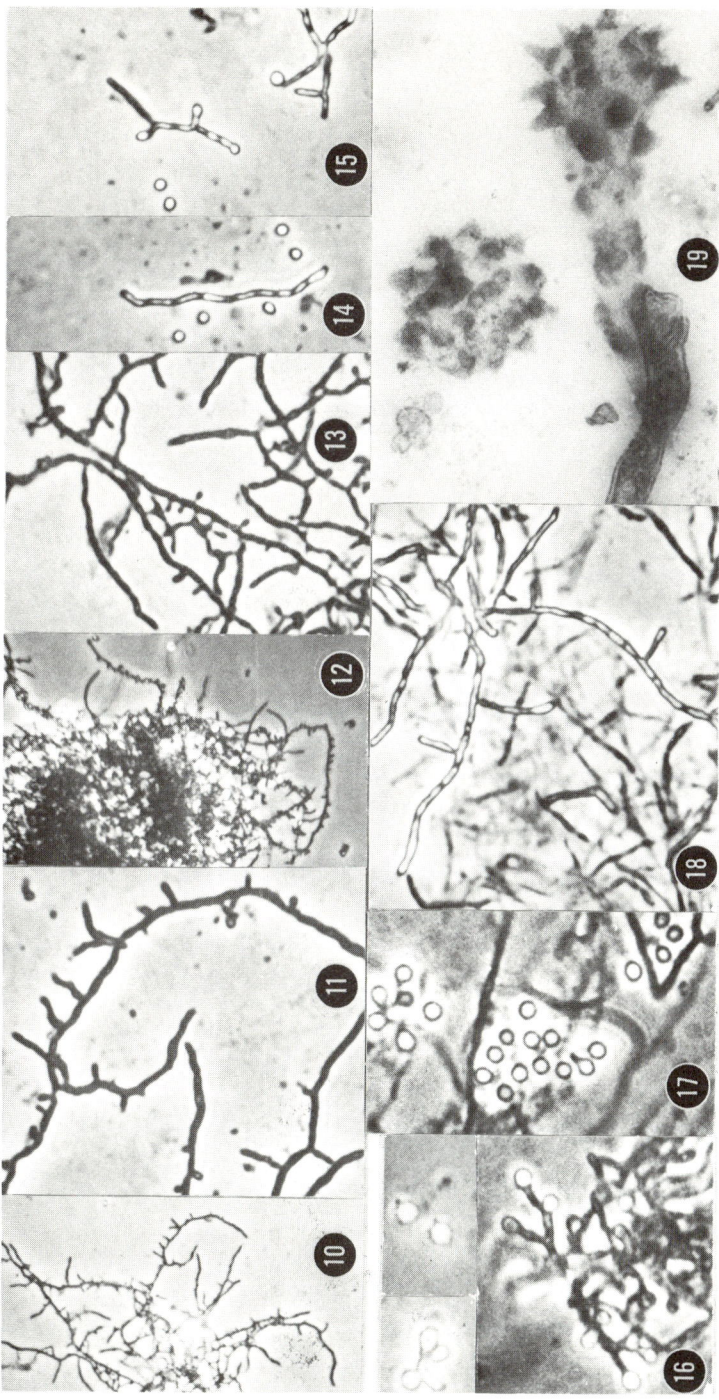

Plate I. (10)–(19). For legend see p. 115.

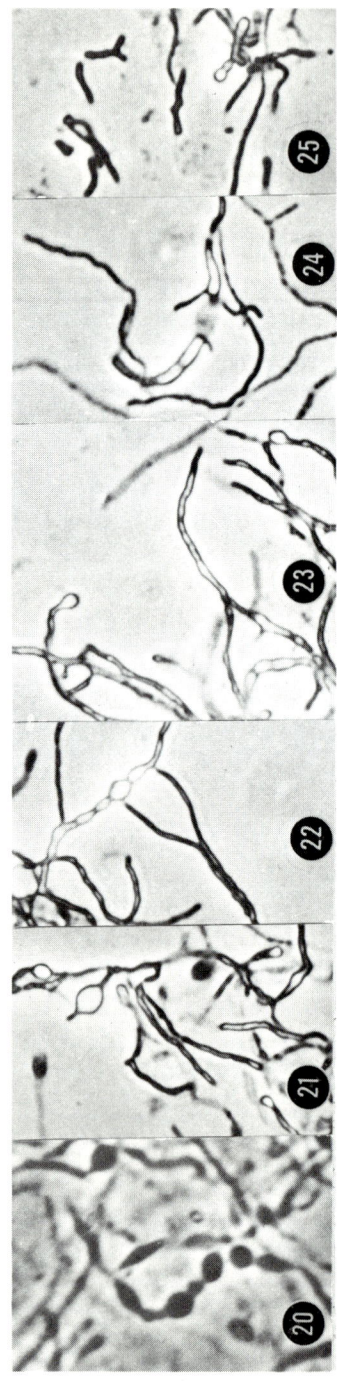

Plate I. (20)–(25). For legend see p. 115.

species producing the gentamicin complex and differed from *M. echinospora* in respect to what was believed to be aberrant spore forms and failure to grow on *l*-rhamnose. Carbohydrate utilization patterns were found to be the most reproducible and capable of giving differential diagnostic information of the physiological tests studied. Yeast extract was substituted for the inorganic nitrogen source used by Pridham and Gottlieb (1948) in carbohydrate utilization studies due to the inability of many micromonosporae to grow on inorganic nitrogen. Spore morphology as revealed by the electron microscope was perhaps the most significant characteristic of *M. echinospora* but in the absence of spores the carbohydrate pattern was of aid along with the biochemical property of gentamicin production. Sensitivity to an acidic environment was a secondary physiological property determined by the use of potato plugs with and without the addition of $CaCO_3$ to neutralize the natural acidity of the potato. In this publication were included comparative descriptions of *Micromonospora chalcea* ATCC 12452, *M. fusca* NRRL B-943, *M. fusca* CBS, and *Micromonospora* sp. ATCC 10026.

Luedemann and Brodsky (1965) described *Micromonospora carbonacea* and *M. carbonacea* subsp. *aurantiaca,* micromonosporae which were found to produce the antibiotic everninomicin (Weinstein *et al.*, 1965). *Micromonospora carbonacea* was characterized by the production of a rather well-developed sympodial type of sporophore best observed in dilute agitated broth cultures. The carbohydrate utilization pattern appeared as good growth on α-melibiose but poor growth on *d*-mannitol, *l*-rhamnose, and raffinose. *Micromonospora carbonacea* subsp. *aurantica* differed from *M. carbonacea* by its inability to reduce nitrate to nitrite, production of a pale yellowish diffusible pigment when grown on mannose and xylose agar in the carbohydrate utilization study, and sparsity of spore production. Biochemically both organisms could be characterized by the production of the everninomicin antibiotic complex. In this publication there appeared a number of drawings characterizing the morphology of several micromonospora species.

Micromonospora halophytica and *M. halophytica* subsp. *nigra* were described by Weinstein *et al.* (1968). These organisms perhaps are best characterized by their biochemical property of producing the halomicin antibiotic complex. Morphologically *M. halophytica* is similar to *M. chalcea,* and is not distinguished from *M. chalcea* by its carbohydrate utilization pattern. Differences between *M. halophytica* and *M. chalcea* may be found in the so called secondary importance criteria. *Micromonospora halophytica* is often found to produce a light reddish brown diffusible pigment on agar media. It produces fair growth on Czapek's sucrose agar and is acid sensitive, growing on potato plug only if $CaCO_3$ is added. The significant characteristics separating *M. halophytica* from the broad *M. chalcea* species group are not readily recognized at this time. As mentioned earlier, when dealing with morphologically similar forms physiological and biochemical properties attain significant proportions. The significant biochemical property of halomicin production and its concomitant characteristic insensitivity to halomicin as an antibiotic results in a simple diagnostic procedure whereby *M. chalcea* is found sensitive to antibiotic sensitivity discs containing the halomicin complex whereas *M. halophytica* strains are found insensitive.

N. *Micromonospora narashinoensis*

The antibiotic nocardorubin was described by Aiso *et al.* (1954). The producing organism was assigned to the genus *Nocardia* based upon the observation that a true branching mycelium after two weeks of culture was found to break up into numerous coccoid cells. The producing culture was later named *Nocardia narashinoensis*. Index Bergeyana (Buchanan *et al.*, 1966) cites Endo 1956, 228 as the author, referring also to Endo 1955, 168. Arai and Kuroda (1965) cite Shidara (1955) as the author of the species in their paper which proposed the new combination *Micromonospora narashino* (Aiso et Arai) Arai and Kuroda [which it appears should be cited: *Micromonospora narashinoensis* (Shidara) Arai and Kuroda]. The Shidara publication (1955) describes the nocardorubin organism and states: "Therefore strain No. 76 was considered to be a new strain belonging to the genus *Nocardia* and was named *Nocardia narashinoensis* nov. sp." I have seen both Endo references (1955 and 1956) and am in agreement with Arai and Kuroda (1965) that Shidara (1955) is the correct citation of author.

Although Arai and Kuroda made an official transfer of *Nocardia narashinoensis* to *Micromonospora narashinoensis* in their 1965 publication an earlier paper of theirs (Arai *et al.*, 1964) used the combination *Micromonospora narashinoensis* and partially redescribed the organism and illustrated it. Electron micrographs of ultrathin sections of spores and mycelium are of interest and demonstrate distinct hyphal septation (Arai *et al.*, 1964). Kenner *et al.* (1968) also noted frequent septation in electron microscope thin sections of micromonospora mycelium.

Arai and Kuroda (1965) presented photomicrographs of the mycelium and spores of *M. narashinoensis*, *M. chalcea* ATCC 12452, and *Micromonospora* sp. ATCC 10026 (Waksman's micromonosporin producer). They pointed out that "With strain No. 76 (*M. narashinoensis*) and *M. chalcea* the spore bearing hyphae is often very short and the growth has an appearance of knobbed mycelia. In most cases, spores are spherical, but elongated forms may be seen especially with *Micromonospora* sp. ATCC 10026. The branchings are more profuse with *Micromonospora* sp. ATCC 10026 than with the other two strains." T'ao (1958) has drawn attention to the botryoid (clustered) appearance of the mycelium in *M. fusca* (CBS) in Figs. 1 and 2 of her thesis. In T'ao's drawings of *M. chalcea* however, she pointed out that the mycelium was long and slightly curved, appearing loosely woven. T'ao did not study *M. chalcea* ATCC 12452.

From my own observations I believe there is a definite mycelial characteristic recognizable in *Micromonospora* sp. ATCC 10026, *M. fusca* (CBS) and aquatic and soil isolates we have made that produce a "micromonosporin-like" antibiotic. These organisms in broth produce usually a tightly woven mycelial colony with characteristic short spikes or knobs (barbed wire appearance) on which spores occasionally are produced. *Micromonospora narashinoensis* IFM 1110 (strain no. 76) produces a similar tightly woven colony with these characteristic short lateral spikes. I have not seen this type of colony in *M. chalcea* ATCC 12452 but have seen loosely woven colonies in this culture similar to those pictured by T'ao (1958) for the *M. chalcea* she studied. The characteristic used by Waksman (1961) separating *M. chalcea* from *M. fusca* by "sporophores form extensive branching" for the latter species is worth further investigation in view of the observations made concerning the clustered, knobbed or spiked hyphae by the above investigators. Arai and Kuroda were not able to demonstrate antibiotic activity in *Micromonospora* sp. ATCC 10026 (micromonosporin producer) and believed the culture probably had lost this property due to a long period of subculturing. They believed that some differences existed between *M. narashinoensis* and *Micromonospora* sp. ATCC 10026 in respect to cultural and physiological characteristics. My observations on *M. narashinoensis* IFM 1110 (strain no. 76) are that it differs principally at present from *Micromonospora* sp. ATCC 10026 in respect to mannitol utilization, 10026 being mannitol positive while no. 76 is mannitol negative. In other respects these two cultures are exceedingly similar.

III. Relationships

A. Relationships within the Genus

What type of information can one consider when speculating about the relationships of the various species of *Micromonospora*? The interesting point about relationships, right or wrong, is that it starts us thinking, thinking about a system which will bring some type of order to the disorder or randomness that exists. The discovery of the periodicity of the chemical elements was a magnificent improvement in organizing information into a retainable form. What the significance of the enzyme systems responsible for carbohydrate utilization patterns is, if any, remains for some inquiring researcher to discover. The positive sucrose and generally negative glycerol utilization pattern of the micromonosporae seems to indicate a fundamental difference between these organisms and the streptomycetes. What relationships may be implied in regard to melibiose, rhamnose, mannitol, and raffinose utilization among the species of *Micromonospora* (see Table I)? What about morphological relationships? Is the clublike echinulate spore development found in *Micromonospora echinospora* a more primitive feature than the botryoid, multibranched systems found in *Micromonospora globosa, M. fusca,* or *M. chalcea* (see Plate I)? The specialized development of a sympodial type of sporophore (sympodula) (Kendrick, 1962) and colonial peripheral sectoring into sporulating and nonsporulating fans (dual phenomenon) in *Micromonospora carbonacea* leads one to suspect a degree of greater specialization and advancement. Do we find certain species of *Micromonospora* adapted to the aquatic environment while other species are found to be more frequent terrestrial inhabitants? My suspicion is that we probably do; *M. chalcea, M. halophytica,* and *Micromonospora* sp. ATCC 10026 seem to be frequently isolated from aquatic samples. These species also appear to sporulate more rapidly and readily in an aqueous medium than *M. carbonacea* and *M. echinospora*. The relationship of species of *Micromonospora* is a problem for investigation and has invited few hypotheses due to the primary problem of not having methods for differentiating species.

Recently a potentially useful method for differentiating streptomycetes based upon their tolerance of NaCl in an agar growth medium has been reported by Tresner *et al.* (1968). Most streptomycetes (70%) appeared able to tolerate 7% NaCl, some 10%, a very few up to 13% and also a few were not able to tolerate 4%. Sodium chloride tolerance appeared characteristic for species groupings rather than simply strain specific. Using a yeast (1%)–starch (2%) medium, I have found that among 40 diverse strains of micromonosporae average NaCl tolerance is 3%, maximum 5% (for one species group it is remarkably diagnostic), and no micromonosporae have yet been found to grow at 7% NaCl (unpublished data). A streptomycete control culture grew nicely in this study at a 7% NaCl concentration. One micromonospora group was found to grow at 1% but not at 2% NaCl concentration. Tolerance of NaCl in micromonosporae appears related to species groups and also may have a more profound consequence in terms of generic relationships.

B. Relationship to Other Actinomycete Genera

The relationship of the genus *Micromonospora* to other genera of actinomycetes has drawn a number of hypotheses from investigators. Hungate (1946) considered the micromonosporae to represent the most primitive form of actinomycete because some still carried on a typical propionic acid fermentation thought to relate them to the propionic acid bacteria. Hungate's species, *Micromonospora propionici*, was an obligate anaerobe and an organism requiring further investigation to determine its relationship to the aerobic micromonosporae. Pridham (1959) considered the micromonosporae simple in terms of their morphology when compared to *Microbispora*, *Actinoplanes*, and *Streptosporangium* and would thus place them lower in a morphologic scheme developing from simplicity to complexity of fruit-bearing structure. Erikson (1953) considered the micromonosporae at the head of a classificatory system. Waksman (1950) thought the micromonosporae might be considered the most highly developed group among the actinomycetes. Hesseltine (1960) placed the micromonosporae at the head of a possible evolutionary tree in the Actinomycetales based upon a concept of reduction of spores in a sporangium to a single spore as being considered advanced, a theory held in the fungal order Mucorales. Hütter (1965) on morphological considerations placed the *Micromonospora* in the family Streptomycetaceae above *Promicromonospora* and below *Microbispora*. Jones and Bradley (1964) primarily upon physiological considerations placed the *Micromonospora* adjacent to *Actinoplanes*, above the *Nocardia* and below the *Streptomyces*. The relationship of *Micromonospora* and *Actinoplanes* was further implicated by cell wall analysis (Becker *et al.*, 1965; Yamaguchi, 1965). The Type II *Micromonospora* cell wall of Becker *et al.* (1965) was characterized by the presence of *m*-diaminopimelic acid and glycine. These latter authors pointed out

TABLE I
Diagnostic Carbohydrate Utilization Patterns [a,b]

Organisms representing known patterns of utilization	l-Arabinose	Sucrose	α-Melibiose	d-Mannitol	Raffinose	Glycerol	l-Rhamnose	Inositol
Micromonospora purpurea ATCC 15835	▨	▨						
Micromonospora echinospora ATCC 15837	▨	▨					▨	
Micromonospora carbonacea NRRL 2972 *Micromonospora narashinoensis* IFM 1110	▨	▨	▨					
Micromonospora sp. ATCC 10026	▨	▨	▨	▨				
Micromonospora chalcea ATCC 12452 *Micromonospora halophytica* NRRL 2998	▨	▨	▨		▨			

#36 (*Micromonospora coerulea?*)		
#46 (*Micromonospora fusca?*)		
Actinomonospora lusitanica CBS		

[a] Major differences in carbohydrate utilization patterns have been assembled in Table I. As many as 24 carbohydrates (Luedemann and Brodsky, 1965) are used in testing for utilization, however, only a portion of these are of differential value. As our knowledge of *Micromonospora* species increases so should the number of differentially useful carbohydrates. Some utilization patterns accommodate a single species while other patterns accommodate a number of species. The blue-green isolate resembling *M. coerulea* and the black diffusible pigment-producing isolate resembling *M. fusca* are at present unique in their utilization patterns. *Actinomonospora lusitanica* has been included as a representative of a nonmicromonosporoid utilization pattern. *Micromonospora* with new utilization patterns are interesting and often lead to the study of cultures which differ morphologically and physiologically from known species. Table I may aid in indicating where new species may be found and what to look for when studying a large group of isolates. It is desirable to base a utilization pattern for an organism on a least three separate trials in order to minimize errors in media preparation, inoculum vigor, and admittance of air to the culture tube or plate. "One run" experiments are potentially misleading.

[b] Positive utilization indicated by darkened squares.

that Couch (1954) had for other reasons noted the similarity of these two genera.

The relationship of *Promicromonospora* to *Micromonospora* by cell wall analysis shows a wide divergence of these two genera (Yamaguchi, 1965); *Promicromonospora* not relating to any of the well known actinomycete genera (M. P. Lechevalier, personal communication, 1969). *Actinomonospora* is presently being investigated by the technique of cell wall analysis by M. P. Lechevalier and the data should be helpful to our understanding of the generic relationship. If as Cummins and Harris (1958) have stated "Cell wall composition may be regarded as an extension of morphology at the biochemical level," this method should be an important tool for establishing generic relationships.

IV. Impressions and Reflections

A. BACKGROUND

This is not meant to be a sentimental or nostalgic digression but an area I would like to think of as impressions — an indulgence frowned upon in past scientific writing, however, an area in which feelings leading up to discoveries can be explored as was so ably demonstrated in Watson's book "The Double Helix."

My work with the micromonosporae came about through a grant I helped obtain for a college acquaintance of mine, A. Woyciesjes, who had become fascinated with these organisms and thought there might be interest in looking at some of them in terms of their biochemical potentialities. It required a while to familiarize myself with the technique for handling these organisms as they did not culture with the same degree of ease as most fungi, streptomycetes, or bacteria. Cultures often died out on agar to agar transfers. Fortunately the cultures proved amenable to lyophilization as a means of maintenance.

B. EARLY SCHEME FOR DIFFERENTIATION

Superficially most micromonosporae appear similar macroscopically and microscopically, a point noted also by other workers. After working rather unproductively with these organisms for better than a year I became interested, as others have, in the pigments produced by these cultures and attempted a simple classification of cultures based upon the extraction of pigments in various solvents. A further characterization of these pigment extracts was made by testing to see if they contained antibacterial or antifungal activity. Some of the extracts were found active and a further step was to chromatogram these extracts and observe the active components by bioautography. Several different patterns of activity were determined and the producing organisms grouped for further study.

C. Traditional Taxonomic Methods

A means for taxonomically describing these organisms by more traditional methods was sought. The methods used by Gordon and Smith (1955) and Gordon and Mihm (1957) for describing species of *Nocardia* and methods used for *Streptomyces* identification (Shirling and Gottlieb, 1966) were applied with results that confirmed original impressions that micromonosporae did not respond well to the media and reactions used for *Nocardia* and *Streptomyces* differentiation. Modifications of these media were made with relatively good success. Substitutes for beef extract, peptone, or inorganic nitrate were found in order to obtain growth. Nitrate nitrogen did support growth in some cultures, but among groups of organisms producing the same antibiotic complex some strains were capable of utilization and others were not (Luedemann and Brodsky, 1964). Organisms producing the same antibiotic complex were found to have similar morphology and similar carbohydrate utilization patterns. It was found that 0.5% yeast extract concentration incorporated into agar produced a minimal growth which could be used as a negative carbohydrate control. Carbohydrates in 1% concentration could be added to this base to obtain a relatively luxuriant growth if the carbohydrate was utilized. The micromonosporae tested for carbohydrate utilization by this method appear to differ from streptomycetes in respect to the utilization of the following carbohydrates. Sucrose is utilized by most micromonosporae. With streptomycetes approximately 50% are able to utilize sucrose. Very few micromonosporae utilize glycerol; most streptomycetes utilize glycerol. Inositol, dulcitol, and D(+)-melezitose have not been utilized by the micromonosporae I have tested; mannitol is utilized by only a few micromonosporae. α-Melibiose is utilized by most micromonosporae, the exception being the gentamicin-producing organisms *M. echinospora* and *M. purpurea* (Luedemann and Brodsky, 1964).

Micromonosporae vary in their sensitivity to acid pH; none that I have observed grow below pH 5.5. One of the substrates used in the older literature as a descriptive medium was growth on potato slice. In dealing with the micromonosporae I believe this is a good test medium, but that it demonstrates a strain's ability to tolerate pH

5.8–6.0 (pH of the raw potato used). The micromonosporae that do not grow on autoclaved potato slice will usually grow luxuriantly on this substrate if a small amount of reagent grade $CaCO_3$ is added to the potato slice before autoclaving, providing, of course, the amount is sufficient to neutralize the weakly acidic potato. Too much $CaCO_3$ applied to the potato slice can prevent the organism reaching the substrate. *Micromonospora chalcea* ATCC 12452 and *Micromonospora* sp. ATCC 10026 tolerate the acid potato slice and grow quite nicely, whereas *M. carbonacea* (Luedemann and Brodsky, 1965) grows rather poorly and *M. echinospora* and *M. purpurea* do not grow at all unless $CaCO_3$ is added (Luedemann and Brodsky, 1964).

Nitrate reduction appears to be a variable reaction with micromonosporae as with streptomycetes (Shirling and Gottlieb, 1966). Micromonosporae hydrolyze starch, peptonize milk, and liquefy gelatin and reactions to these tests appear to be more a generic characteristic than one useful for species differentiation (see Krasilńikov, 1960). Cellulose decomposition may possibly be of diagnostic value but the test requires 3–6 months for evaluation when cellulose strip disintegration is used as the test method.

The tyrosine reaction (Gordon and Smith, 1955) using beef extract as a component of the medium has given mostly negative reactions with the micromonosporae. Recently I have substituted yeast extract for the beef extract and found most test micromonosporae produce a beautiful dark reddish brown to black positive reaction.

D. Transfer to Transfer Variation

My impression after performing many physiological tests on the micromonosporae is that there is no simple solution to speciation by these methods. The carbohydrate utilization test is the most diagnostic of these methods. It requires however, a clearly positive or negative reaction. To base a new species upon a single difference in carbohydrate utilization *alone* is not sufficient criteria for speciation. I have seen a case in my own laboratory where *Micromonospora* sp. (Waksman 3450, ATCC 10026) and its counterpart in the CBS, *M. fusca* (Waksman 3450) both were mannitol positive. *Micromonospora fusca* CBS was reported (Luedemann and Brodsky, 1964) by us to be a poorly sporulating orange colored, nonantibiotic-producing culture. This organism had been carried through a number of subcultures and became a highly sporulating culture probably through transfer to transfer selection. This culture had lost its ability to utilize mannitol but regained its ability to produce an antibiotic complex which chromatographically was identical to the antibiotic complex produced by *Micromonospora* sp. ATCC 10026. Morphologically these two cultures are similar and are characterized by tightly woven mycelial clusters bearing many short lateral spikes (barbed wire appearance). This mannitol negative varient of *M. fusca* CBS is very similar to the culture I recently received as *M. narashinoensis* No. 76, IFM 1110. Our lyophilized *M. fusca* CBS culture is still mannitol positive, poorly sporulating, and not producing demonstrable antibiotic activity.

E. A Method for Studying Morphology

My impression concerning micromorphology is that it is the most reliable method for the characterization of species if morphologic differences exist. However, it is one of the most difficult methods, requiring patience and skill in developing the microscopist's techniques. I enjoy studying microscopic morphology and marvel at the way nature has put these organisms together much the way, I am certain, a physiologist derives satisfaction from determining the metabolic pathway in which a compound is broken down and reassembled. Each specialist develops his own techniques much like an artist attempting to communicate his impressions in lines, shades, and colors. In attempting to understand micromonospora morphology the techniques used in studying streptomycete morphology cannot be borrowed. Streptomycete aerial morphology is best observed on an agar surface (Lechevalier, 1964). Because micromonosporae do not have an aerial morphology other methods must be devised. Agar grown colonies of micromonosporae when viewed by a slide smear technique commonly used in bacteriology are disappointing because the structure of the spore and sporophore are generally disrupted. Impression slides also suffer due to disruption of the mycelium and spores. Direct examination of a micromonospora colony on water agar was used by Ørskov (1923) with some success but he also stated that the organism was found to develop in broth culture in an identical manner. My own observations have met with most success by modifications of this latter method using agitated broth cultures. In respect to a standard broth for morphological study there does not seem to be one. Different species of *Micromonospora* require different techniques in order to gain an insight into the morphological development of the sporulation apparatus. *Micromonospora chalcea* ATCC 12452 and *Micromonospora* sp. ATCC 10026 generally sporulate readily and rapidly. A lean medium is often successful for observing these organisms in the sporulation process. It requires microscopic observations at 24-hour intervals for often sporulation occurs rapidly and a slide of the culture made too late will find only the coccoid spores and fragments of mycelium. Other cultures like *M. carbonacea* and *M. echinospora* require richer media and longer incubation periods, 7–21 days or longer. With the richer media often 1:10 dilutions must be made in sterile tap water so that a thin, even film of mycelium may be obtained. The best preparations I have made have been to take a sterile capillary tube and remove a drop or two of the mycelial suspension from the broth or water culture. This material is spread evenly over the surface of a cover slip. Sometimes several cover slips must be used before one is found which allows the suspension to be evenly spread. The material is air dried and/or is gently heated over an alcohol lamp in order to fix the material to the cover slip but not to cook and distort the mycelium. Once the mycelium has been fixed to the cover slip a drop of dilute crystal violet stain is placed on a slide with a drop of water and the cover slip, face down, is lowered into contact with the stain solution. A phase contrast microscope has been found to give excellent results under oil immersion with this method. Often the spores will be found to fluoresce a brilliant yellow and the mycelium is either blue or sometimes a pale orange. Very little distortion is observed through the thickness of the cover slip whereas material on the slide is not sharply in focus. Direct oil observations of the dried material have been disappointing. By the above technique I have been able to observe the roughened outline caused by the spines on the spores of *M. echinospora* which by electron microscopy measure approximately $0.1~\mu m$ in length (Luedemann and Brodsky, 1964). Ideally the crystal violet should be dilute enough so that it is taken up almost entirely by the spores and mycelium on the slide, leaving a clear background. In practice certain portions of a slide may be found to be stained perfectly, other areas less so. There are two areas of the micromorphology of these organisms as viewed by the light microscope which are worth considering. One is the mycelium in terms of its length, amount of branching, length of branches, segmentation, diameter, and compactness of the mycelium in a colony —whether the colony is loosely or densely woven. The second is the characteristics of the spores and sporophores: Are the spores randomly distributed throughout the mycelium or do they appear in clusters? Are the clusters arranged in a sympodially branched system? What are the general characteristics of the spores, smooth or rough?

Once good spore and sporophore preparations have been obtained and confirmed by light microscopy, electron microscopy must be employed to answer the question: How is the spore produced and attached to its sporophore? Can we determine whether the micromonospora spore is a true conidium, which it appears to me to be, or is it more like a terminal chlamydospore in its formation? Terminal chlamydospores do appear to occur with some frequency in the micromonosporae but their irregular diameter and shape can most often aid in distinguishing them. The electron microscopist will, I believe, resolve a number of the problems of speciation in the micromonosporae once good spore and sporophore preparations are obtained.

REFERENCES

Acton, H. W., and McGuire, C. (1931). *Indian Med. Gaz.* **66**, 65–70.
Aiso, K., Arai, T., Shidara, I., and Ogi, K. (1954). *J. Antibiot. (Tokyo)* **A8**, 1–6.
Arai, T., and Kuroda, S. (1965). *Ann. Rept. Inst. Food Microbiol. Chiba Univ.* **18**, 33–38.
Arai, T., Koyama, Y., Kuroda, S., and Honda, H. (1964). *Kogyo Kagaku Zasshi* **67**, 728–733.
Baldacci, E. (1938). *Atti Ist. Botan. Lab. Crittogam. Univ. Pavia* **10**, 321–329.
Baldacci, E. (1958). *Giorn. Microbiol.* **6**, 10–27.
Baldacci, E., and Locci, R. (1961). *Ann. Microbiol. Enzimol.* **11**, 19–30.
Baldacci, E., Comaschi, G. F., Scotti, T., and Spalla, C. (1953). *Intern. 6th Congr. Microbiol. Symp. Actinomycetales, Rome*, pp. 20–39.
Becker, B., Lechevalier, M. P., and Lechevalier, H. A. (1965). *Appl. Microbiol.* **13**, 236–243.
Bisset, K. A. (1957). *J. Gen. Microbiol.* **17**, 562–566.
Breed, R. S., Murray, E. G. D., and Parker-Hitchens, A. (1948). "Bergey's Manual of Determinative Bacteriology," 1529 pp. Williams & Wilkins, Baltimore, Maryland.
Buchanan, R. E., Holt, J. G., and Lessel, E. F., Jr. (1966). "Index Bergeyana," 1472 pp. Williams & Wilkins, Baltimore, Maryland.
Colmer, A. R., and McCoy, E. (1943). *Trans. Wisconsin Acad. Sci.* **35**, 187–220.
Couch, J. N. (1954). *Trans. N. Y. Acad. Sci.* **16**, 315–318.
Cowan, S. T. (1962). *In* "Microbial Classification" (G. C. Ainsworth, and P. H. A. Sneath, eds.), Symp. Soc. Gen. Microbiol. No. 12, pp. 433–455. Cambridge Univ. Press, Cambridge.
Cummins, C. S., and Harris, H. (1958). *J. Gen. Microbiol.* **18**, 173–189.
Endo, T. (1955). *J. Antibiot. (Tokyo)* Ser. **B8**, 163–169.
Endo, T. (1956). *J. Antibiot. (Tokyo)* Ser. **A9**, 228.
Erikson, D. (1935). *Med. Res. Council Spec. Rept. Ser.* **203**, 5–61.
Erikson, D. (1941). *J. Bacteriol.* **41**, 277–300.
Erikson, D. (1949). *Ann. Rev. Microbiol.* **3**, 23–54.
Erikson, D. (1953). *Intern. 6th Congr. Microbiol. Symp. Actinomycetales*, Rome, pp. 102–121.
Fisher, W. P., Charney, J., and Bolhofer, W. A. (1951). *Antibiot. Chemotherapy* **1**, 571–572.
Foulerton, A. (1905). *Lancet* **i**, 1199–1200.

Gordon, R. E., and Mihm, J. M. (1957). *J. Bacteriol.* **73**, 15-27.
Gordon, R. E., and Smith, M. M. (1955). *J. Bacteriol.* **69**, 147-150.
Harden, M. E. (1941). B. A. Thesis, University of Wisconsin, Madison, Wisconsin.
Hesseltine, C. W. (1960). *Mycologia* **52**, 460-474.
Hütter, R. (1965). *Pathol. Microbiol.* **28**, 567-574.
Hungate, R. E. (1946). *J. Bacteriol.* **51**, 51-56.
Ilavsky, J., *Paper presented Spolok Slovenskych Lekarov Bratislave, Czechosolvakia, June, 1968* (to appear in *Lekarsky Obzor* **18**, in press).
Jensen, H. L. (1930). *Proc. Linnean Soc. N.S. Wales* **55**, 231-248.
Jensen, H. L. (1932). *Proc. Linnean Soc. N. S. Wales* **57**, 173-180.
Jones, L. A., and Bradley, S. G. (1964). *Mycologia* **56**, 505-513.
Kendrick, W. B. (1962). *Can. J. Botany* **40**, 771-797.
Kenner, D. D., Hohl, H., and Baker, G. E. (1968). *Pacific Sci.* **22**, 52-55.
Krainsky, A. (1914). *Zentr. Bakteriol. Parasitenk. Abt. II*, **41**, 649-688.
Krasilńikov, N. A. (1938). *Akad. Nauk SSSR, Moscow*, pp. 1-328.
Krasilńikov, N. A. (1941). *Izv. Akad. Nauk. SSSR*, pp. 1-147. (Engl. Transl. by S. Nemchonok, 1966, Federal Sci. Tech. Inform.)
Krasilńikov, N. A. (1949). *Akad. Nauk. SSSR, Moscow*, pp. 1-830. [Engl. Transl. (in part) by J. B. Routien (ed.), published by Chas. Pfizer & Co., New York, 1957.]
Krasilńikov, N. A. (1960). *J. Bacteriol.* **79**, 75-80.
Kriss, A. E. (1939). *Mikrobiologiya* **8**, 178-185.
Küster, E., and Locci, R. (1964), *Intern. Bull. Bacteriol. Nomenclature Taxonomy* **14**, 109-114.
Lechevalier, H. (1964). *In* "Principles and Applications in Aquatic Microbiology" (H. Heukelekian and N. C. Dondero, eds.), pp. 230-253. Wiley, New York.
Lechevalier, H., and Lechevalier, M. P. (1967). *Ann. Rev. Microbiol.* **21**, 71-100.
Luedemann, G. M., and Brodsky, B. C. (1964). *Antimicrobial Agents Chemotherapy–1963*, pp. 116-124.
Luedemann, G. M., and Brodsky, B. C. (1965). *Antimicrobial Agents Chemotherapy–1964*, pp. 47-52.
Morquer, R., and Comby, L. (1943). *Bull. Soc. Hist. Nat. Toulouse* **78**, 23-28.
Ørskov, J. (1923). "Investigations into the Morphology of the Ray Fungi," 171 pp. Levin and Munksgaard, Copenhagen.
Potter, L. F., and Baker, G. E. (1956). *Ecology* **37**, 351-355.
Pridham, T. G. (1959). *Rev. Latinoam. Microbiol. Suppl.* **3**, 1-22.
Pridham, T. G., and Gottlieb, D. (1948). *J. Bacteriol.* **56**, 107-114.
Sebald, M., and Prévot, A. R. (1962). *Ann. Inst. Pasteur* **102**, 199-214.
Shidara, I. (1955). *J. Chiba Med. Soc.* **30**, 551-559.
Shirling, E. B., and Gottlieb, D. (1966). *Intern. J. System Bacteriol.* **16**, 314-340.
Shull, G. M. (1959). U.S. Patent 2,890,153.
Simmons, E. G. (1966). *Quart. Rev. Biol.* **41**, 113-123.
Taira, T., and Fujii, S. (1951). *J. Antibiot. (Tokyo)* **5**, 185-187.
T'ao, L. H. (1958). M.S. Thesis, Vassar College, Poughkeepsie, New York, 190 pp.
Tresner, H. D., Hayes, J. A., and Backus, E. J. (1968). *Appl. Microbiol.* **16**, 1134-1136.
Umbreit, W. W., and McCoy, E. (1941). "Symposium on Hydrobiology," pp. 106-114. Univ. of Wisconsin, Madison, Wisconsin.
Waksman, S. A. (1950). "The Actinomycetes: their Nature, Occurrence, Activities, and Importance," 230 pp. Chronica Botanica Co., Waltham, Massachusetts.
Waksman, S. A. (1957). *Bacteriol. Rev.* **21**, 1-29.

Waksman, S. A. (1959). "The Actinomycetes, Vol. I: Nature, Occurrence and Activities," 327 pp. Williams & Wilkins, Baltimore, Maryland.

Waksman, S. A. (1961). "The Actinomycetes, Vol. II: Classification, Identification and Descriptions of Genera and Species," 363 pp. Williams & Wilkins, Baltimore, Maryland.

Waksman, S. A. (1967). "The Actinomycetes: A Summary of Current Knowledge," 280 pp. Ronald Press, New York.

Waksman, S. A., Geiger, W. B., and Bugie, E. (1947). *J. Bacteriol.* **53**, 355–357.

Waksman, S. A., and Lechevalier, H. A. (1962). "The Actinomycetes, Vol. III: Antibiotics of Actinomycetes," 430 pp. Williams & Wilkins, Baltimore, Maryland.

Wang, C. J. K. (1965). *N.Y. State College Forestry, Syracuse, N.Y. Tech. Publ. No.* **87**, 115 pp.

Weinstein, M. J., Luedemann, G. M., Oden, E. M., Wagman, G. H., Rosselet, J. P., Marquez, J. A., Coniglio, C. T., Charney, W., Herzog, H. L., and Black, J. (1963). *J. Med. Chem.* **6**, 463–464.

Weinstein, M. J., Luedemann, G. M., Oden, E. M., and Wagman, G. H. (1965). *Antimicrobial Agents Chemotherapy–1964*, pp. 24–32.

Weinstein, M. J., Luedemann, G. M., Oden, E. M., and Wagman, G. H. (1968). *Antimicrobial Agents Chemotherapy–1967*, pp. 435–441.

Welsch, M. (1942). *J. Bacteriol.* **44**, 571–588.

Yamaguchi, T. (1965). *J. Bacteriol.* **89**, 444–453.

Dental Caries and Periodontal Disease Considered as Infectious Diseases*

WILLIAM GOLD

*Department of Microbiology,
College of Dentistry,
New York University, New York*

I.	Introduction	135
II.	Caries in Man and Animals	136
	A. Methods of Study	136
	B. Interactions of Host, Diet, and Microbes	138
	C. Special Role of Sucrose, and Microbial Dextran and Levan Formation	144
	D. Possible Connection with Other Diseases	145
	E. Treatment by Antimicrobial Agencies and Other Means	146
III.	Periodontal Disease in Man and Animals	148
	A. Methods of Study	148
	B. Interactions of Microbes, Host, and Diet	149
	C. Treatment with Antimicrobial Agencies	151
IV.	Perspectives	152
V.	Summary	153
	References	153

I. Introduction

Despite the early recognition of the relationship of microbes with tooth decay processes (Miller, 1883), and the more recent general acceptance of a theory of microbial etiology of the disease (Teuscher et al., 1948), there were still sufficient reservations about the causes of dental caries. Orland and co-workers (1954) defined the disease only descriptively in its progression from enamel "white spot" to large, open cavity, without an etiological reference. They then described their experiments with germfree rats which showed that microorganisms were indeed indispensable to the process.

Periodontal disease is a progressive disease syndrome which may involve inflammation and retraction of the gums, resorption of the bony support of the teeth, and destruction of the periodontal ligament, leading to formation of deep gingival pockets and loosening of the teeth, with their eventual loss. It is the major cause of tooth loss in countries where tooth decay is uncommon (Sinrod, 1965). Although it is recognized that tooth cleanliness and bacterial plaque accumula-

---

*Supported in part by a grant from the John Hartford Foundation.

tions along the gums are correlated with the occurrence and severity of the disease (Dunbar *et al.*, 1968; Littleton, 1963), the nature of this relationship and whether it is one of cause and effect is still an open question among periodontologists (Mandel, 1966, 1967).

Recent research reports have given such a broad understanding of tooth decay processes and their expression under various conditions, that caries will be described as an infectious disease, in considerable detail, in this article. Periodontal disease research, however, has not gone so far that one can claim that the disease in man stems directly from specific microbial infections. Still, as will be shown, progress has been made in this direction.

II. Caries in Man and Animals

A. METHODS OF STUDY

Research on tooth decay in man is limited by the slow rate of disease development, the problems of controlling the diets and environments of large numbers of heterogeneous subjects, and the great expense of adequate clinical trials. Investigators turned to readily available laboratory animals with teeth similar in structure to those of man. The Norwegian white rat, the Syrian hamster, the cotton rat, gerbil, and monkey have been used successfully (Bowen and Cornick, 1967; Fitzgerald and Fitzgerald, 1965; Keyes, 1946; Madsen and Edmonds, 1962; Shibata, 1929). The smaller size of the teeth and the use of highly cariogenic diets lead, in months, to extensive decay. Large numbers of animals may be used at small cost.

On the assumption that acid production alone was involved in clinical caries etiology, the mere presence of acidogenic microorganisms together with an available fermentable carbohydrate substrate in the mouth was considered a sufficient condition for production of decay; any fermentation inhibitor was therefore a possible cure for caries. The genus *Lactobacillus* was thought by many to be a prime cause of the disease because its species are commonly found in mouths with active caries and are able to grow at low pH; the infection was believed to be nonspecific (Shaw, 1950; Teuscher *et al.*, 1948). Most experiments with man or animals involved tests that showed that when caries was found, acidogenic bacteria and carbohydrates were both present in the mouth. Attempts were also made to produce caries in animals so that the process might be readily studied. This phase of caries research has been reviewed by Shaw (1950) and therefore will not be discussed in detail. The same review, however, points out many experiments that failed to produce caries in animals that harbored or were superinfected with acidogenic bacteria, or with debris from carious lesions. Individual case studies showed instances of subjects with extensive caries but low *Lactobacillus* counts, as well as caries-resistant subjects with high counts and high sugar consumption. Clearly, despite the logic of the theory, the etiology of caries was not proved as it might had Koch's postulates been applied.

Direct tests of the acidogenic theory were made in animals, and have provided powerful new methods of investigation. Kite *et al.* (1950) fed a cariogenic diet to normal rats and to rats with their salivary glands removed to make them more susceptible to decay. A second set of normal and desalivated rats was given the same diet by intubation, so that the food was never in the mouth. The normal rats that ate the cariogenic diet developed moderate decay; the desalivated rats on the same diet showed rampant decay. In contrast, both groups of rats fed by stomach tube had no decay.

Rats taken by Caesarian section and reared germfree were fed the same cariogenic diet as conventional rats (Orland *et al.*, 1954). Decay was found only in the conventional animals. It was not found in any of the germfree rats, even by microscopic examination of ground-down teeth. This demonstration unequivocally established the necessity of bacteria to the decay process, since diet alone caused no disease. It also provided a tool by which pure cultures of bacteria could be tested for cariogenic potential (Orland, *et al.*, 1955). The method has the disadvantage that some pure cultures may cause decay only in the absence of competition with other microbes normally found in the mouth, but presumably even this could be tested by using mixtures of known cultures of which only one is cariogenic.

A means of overcoming this difficulty was found when Keyes (1960) discovered that one line of hamsters, descended from a caries-active female, that had been treated with penicillin to depress cariogenic flora, was resistant to caries. The animals failed to develop extensive decay unless they were caged with hamsters having caries, or were infected with fecal material from such hamsters. The caries-resistant strain of hamsters was then used to screen pure cultures isolated from caries-active hamsters (Fitzgerald and Keyes, 1960), and from man (Krasse, 1966; Zinner *et al.*, 1965). The task of showing the establishment of the bacteria among the mixed flora of the hamster's mouth was made easy by using antibiotic-resistant cultures to infect the animals, and then isolating them on media with the antibiotic added (Fitzgerald and Keyes, 1963).

Another method has been used (Guggenheim *et al.*, 1965; 1966a,b), in which conventional rats are continually given erythromycin in their drinking water to suppress streptococci. The animals are then infected with strains of streptococci made resistant to very high doses of the antibiotic.

These methods have been used to determine whether caries in animals results from a specific or nonspecific infection, and to investigate interactions between the cariogenic flora, and other factors such as the composition of the diet and the condition of the host. Parallel studies in man and in monkeys so far show that the disease is similar in all species, as discussed in the following sections, so that the animal experiments are being used to pilot the development of methods for caries reduction and prevention (Keyes *et al.*, 1966).

B. Interactions of Host, Diet, and Microbes

Caries incidence and severity are influenced by a variety of conditions (Teuscher *et al.*, 1948). All of these modify the host and his resistance, or the microbe and its virulence.

Host susceptibility is partly determined by conditions such as the spacing of teeth and the depth and arrangement of pits and fissures in the surfaces of molars. These lead to the impaction of food particles in enclosed spaces where bacteria can act on carbohydrate close to the tooth surface. Accordingly, Hoppert *et al.* (1932) found that certain grain diets were cariogenic to rats only if the grain particles were large enough to be trapped in the fissures of the molars by chewing; caries developed only in the fissures. Rats have unusual molars with very deep, narrow, angled grooves or sulci in the occlusal surface, and much of the work with rats is characterized by crevice decay with little or no decay of the smooth surfaces of the teeth (Fitzgerald *et al.*, 1960; Hoppert *et al.*, 1932; Orland *et al.*, 1954; Rosebury *et al.*, 1933).

Early decay in the human mouth is also due to food impaction in grooves of the molars. The fissures are hard to clean by brushing, and are the chief loci of decay in populations drinking fluoridated water (Reid and Grainger, 1955). Caries of crevices has been prevented experimentally by filling the grooves with adhesive plastics (Cueto and Buonocore, 1967).

The rate and volume of saliva flow modify the susceptibility of the host, since saliva can wash away or dilute food debris or the acid formed from it. In man, the condition of xerostomia or "dry mouth" is accompanied by rampant caries. A standard procedure for producing rampant caries in animals is surgical removal of the salivary glands (Haldi et al., 1967; Klapper and Volker, 1963). In a survey of thousands of inductees in the armed services, a correlation has been found between the unstimulated flow rate of parotid gland saliva and total caries experience (Shannon and Terry, 1965). A significant difference in flow rate was found between those having little or no decay, and those having large numbers of decayed, missing, or filled teeth. A similar significant difference in the rate of flow of saliva stimulated by chewing paraffin was found by Wright (1964). Groups of students with little or no decay were compared with other groups with large numbers of missing and filled teeth, and teeth with active, progressive decay at the time they were selected. This gave assurance that the latter group was still highly susceptible to the decay process. Students with high stimulated saliva flow rates had little or no decay, while low saliva flow rates were found in students with extensive decay. Ahrens (1965) examined a group of patients at the beginning and end of a year to determine, at each examination, total caries and the rate of flow of the whole unstimulated saliva. He found that instances of more than two new cavities a year were associated with low flow rates of unstimulated saliva. Other conditions change saliva flow, and thus, can affect caries development. Schmidt-Nielsen (1946), in a study of chemical composition of unstimulated saliva of a group of pregnant women, found that she could not get saliva samples *post partem* because of the low flow rates. Kullander and Sonesson (1965) have found decreases in the saliva flow rates of women in the late months of pregnancy and in the postmenopausal period.

The eating habits of the host also influence the course of the disease by controlling the availability of the diet to the oral microbes. Rodents on *ad libitum* feeding normally eat small amounts of a diet at a time with considerable frequency. This aids in the rapid development of destructive decay when it is desired for experimental purposes. In both the rat (König et al., 1968; Larson et al., 1962) and in the hamster (Gustafson et al., 1959) the restriction of feeding to a few larger meals a day decreases the amount of decay inversely to the frequency of feeding. The same pattern has been found true in man in the Vipeholm studies (Lundqvist, 1952; Gustafsson et al., 1953). Increased caries was not so much the result of the total sugar intake as it was related to the frequency with which high levels of sugar could be found in the saliva during the day. Sticky or chewy sweets, given frequently, were more cariogenic than sweet drinks between meals or sweets at meals only.

Rodents chew bedding materials, which stimulates saliva flow and rinses the mouth. Crude bedding materials also could contain extractives which modify the oral flora. Shibata (1929) found that rats in cages with straw bedding developed caries more slowly and of less severity than rats without bedding. Orland et al. (1954) mention reduced decay in germfree rats with bedding, but the result is complicated by a simultaneous elimination of sugar from the drinking water. Most investigators use screen-bottom cages, avoiding the problems with bedding (personal inquiries).

A discussion of the microbes causing caries must include the contributions of the diet to the process. In addition to the requirement that an experimental diet be consumed orally, it must also contain carbohydrate, or decay will not develop (Shaw, 1954), as would be expected from the acid theory of the disease. Not only has plaque from human teeth been shown to produce acid from carbohydrates *in vitro* (Miller, 1883), but also *in situ*, after rinsing the mouth with sugar solutions (de Crousaz et al., 1967; Kleinberg, 1961; Stephan, 1940, 1944). Most convincing is the work of Dirksen et al. (1962, 1963), who showed that the deep, advancing edges of active carious lesions are always acid, whereas the shallower layers and top surfaces are closer to the pH of the saliva. In enlarged and exposed cavities, which have emptied out their contents, the carious layer is shallow and more alkaline, probably because of saliva washing, and the decay process becomes arrested. A number of studies have shown that sucrose is more cariogenic for experimental animals than other sugars, starch, flour, white or whole grain breads (Grenby, 1963; König, 1967; Krasse, 1965a,b). This is paralleled in man by the observation that caries is rare in countries where the major part of the diet is grain (MacGregor, 1964; Sinrod, 1965), but that introduction of sweets is followed by the appearance of extensive decay (MacGregor, 1964).

Phosphates, both inorganic and organic, added at levels of about 1.5% to the diets of experimental animals, reduce caries by means not as yet explained. The soluble salts are more active than the insoluble calcium phosphates, and some of the polymerized phosphates are as active or more active than orthophosphates (Nizel and Harris, 1964). It is not clear whether the effect is one of reducing the solubility of enamel, or whether it is exerted on the bacteria alone.

Attempts to reduce caries in man by addition of phosphate to the diet have not been generally successful in prolonged trials (Averill et al., 1966; Ship and Mickelsen, 1963; Strålfors, 1963). However, these studies were complicated by the need to use insoluble calcium phosphate to reduce the objectionable taste of the high levels of additive, which may have reduced the effectiveness of the salts. More recent studies used soluble calcium sucrose phosphate added to all the carbohydrate in the diet of institutionalized children. An unusually high caries rate of about 7 new decayed, missing, or filled surfaces per year was reduced to the still somewhat high figure of about 5 new surfaces a year (Harris et al., 1967). Stookey and co-workers (1967) claim to have reduced caries in clinical trials with sodium phosphate added only to breakfast cereals, a very small part of the total diet, eaten usually once a day.

A possible reason why dietary phosphates protect animals better than man is the low phosphate content of rat and hamster saliva, which is about one-tenth that in human saliva (Ericsson, 1962; Holloway and Williams, 1965). At such low levels of phosphate, plaque microbes will grow partly by extracting phosphate from the teeth, especially if they can attach themselves firmly to the enamel and lower the pH to solubilize tooth mineral. This would be true particularly with high sugar diets since plaque organisms have been shown to take up greater amounts of phosphate after sugar intake (Dawes and Jenkins, 1962; Luoma, 1964). Conversely, the presence of much soluble phosphate in the saliva will satisfy their requirement thereby sparing the tooth mineral. Adding soluble phosphate to the animal diet will produce a great change in available saliva phosphate, but this does not occur in man.

Fluorides added to the diet of man and of animals reduces decay, but it is not certain how much of the effect is due to a direct action of the fluoride on microbial metabolism. Hardwick and Leach (1962) showed that the plaque may accumulate fluoride to levels which can inhibit fermentation, and it has been shown that concentrations of fluoride too low to inhibit fermentation will reduce microbial synthesis of intracellular storage polysaccharides (Weiss et al., 1964). The topical application of high concentrations of fluoride in hamsters (Keyes et al., 1966) reduced or prevented plaque formation in experiments where decay was halted and periodontal disease prevented. Reid and Grainger (1955) found that fluoride in drinking water reduced decay in man only on smooth surfaces of the teeth, but not in the occlusal pits of the molars. The effects may be exerted primarily on certain bacteria that are discussed below, and which are especially active in decay of smooth surfaces.

The demonstration that microbes were indispensable to caries was followed by tests, in germfree rats, of an enterococcus culture together with either a proteolytic rod or an anaerobic rod. The combinations produced decay, but it was not shown what part any one of the cultures played in the process, or whether all were needed (Orland et al., 1955). Since then Fitzgerald et al. (1960) produced caries in germfree rats with a single streptococcus strain which was typically acidogenic, and did not digest gelatin, collagen, or casein, and which was isolated from carious matter from conventional rats. More recently, lactobacilli alone have been able to cause caries in germfree rats (Fitzgerald et al., 1966; Rosen et al., 1968). It has been demonstrated that many acidogenic microbes have the potential to participate in the decay process in the rat, if they can establish themselves in the mouth. The lesions are always fissure-type, and never are found between the teeth, or along the gums and on the smooth surfaces of the crowns, where they are frequently found in man. Further difficulties with this model system occur because the causative bacteria have no competitors. Lactobacilli, which normally comprise a minute fraction of the plaque flora of man (Gibbons, 1964), assume an unrealistic cariogenic potential in the otherwise germfree rat.

The model becomes even less satisfactory when results in germfree rats are compared with those in caries-resistant, but otherwise conventional hamsters. Fitzgerald and Keyes (1960) reinfected such hamsters with a number of different *Streptococcus* and *Lactobacillus* species isolated from hamsters with decay, but were unable to induce caries in the resistant animals with most of them. Only certain streptococcus isolates could cause decay in the resistant animals, all of a type able to synthesize a tough, tenacious dextran from sucrose. Decay in the hamsters was characterized by the production of an adhering microbial mass on smooth surfaces of the molars, which was followed by decay. In resistant animals, free of such streptococci, the film failed to form on the molars despite the feeding of a high sucrose diet, and decay did not develop though other streptococci were present.

It has since been shown that similar streptococci are found in large numbers in the dental plaque of man, and that they will cause decay in the resistant hamsters (Krasse, 1966; Zinner et al., 1965), in germfree rats (Gibbons et al., 1966), and in conventional rats in which resident streptococci are suppressed with erythromycin (Guggenheim et al., 1965). Monkeys on a high sucrose diet also develop decay together with a plaque consisting of dextran-producing streptococci (Bowen and Cornick, 1967), and are susceptible to several cultures from man (Bowen, 1968).

Similar conditions prevail in man as well. The pitted surfaces of the molars are very susceptible, and decay usually appears there first probably because of food retention, before any extensive damage to smooth surfaces (König, 1963; Reid and Grainger, 1955). The presence of sucrose in the diet of man hastens decay; its absence reduces decay, even when the diet is primarily carbohydrate (MacGregor, 1964). Addition of sugar increases the caries rate among persons on controlled diets when it is given frequently in sticky forms (Lundqvist, 1952; Gustafsson et al., 1953). Extensive decay, especially around the necks of the lower incisors, is common among confectioners and cake bakers (Gerke and Klemt, 1955).

More recently, persons suffering from hereditary fructose intolerance and not able to eat any sucrose-containing foods, have been found to have little or no decay despite poor mouth care (Cornblath et al., 1963; Froesch et al., 1963; Marthaler and Froesch, 1967). Aside from the absence of sweets, their diets are the same as those of their families and neighbors, as high in starches and as "refined." Members of their families not suffering from the genetic disease and eating sweets have the expected decay experience.

As mentioned above, the dextran-forming bacteria that have been isolated from carious teeth in man, are capable of causing decay in animals on sucrose diets. The polysaccharides, dextran and levan, have been found in the microbial plaque of human teeth, and have been shown to increase when sucrose is eaten (Gibbons and Banghart, 1967; McDougall, 1964; Wood, 1967). Carlsson and Egelberg (1965) found that sugar-free, and glucose or fructose diets produce slight to moderate amounts of plaque on human teeth, whereas sucrose diets produce voluminous, tough plaque resembling the colonies produced on sucrose agar by dextran-producing streptococci. *In vitro*, these cultures stick to the walls of the culture vessels (Gibbons and Banghart, 1967), to wires or to other smooth objects suspended in them (Jordan and Keyes, 1966). Cracking, by colony growth, occurs on the agar surfaces on which they grow (Edwardsson, 1968). They produce extracellular enzymes that form dextran gels from the sucrose in which the cells are imbedded (Dahlqvist et al., 1967; Guggenheim and Schroeder, 1967). Some of the cultures also produce a fructose-containing polysaccharide from sucrose, probably a levan, in addition to the dextran (Dahlqvist et al., 1967). Others may form small amounts of dextran even in the absence of sucrose (Gibbons and Banghart, 1967).

None of the streptococci so far tested fall into Lancefield serological groups, but seem to be related, serologically, either to the hamster strain of Fitzgerald and Keyes (1960), the rat strain of Fitzgerald *et al.* (1960), or to both (Jablon and Zinner, 1966; Zinner *et al.*, 1965). Biochemically, they are almost indistinguishable, and resemble *Streptococcus sanguis*, *S. mutans*, and *S. salivarius* in their sugar fermentations, hemolysis patterns, and colonial morphology (Edwardsson, 1968; Jablon and Zinner, 1966).

The levan produced from sucrose by some of the cultures is readily fermentable by dental plaque with the resultant formation of acid (Manly and Dain, 1963; Wood, 1964). In addition, an intracellular, iodine-staining, glycogenlike polysaccharide is formed by many oral microbes, including the dextran-forming streptococci, during growth on various sugars. This material accumulates when excess sugar is present, and is metabolized resulting in acid formation when sugar in the medium is depleted (Gibbons, 1964; Gibbons and Socransky, 1962). These are possibly the methods by which the plaque maintains a low pH for prolonged periods after eating of sugar (Stephan, 1940; de Crousaz *et al.*, 1967).

C. Special Role of Sucrose, and Microbial Dextran and Levan Formation

From the preceding discussion, it is evident that sucrose is a major cause of tooth decay in civilized communities, probably because of its role in dextran formation. However, microbial dextrans differ from one another in many properties such as solubility, degree of branching, and digestibility with enzymes. In addition, not all dextran-producing streptococci are cariogenic in experimental animals (Tsuchiya *et al.*, 1952; Bowen, 1968). The dextrans produced by cariogenic strains are primarily α-1,6-linked glucans, with branches at the 1,2-, 1,3-, and sometimes 1,4-positions (Dahlqvist *et al.*, 1967; Guggenheim and Schroeder, 1967). They are produced by extracellular enzymes that are partly free in the medium, and partly bound to the cell surface.

Dextrans have a number of properties that help them to build plaque. They are strongly adsorbed to hydroxyapatite, the chief tooth mineral, and they can form insoluble precipitates with some saliva proteins (Gibbons and Banghart, 1967). As they are formed on agar plates and on teeth, they are tough and difficult to remove, and are a means by which a strain can crowd out competitors by simple displacement. They are not degraded readily by other organisms in the plaque (Wood, 1967).

Levans, on the other hand, are water soluble in high concentration, and cannot act as strong structural components of plaque matrix. However, their large molecular size prevents them from diffusing from the plaque once they are formed, and they remain as a storage carbohydrate which is fermented readily, once other carbohydrates are exhausted.

Dextrans and levans also have immunological properties which will be discussed here, although their application is related more to the development of periodontal disease. When injected intravenously into various experimental animals, levans and dextrans promote the peritoneal infection of animals by very small inocula of pathogens, the effect being greater with the larger polymers. They prevent the diffusion of antibodies and diapedesis of leukocytes into experimental skin infections, as well as diffusion of trypan blue from the blood into injured skin. They act as a blockade in the skin against the Schwartzman reaction, interfere with the formation of granulation tissue, and cause withdrawal of leukocytes from the circulation (Davies *et al.*, 1955; Hestrin, 1956; Hestrin and Davies, 1956; Hestrin *et al.*, 1954a,b; Wolman and Wolman, 1956). These properties were displayed primarily by intravenous injection, but diffusion of antibodies and trypan blue in the injured skin was not stopped by local injection of levan. However, we must consider the possibility that local presence of the polysaccharides promote the invasiveness of some plaque microbes, just as other gums and gastric mucin do.

D. Possible Connection with Other Diseases

The production of dextrans from sucrose is a property shared by other medically important streptococci. Hehre and Neill (1946) found that 22 out of 48 strains of streptococci, isolated from patients with subacute bacterial endocarditis, produced dextran from sucrose, and one strain also produced levan. Porterfield (1950) found that 13 out of 29 strains of streptococci from cases of subacute bacterial endocarditis produced dextran from sucrose, as did also 3 type cultures of *Streptococcus sanguis*, and 10 out of 80 cultures of streptococci isolated from the blood of persons after dental extractions. Conner *et al.* (1967) found that periodontal scaling of the teeth in patients with healthy or diseased gums resulted in bacteremias in 22 to 50% of the patients, respectively.

Gibbons and Banghart (1968) showed that a dextran-forming streptococcus isolated from a case of subacute bacterial endocarditis was able to cause tooth decay in germfree rats.

It thus appears likely that tooth decay and subacute bacterial endocarditis are caused, at least in part, by the same kinds of streptococcus, which may gain entrance to the body through various dental operations or possibly by the loss of deciduous teeth in children.

E. Treatment by Antimicrobial Agencies and Other Means

From the present understanding of the interactions of host, microbes, and diet in caries development, control might be exerted over the whole process through control of any one of the interactions. Decay in the occlusal crevices of molars has already been controlled experimentally by filling them prophylactically with an adhesive plastic to prevent food impaction and microbial growth (Cueto and Buonocore, 1967). Increasing the saliva flow or rinsing the mouth, has been shown to flush food particles from the mouth, whether with water (Lundqvist, 1952), chewing gum (Volker, 1948), chewing apple slices (Slack and Martin, 1958), or by sucking acidulated sorbitol tablets (Clark et al., 1961). In two-year clinical trials with the sorbitol tablets taken after every meal and snack, small but significant reductions in caries were obtained (Slack et al., 1964; Goose et al., 1964). There was difficulty in maintaining cooperation of test subjects for 2 years, because the tablets were not completely acceptable, and many subjects dropped out. Chewing gum taken only after meals did not result in increased or decreased decay (Volker, 1948; Toto et al., 1960). A pack of gum a day for 30 months chewed between meals, after school, and after supper for a total of 100 minutes a day by institutionalized children had little effect on decay (Finn and Jamison, 1967). These children were on a diet with low sugar and had limited access to sweets. They exhibited a very low caries rate of about 2 new cavities a year not only in this study, but also in a previous 2-year dentifrice study (Finn and Jamison, 1963).

Since it is not the sugar taken at meals, but as snacks between meals which is responsible for high decay rates (Lundqvist, 1952; Rosenstein, 1966) it would be expected that timing these saliva-stimulating methods to snack periods, and selection of test groups that eat frequently between meals would show the greatest effect and utility of the method for confirmed between-meals nibblers. Gerke and Klemt (1955) were able to reduce the decay rate of confectioners and bakers from 4 to less than 2 new cavities a year, by having them chew gum constantly during working hours, when they might be tasting their products. The greatest reduction was in the incidence of cervical decay. From the reports of de Crousaz et al. (1967), such results might be expected. Using a pH probe built into a dental prosthetic device, they were able to monitor pH changes in interproximal spaces in vivo, and found that the profound drop in pH that follows the eating of carbohydrates was reversed in minutes upon the chewing of a neutral tablet. The work demonstrated the ability of chewing to rinse poorly accessible surfaces, and the buffering capacity of stimulated saliva.

Elimination of sucrose from the diet has been effective in reducing or eliminating decay in persons with hereditary fructose intolerance. Although it is unlikely that sweets can be eliminated from the diet, they can be made less harmful by restricting them to mealtimes, or by substituting less cariogenic snacks. In this country, sorbitol-based candies are sold as noncariogenic sweets, and in Sweden, candies are sold which are made from a starch syrup partially reduced, so that sorbitol replaces the free glucose and the reducing end groups of the dextrins. The syrup and candy made from it is much less fermentable by microbes of the saliva and plaque, and when it is fed to rats and hamsters, it is less cariogenic than sucrose (Frostel, 1963; Frostel et al., 1967).

Removing the bacteria or blocking their activies can bring some reduction in decay. Children taking high doses of penicillin to control subacute bacterial endocarditis have been found to have progressively less decay the longer they have been under treatment (Handleman et al., 1966). Addition of penicillin to dentifrices has had mixed results (Hill et al., 1953a), and the development of penicillin resistance by the streptococci of subjects using penicillin dentifrices made this approach undesirable (Hill et al., 1953b).

As yet, no inhibitor has been found to prevent dextran synthesis by plaque organisms, but dextrans have been dissolved enzymatically (Bowen, 1968; Fitzgerald et al., 1968a; Guggenheim and Schroeder, 1967). From in vitro work, it appears that dextrans of cariogenic bacteria, once formed, are only partly degraded by *Penicillium funiculosum* dextranase. By adding the enzyme in drinking water, Fitzgerald et al. (1968b) were able to reduce plaque formation to some extent in hamsters on a sucrose diet, and also the severity of decay.

Carlsson and Krasse (1968) found that antisera of animals immunized with dextran-forming streptococci are inhibitory to dextran-forming enzymes. They propose possible immunization against caries by using a similar vaccine in man. Some support is given to the idea by Rovelstad (1967) who found that the blood sera of naval recruits who remained caries-free for 1 year in the service had opsonins against some dextran-forming streptococci. This apparently is rare, because de Crousaz and Guggenheim (1966) were unable to find, using fluorescent antibody techniques, antibodies against cariogenic dextran-forming streptococci in over 200 sera. Presumably, such protection would be primarily against smooth surface lesions, and might have little effect on crevice lesions.

Prevention of new decay by direct antimicrobial action of fillings was shown by Lind *et al.*, (1964). Decay on one of two contacting tooth surfaces is usually followed soon after by decay of the second surface, even if the first has meanwhile been filled. Silver fillings do not, apparently, delay this process, but porcelain fillings, which contain fluoride, and copper amalgam fillings were found to reduce significantly the numbers of new cavities in neighboring teeth. Since direct fluoride applications have been found clinically useful in decay reduction, some similar use might be found for dilute copper solutions, especially since copper is known to complex with carbohydrates and proteins, and to precipitate with phosphates. This would tend to fix it in plaques, where small amounts dissolved from copper amalgam fillings have been effective.

III. Periodontal Disease in Man and Animals

A. METHODS OF STUDY

Periodontal disease in man and animals has long been studied by microscopic examination of sections from diseased tissues, but the sections usually are not stained to show bacteria, and are examined at powers too low to see microbes, except in large masses (Haberman, 1959). Primarily, they have been stained to show inflammation types of tissue reaction. Biopsy and post-mortem samples of normal and inflamed gingivae, freshly extracted, periodontally involved teeth, and portions of the jaws of cadavers, dogs, and of smaller animals have been examined to determine the sequence of events and causal relationships among them (Held, 1950).

Clinical observations have linked severity of disease with the degree of accumulation on the teeth of microbial plaque and calculus, usually called "debris" (Littleton, 1963). Samples of this microbial mass, taken from diseased and from clinically healthy areas in the gingival crevice, have been examined microscopically, and viable counts of the microbial varieties have been made to determine differences in their qualitative or quantitative composition.

Some investigators have used small animals in the same way as they have used them in caries research. Experimental diets have been devised which produce periodontal disease in the animals, and specific microbes from diseased lesions have been tested in animals made resistant by controlling the naturally occurring oral flora. Antibiotics have been tested to control infection, and to establish the bacterial causation of the disease states. Examples of these clinical and experimental approaches are given below.

B. Interactions of Microbes, Host, and Diet

The examination of large numbers of extracted teeth and sections of human jaws led Waerhaug (1952) to conclude that all periodontal disease is a response to microbial irritation. Even when calculus is present, he considered the irritation to be not from the hard accretions, but from the soft plaque overlying them (Waerhaug, 1956). On teeth with no loss of periodontal ligament, microbial and calculus accumulations adhere to the enamel about 2 mm. from the ligament, and are separated from it only by a collar of epithelial cells lining the gingival crevice and adhering weakly to the enamel. As the disease progresses, the plaque and calculus approach the ligament and move down onto the root, while the ligament disintegrates, so that they are about 1 mm. from the ligament in deep lesions. Observations made in the present author's laboratory show that most teeth with deep pockets all around the root before extraction have less than 0.25 mm. between the plaque and remaining ligament at one or more points (Hoffman, 1968). As the microbial plaque approaches the crest of bone supporting the teeth, resorption of the bone begins. Sections show leukocytes moving out of the gingival tissue into the space between the epithelial lining of the pocket and the microbial plaque. Waerhaug interprets the series of events as an irritation of gingival and bony tissue by the bacteria, followed by an inflammatory tissue response that leads to detachment of the ligament and bone resorption. In this concept, the microbes and irritation are nonspecific, and the damage results from an autolytic response of the host tissue. Waerhaug's observations, however, are incomplete in that no specific bacterial stains were used, and the sections were examined under low magnifications. The masses of plaque were visible, but it was not determined whether a few bacterial cells preceded the margin of the plaque mass and were located at the edge of the periodontal ligament, digesting and loosening the attachment to the tooth. It was also impossible to see whether there was any microbial invasion of the gum tissues.

Most sections of inflamed gingival tissue stained for bacteria do not show evidences of invasion according to Gibson and Shannon (1964). However, by means of fluorescent antibody techniques, Tsutsui *et al.* (1968) have shown both *Staphylococcus aureus* and *Streptococcus salivarius* to be contained inside macrophages and other leukocytes in the subepithelial connective tissue of inflamed gingival biopsy samples. In five clinically healthy samples, there were no fluorescent cells.

The microscopic and viable components of the gingival plaque from man include streptococci, filamentous and diphtheroid organisms, fusobacteria, gram-negative cocci, spirochetes, and others. Their relative numbers do not appear to differ in plaque from healthy or diseased sites, but the amount of plaque removable from the gingival crevice is increased severalfold in the diseased condition (Gibbons *et al.*, 1963; Socransky *et al.*, 1963).

Löe *et al.* (1965) studied the succession of microbial forms in the surface of the plaque when brushing and oral hygiene are stopped. They found that the plaque consists first of gram-positive cocci and small rods, but after a few days contains many filamentous forms. Finally, vibrios and spirochetes appear and gingivitis follows shortly. As cleansing and brushing are reinstituted, the organisms disappear in reverse order; however, some gram-positive cocci are always found at the gingival margin.

Streptococcus mitis has been found in large numbers in the human gingival crevice (Dwyer and Socransky, 1968), and some of its biochemical activities fit the species to a possible role in periodontal disease. Some strains produce collagenase (Schultz-Haudt *et al.*, 1954), and some produce hyaluronidase (Parikh *et al.*, 1965). The species is also described as usually forming dextran on high sucrose-containing media (Breed *et al.*, 1957).

There is much evidence that the bone loss and destruction of the periodontal ligament in experimental animals results from a microbial infection. This is not to say that metabolic derangements of the host or physical damage to the gums play no part in bone loss; aging germfree mice and rats may undergo loss of alveolar bone in the absence of infection (Baer and Fitzgerald, 1966; Baer and Newton, 1960; Baer *et al.*, 1964). However, in young animals, specific infections have been found to induce not only bone loss but also soft tissue damage which do not develop in uninfected controls. For instance, Keyes and Jordan (1964) produced periodontal disease in hamsters on a sucrose diet. The causative agent was shown to be a filamentous, actinomyces-like organism which forms a levan from sucrose, and produces a tenacious plaque that invades the gingival crevice (Howell and Jordan, 1967; Jordan and Keyes, 1965). Gibbons et al., (1966) reported the production of both dental caries and also alveolar bone loss in germfree rats infected with dextran-forming streptococci of human origin. In 1968, Gibbons et al. also produced caries of the roots of the molars in germfree rats infected with a levan-forming streptococcus, but only with a high sucrose diet, not with a low sucrose diet.

Auskaps et al. (1957) produced periodontal disease in rice rats only on carbohydrate-containing diets. The agent or agents responsible do not appear to be streptococci, but may be one or more of several different rods isolated from diseased animals and inoculated into rice rats whose normal flora were suppressed with antibiotics (Dick et al., 1968).

Sucrose may play a similar role in the diet of man, but periodontal disease is found universally in places like Ethiopia where sucrose is scarce in the diet and caries rare (Littleton, 1963). Dextrans have not yet been shown to be a major constituent of the dental plaque under these conditions, but some streptococci form dextrans even in glucose media (Gibbons and Banghart, 1967) so that this possibility may exist. Specific roles for the filamentous and diphtheroid microbes in the gingival plaque of man have not yet been demonstrated.

C. Treatment with Antimicrobial Agencies

The periodontal disease produced with rice rats on carbohydrate diets could be completely controlled by introduction of penicillin into the diet, water, or by injection (Shaw and Dick, 1966). Keyes et al. (1966) tested a number of treatments in hamsters infected both with dextran-forming streptococci and with the levan-forming filamentous culture, Odontomyces viscosus. He found that acidulated fluoride gels, and to some extent neutral fluoride gels, applied topically, and the antibiotics spiramycin and vancomycin were able to reduce plaque formation, tooth decay, and alveolar bone loss. Cessation of treatment allowed resumption of active disease with cavitation and bone resorption.

Spiramycin has already been used in the clinical control of periodontal disease in man (Harvey, 1961; Hess, 1958), and vancomycin, applied to the teeth of mentally retarded patients as an adhesive paste, reduced both the accumulation of plaque and the incidence of gingivitis (Mitchell and Holmes, 1965). Both antibiotics, as well as penicillin, have a narrow spectrum of activity primarily against gram-positive bacteria. Their activity indicates that the primary incitants of gingivitis and of periodontal disease in man are similar to those in experimental animals.

Physical removal of the plaque by brushing and cleaning methods reduces gingivitis in clinical studies in man. Brandtzaeg and Jamison (1964a,b) reported improved health of the gums in Norwegian army recruits after instruction in and institution of a program of proper oral hygiene.

IV. Perspectives

The chief interactions between host, microbe, and diet in the causation of dental caries are sufficiently well established that rational rather than empirical research programs for caries prevention can be and have been formulated. Some approaches would involve modification of the host, such as methods of filling occlusal fissures of the molars with permanent materials without drilling to prepare a base, or immunizing the host with an effective vaccine. Others would try to control the activities of dextran-forming bacteria by methods such as improved physical or chemical removal of plaque from the tooth surface, or by specific means of inhibiting dextran formation. Conventional sweets such as candy and cookies can be made without sucrose, to reduce their cariogenicity, although some technical problems arise in substituting invert sugar or dextrose. The tools for testing cariogenicity in the laboratory are already available and sufficiently trustworthy to predict the probability of success in clinical trials.

On the other hand, the causes of periodontal diseases in man are still clouded, and much has to be done to show that specific microbes from human periodontal lesions will cause similar injuries to the soft tissues and alveolar bone in conventional animals, as has been done in the case of dental caries. This will be a necessary prelude to the testing of vaccines against the invading organisms, which would be preferred for preventive or curative measures. Although the prolonged use of antibiotics is not popular because of the possibility of developing resistant strains of pathogens, the effectiveness of narrow-spectrum antibiotics in controlling periodontal disease in man and animals makes an exception possible. Antibiotics can be used which have little or no commercial application because of their restricted activity, and which do not produce cross resistance to those commonly used. In conjunction with present surgical and other manipulative procedures, somewhat prolonged antibiotic treatment will not only reduce possibility of damage from bacteremia, but also promote more rapid healing and a faster return to unirritated conditions in cases requiring more drastic treatment.

V. Summary

Tooth decay has been studied recently by experimental methods which permit the satisfactory application of Koch's postulates in defining caries as an infectious disease. Germfree or antibiotic-treated animals, fed cariogenic diets, have been found to develop tooth decay with specific infections, which can be transmitted to other animals.

Susceptibility of the host is modified by a number of conditions, such as the depth and shape of tooth crevices, spacing of teeth, rate of saliva flow, and frequency of eating.

The diet must contain carbohydrate for the bacteria to cause decay. Sucrose is much more cariogenic than other sugars, both in animals and in man, because of the ability of certain streptococci to form tough, insoluble dextrans from it. The polysaccharide anchors the bacteria to the smooth surfaces of teeth promoting decay. Many of the same streptococci also form levans from sucrose, which are then fermented to form acid after other fermentable carbohydrates are used up.

The discovery of the special connection of sucrose with tooth decay, and the demonstration of the interactions of modifying conditions have made possible rational approaches to the control of caries.

Periodontal disease, particularly the destruction of the periodontal ligament and the resorption of the bony support of the teeth, has been produced experimentally in animals, and controlled with antibiotics. The infections in some cases are very specific, and require sucrose in the diet. As yet, the disease has not been produced in animals with microbes from the human mouth other than with streptococci. Some cases of periodontal disease in man have been managed clinically with the antibiotic spiramycin.

REFERENCES

Ahrens, G. (1965). *Advan. Fluorine Res.* 3, 227–231.
Auskaps, A. M., Gupta, O. P., and Shaw, J. H. (1957). *J. Nutr.* 63, 325–343.
Averill, H. M., Freire, P. S., and Bibby, B. G. (1966). *Arch. Oral Biol.* 11, 315–322.
Baer, P. N., and Fitzgerald, R. J. (1966). *J. Dental Res.* 45, 406.
Baer, P. N., and Newton, W. L. (1960). *Oral Surg. Oral Med. Oral Pathol.* 13, 1134–1144.
Baer, P. N., Newton, W. L., and White, C. L. (1964). *J. Periodontol.* 35, 388–396.
Bowen, W. H. (1968). *Brit. Dental J.* 124, 347–349.
Bowen, W. H., and Cornick, D. E. (1967). *Helv. Odontol. Acta* 11, 27–31.
Brandtzaeg, P., and Jamison, H. C. (1964a). *J. Periodontol.* 35, 302–307.

Brandtzaeg, P., and Jamison, H. C. (1964b). *J. Periodontol.* **35**, 308-312.
Breed, R. S., Murray, E. G. D., and Smith, N. R. (1957). "Bergey's Manual of Determinative Bacteriology," 7th ed. Williams & Wilkins, Baltimore, Maryland.
Carlsson, J., and Egelberg, J. (1965). *Orontol. Rev.* **16**, 112-125.
Carlsson, J., and Krasse, B. (1968). *Arch. Oral Biol.* **13**, 849-852.
Clark, R., Hay, D. I., Schram, C. J., and Wagg, B. J. (1961). *Brit. Dental J.* **111**, 244-248.
Conner, H. D., Haberman, S., Collings, C. K., and Winford, T. E. (1967). *J. Periodontol.* **38**, 466-472.
Cornblath, M., Rosenthal, I. M., Reisner, S. H., Wybregt, S. H., and Crane, R. K. (1963). *New Engl. J. Med.* **269**, 1271-1278.
Cueto, E. I., and Buonocore, M. G. (1967). *J. Am. Dental Assoc.* **75**, 121-128.
Dahlqvist, A., Krasse, B., Olsson, I., and Gardell, S. (1967). *Helv. Odontol. Acta* **11**, 15-21.
Davies, A. M., Shilo, M., and Hestrin, S. (1955). *Brit. J. Exptl. Pathol.* **36**, 500-506.
Dawes, C., and Jenkins, G. N. (1962). *Arch. Oral Biol.* **7**, 161-172.
de Crousaz, P. A., and Guggenheim, B. (1966). *Helv. Odontol. Acta* **10**, 38-40.
de Crousaz, P. A., Graf, H., and Bibby, B. G. (1967). *Helv. Odontol. Acta* **11**, 157-162.
Dick, D. S., Shaw, J. H., and Socransky, S. S. (1968). *Arch. Oral Biol.* **13**, 215-228.
Dirksen, T. R., Little, M. F., Bibby, B. G., and Crump, S. L. (1962). *Arch. Oral Biol.* **7**, 49-58.
Dirksen, T. R., Little, M. F., and Bibby, B. G. (1963). *Arch. Oral Biol.* **8**, 91-97.
Dunbar, J. B., Wolff, A. E., Volker, J. F., and Moller, P. (1968). *Arch. Oral Biol.* **13**, 387-406.
Dwyer, D. M., and Socransky, S. S. (1968). *Brit. Dental J.* **124**, 560-564.
Edwardsson, S. (1968). *Arch. Oral Biol.* **13**, 637-646.
Ericsson, Y. (1962). *Advan. Fluorine Res.* **1**, 327-336.
Finn, S. B., and Jamison, H. C. (1963). *J. Dent. Children* **30**, 17-25.
Finn, S. B., and Jamison, H. C. (1967). *J. Am. Dental Assoc.* **74**, 987-995.
Fitzgerald, D. B., and Fitzgerald, R. J. (1965). *Arch. Oral Biol.* **11**, 139-140.
Fitzgerald, R. J., and Keyes, P. H. (1960). *J. Am. Dental Assoc.* **61**, 23-33.
Fitzgerald, R. J., and Keyes, P. H. (1963). *Am. J. Pathol.* **42**, 759-772.
Fitzgerald, R. J., Jordan, H. V., and Stanley, H. R. (1960). *J. Dental Res.* **39**, 923-935.
Fitzgerald, R. J., Jordan, H. V., and Archard, H. O. (1966). *Arch. Oral Biol.* **11**, 473-476.
Fitzgerald, R. J., Spinell, D. M., and Stoudt, T. H. (1968a). *Arch. Oral Biol.* **13**, 125-128.
Fitzgerald, R. J., Keyes, P. H., Stoudt, T. H., and Spinell, D. M. (1968b). *J. Am. Dental Assoc.* **76**, 301-304.
Froesch, E. R., Wolf, E. P., Baitsch, H., Prader, A., and Labhart, A. (1963). *Am. J. Med.* **34**, 151-167.
Frostel, G. (1963). *Svensk Tandläkare Tidskr.* **18**, 1-14.
Frostel, G., Keyes, P. H., and Larsen, R. H. (1967). *J. Nutr.* **93**, 65-76.
Gerke, J., and Klemt, W. (1955). *Zahnarztl. Rundschau.* pp. 499-506.
Gibbons, R. J. (1964). *J. Dental Res.* **43**, 1021-1027.
Gibbons, R. J., and Banghart, S. (1967). *Arch. Oral Biol.* **12**, 11-24.
Gibbons, R. J., and Banghart, S. (1968). *Arch. Oral Biol.* **13**, 297-308.
Gibbons, R. J., and Socransky, S. S. (1962). *Arch. Oral Biol.* **7**, 73-80.
Gibbons, R. J., Socransky, S. S., Sawyer, S., Kapsimalis, B., and MacDonald, J. B. (1963). *Arch. Oral Biol.* **8**, 281-289.
Gibbons, R. J., Berman, K. S., Knoettner, P., and Kapsimalis, B. (1966). *Arch. Oral Biol.* **11**, 549-560.

Gibson, W. A., and Shannon, I. L. (1964). *Periodontics* **2**, 119–121.
Goose, D. H., Hartles, R. L., and Tweedie, M. C. K. (1964). *Brit. Dental J.* **117**, 283–286.
Grenby, T. H. (1963). *Arch. Oral Biol.* **8**, 27–30.
Guggenheim, B., and Schroeder, H. E. (1967). *Helv. Odontol. Acta* **11**, 131–151.
Guggenheim, B., König, K. G., and Mühlemann, H. R. (1965). *Helv. Odontol. Acta* **9**, 121–129.
Guggenheim, B., König, K. G., Herzog, E., Mühlemann, H. R. (1966a). *Helv. Odontol. Acta* **10**, 100–113.
Guggenheim, B., König, K. G., and Mühlemann, H. R. (1966b). *Pathol. Microbiol.* **29**, 656–662.
Gustafson, G., Stelling, Em., Abramson, E., and Brunius, E. (1959). *Arch. Oral Biol.* **1**, 42–47.
Gustafsson, B. E., Quensel, C. E., Lanke, L. S., Lundqvist, C., Grahnen, H., Bonow, B. E., and Krasse, B. (1953). *Acta Odontol. Scand.* **11**, 232–363.
Haberman, S. (1959). *J. Periodontol.* **30**, 190–195.
Haldi, J., Law, M. L., and John, K. (1967). *J. Dental Res.* **46**, 739–741.
Handleman, S. L., Mills, J. R., and Hawes, R. R. (1966). *J. Oral Therap. Pharmacol.* **2**, 338–345.
Hardwick, J. L., and Leach, S. A. (1962). *Advan. Fluorine Res.* **1**, 151–158.
Harris, R., Schamschula, R. G., Gregory, G., Roots, M., and Beveridge, J. (1967). *Australian Dental J.* **12**, 105–113.
Harvey, R. F. (1961). *J. Canad. Dental Assoc.* **27**, 576–585.
Hehre, E. J., and Neill, J. M. (1946). *J. Exptl. Med.* **83**, 147–162.
Held, A. J. (1950). *Intern. Dental J.* **1**, 75–100.
Hess, J. C. (1958). *Rev. Franc. Odonto-stomatol.* **5**, 1560–1564.
Hestrin, S. (1956). *Ann. N. Y. Acad. Sci.* **66**, 401–409.
Hestrin, S., and Davies, A. M. (1956). *Brit. J. Exptl. Pathol.* **37**, 235–238.
Hestrin, S., Shilo, M., and Feingold, D. S. (1954a). *Brit. J. Pathol.* **35**, 107–111.
Hestrin, S., Shilo, M., Feingold, D. S., and Wolman, B. (1954b). *Brit. J. Exptl. Pathol.* **35**, 112–117.
Hill, T. J., Sims, J., and Newman, M. (1953a). *J. Dental Res.* **32**, 448–452.
Hill, T. J., Rasch, C., and Wolpert, B. (1953b). *J. Dental Res.* **32**, 453–457.
Hoffman, I. D. (1968). M.S. Thesis, New York University College of Dentistry, New York.
Holloway, P. J., and Williams, R. A. D. (1965). *Arch. Oral Biol.* **10**, 237–244.
Hoppert, C. A., Webber, P. A., and Canniff, T. L. (1932). *J. Dental Res.* **12**, 161–173.
Howell, A., Jr., and Jordan, H. V. (1967). *Arch. Oral Biol.* **12**, 571–573.
Jablon, J. M., and Zinner, D. D. (1966). *J. Bacteriol.* **92**, 1590–1596.
Jordan, H. V., and Keyes, P. H. (1965). *Am. J. Pathol.* **46**, 843–857.
Jordan, H. V., and Keyes, P. H. (1966). *Arch. Oral Biol.* **11**, 793–801.
Keyes, P. H. (1946). *J. Dental Res.* **25**, 341–353.
Keyes, P. H. (1960). *Arch. Oral Biol.* **1**, 304–320.
Keyes, P. H., and Jordan, H. V. (1964). *Arch. Oral Biol.* **9**, 377–400.
Keyes, P. H., Rowberry, S. A., Englander, H. R., and Fitzgerald, R. J. (1966). *J. Oral Therap. Pharmacol.* **3**, 157–173.
Kite, O. W., Shaw, J. H., and Sognnaes, R. F. (1950). *J. Nutr.* **42**, 89–106.
Klapper, C. E., and Volker, J. F. (1963). *J. Dental Res.* **42**, 763–767.
Kleinberg, I. (1961). *J. Dental Res.* **40**, 1087–1111.

König, K. G. (1963). *J. Dental Res.* **42**, 461–476.
König, K. G. (1967). *Brit. Dental J.* **123**, 585–589.
König, K. G., Schmid, P., and Schmid, R. (1968). *Arch. Oral Biol.* **13**, 13–26.
Krasse, B. (1965a). *Arch. Oral Biol.* **10**, 215–221.
Krasse, B. (1965b). *Arch. Oral Biol.* **10**, 223–226.
Krasse, B. (1966). *Arch. Oral Biol.* **11**, 429–436.
Kullander, S., and Sonesson, B. (1965). *Acta Endocrinol.* **48**, 329–336.
Larson, R. H., Rubin, M., and Zipkin, I. (1962). *Arch. Oral Biol.* **7**, 463–468.
Lind, V., Wennerholm, G., and Nyström, S. (1964). *Acta Odontol. Scand.* **22**, 333–342.
Littleton, N. W. (1963). *Publ. Health Rept.* **78**, 631–640.
Löe, H., Theilade, E., and Jensen, S. (1965). *J. Periodontol.* **36**, 177–187.
Lundqvist, C. (1952). *Odontol. Rev.* **3** (Suppl. 1), 1–121.
Luoma, H. (1964). *Advan. Fluorine Res.* **3**, 217–226.
MacGregor, A. B. (1964). *Nutr. Dieta* **5**, 119–142.
McDougall, W. A. (1964). *Australian Dental J.* **9**, 1–5.
Madsen, K. O., and Edmonds, E. J. (1962). *J. Dental Res.* **41**, 405–412.
Mandel, I. D. (1966). *J. Periodontol.* **37**, 357–367.
Mandel, I. D. (1967). *J. Periodontol.* **38**, 721–735.
Manly, R. S., and Dain, J. A. (1963). *Proc. Intern. Assoc. Dental Res.*, Abst. No. 359.
Marthaler, T. M., and Froesch, E. R. (1967). *Brit. Dental J.* **123**, 597–599.
Miller, W. (1883). *Arch. Exptl. Pathol. Pharmacol.* **16**, 291–304.
Mitchell, D. F., and Holmes, L. A. (1965). *J. Periodontol.* **36**, 202–208.
Nizel, A. E., and Harris, R. S. (1964). *J. Dental Res.* **43**, 1123–1136.
Orland, F. J., Blayney, J. R., Harrison, R. W., Reyniers, J. A., Trexler, P. C., Wagner, M., Gordon, H. A., and Luckey, T. D. (1954). *J. Dental Res.* **33**, 147–174.
Orland, F. J., Blayney, J., Harrison, R. W., Reyniers, J. A., Trexler, P. C., Ervin, F., Gordon, H. A., and Wagner, M. (1955). *J. Am. Dental. Assoc.* **50**, 259–272.
Parikh, S. R., Toto, P. D., and Grisamore, T. L. (1965). *J. Dental Res.* **44**, 996–1001.
Porterfield, J. S. (1950). *J. Gen. Microbiol.* **4**, 92–101.
Reid, D. B. W., and Grainger, R. M. (1955). *Human Biol.* **27**, 1–11.
Rosebury, T., Karshan, M., and Foley, G. (1933). *J. Dental Res.* **13**, 379–398.
Rosen, S., Lenney, W. S., and O'Malley, J. E. (1968). *J. Dental Res.* **47**, 358–363.
Rosenstein, S. N. (1966). *N. Y. State Dental J.* **32**, 400–406.
Rovelstad, G. H. (1967). *Proc. Inst. Med. Chicago* **26**, 226.
Schmidt-Nielsen, B. (1946). *Acta Odontol. Scand.* **7** (Suppl. 2), 71.
Schultz-Haudt, S., Bibby, B. G., and Bruce, M. A. (1954). *J. Dental Res.* **33**, 624–631.
Shannon, I. L., and Terry, J. M. (1965). *J. Dental Med.* **20**, 128–132.
Shaw, J. H. (1950). *Intern. Dental J.* **1**, 48–74.
Shaw, J. H. (1954). *J. Nutr.* **53**, 151–162.
Shaw, J. H., and Dick, D. S. (1966). *Arch. Oral Biol.* **11**, 369–371.
Shibata, M. (1929). *Japan. J. Exptl. Med.* **7**, 247–251.
Ship, I. I., and Mickelsen, O. (1963). *J. Dental Res.* **43**, 1144–1149.
Sinrod, H. S. (1965). *Science* **149**, 400–402.
Slack, G. L., and Martin, W. J. (1958). *Brit. Dental J.* **105**, 366–370.
Slack, G. L., Millward, E., and Martin, W. J. (1964). *Brit. Dental J.* **116**, 105–108.
Socransky, S. S., Gibbons, R. J., Dale, A. C., Bortnick, L., Rosenthal, E., and MacDonald, J. B. (1963). *Arch. Oral Biol.* **8**, 275–280.
Stephan, R. M. (1940). *J. Am. Dental Assoc.* **27**, 718–723.
Stephan, R. M. (1944). *J. Dental Res.* **23**, 257–266.

Stookey, G. K., Carroll, R. A., and Muhler, J. C. (1967). *J. Am. Dental Assoc.* **74**, 752–758.
Strålfors, A. (1963). *J. Dental Res.* **43**, 1137–1143.
Teuscher, G. W., McDonald, R. E., Clopper, P. W., Cox, M. A., Gruebbel, A. O., Hoffman, O. E., McLaren, H. R., Noyes, H. J., Permar, D., Polevitsky, K. A., Wertheimer, F., Bibby, B. G., Chase, S. W., Crowley, M. C., Stephan, R. M., Tramp, M. I., Williams, N. B., Zander, H. A., and Fosdick, L. S. (1948). *J. Dental Res.* **27**, 419–421.
Toto, P. D., Rapp, G., and O'Malley, J. (1960). *J. Dental Res.* **39**, 750–751 (Abstr.).
Tsuchiya, H. M., Jeanes, A., Bricker, H. M., and Wilham, C. A. (1952). *J. Bacteriol.* **64**, 513–519.
Tsutsui, M., Utsumi, N., and Tsubakimoto, K. (1968). *J. Dental Res.* **47**, 663.
Volker, J. F. (1948). *J. Am. Dental Assoc.* **36**, 23–27.
Waerhaug, J. (1952). *Odont. Tidsk.* **60** (Suppl. 1), 9–186.
Waerhaug, J. (1956). *J. Dental Res.* **34**, 313–322.
Weiss, S., Schnetzer, J. D., and King, W. J. (1964). *J. Dental Res.* **43**, 745 (Abstr.).
Wolman, M., and Wolman, B. (1956). *A. M. A. Arch. Pathol.* **62**, 74–84.
Wood, J. M. (1964). *J. Dental Res.* **43**, 955 (Abstr.).
Wood, J. M. (1967). *Arch. Oral Biol.* **12**, 849–858.
Wright, D. E. (1964). *Arch. Oral Biol.* **9**, 321–329.
Zinner, D. D., Jablon, J. M., Aran, A. P., Saslaw, M. S. (1965). *Proc. Soc. Exptl. Biol. Med.* **118**, 766–770.

The Recovery and Purification of Biochemicals

VICTOR H. EDWARDS

*School of Chemical Engineering,
Cornell University,
Ithaca, New York*

I.	Introduction	159
II.	Removal of Suspended Solids	163
III.	Liquid Ion Exchange	168
IV.	Gel Filtration	173
	A. Media for Gel Filtration	175
	B. Theory	178
	C. Applications	188
V.	Membrane Separations	190
	A. Dialysis	196
	B. Ultrafiltration	198
VI.	Conclusions	207
	References	207

I. Introduction

The recovery and purification of fermentation products can be difficult and costly. The desired product is often present at low concentrations in an aqueous mixture that contains both suspended and dissolved materials. The product may be sensitive to temperature, extremes of pH, and certain chemicals. Also, because the mixture favors the growth of microorganisms, the product must be protected from degradation by foreign microorganisms. The product may be intimately associated with the cell, as in the cases of intracellular enzymes and of nucleic acids; product recovery then requires rupture of the cell and sometimes partial degradation of cellular components.

Pertinent information on typical reactants (nutrients) and fermentation products is presented in Table I. The nutrients in a fermentation are low molecular weight compounds such as sugars and salts, plus complex mixtures such as soybean meal and corn steep liquor solids that contribute suspended solids and higher molecular weight compounds to the fermentation broth. The latter components may increase the difficulty of product recovery. The nutrient concentrations in Table I represent values typically found at the start of a fermentation, and microbial utilization will result in significantly reduced concentrations at the end of the fermentation process. Fermentation

TABLE I
SOME COMPOUNDS OF INTEREST IN INDUSTRIAL FERMENTATIONS

Compound	Molecular weight	Diffusivity (10^5cm.^{2}/second)	Boiling point (°C.)	Polarity	Typical concentration (gm/liter)	Cost (cents/lb.)	Broth value (cents/liter)
Reactants							
Glucose	180.16	0.52	d.^{a}	Nonpolar	50	10.	1.1
Ammonium sulfate	132.15	ca. 0.8	d. 160°C.	Strong electrolyte, slightly acidic	2	2.	0.01
Urea	60.06	1.06	d.	Very weak electrolyte, very slightly basic	5	4.6	0.05
Potassium dihydrogen phosphate	136.09	ca. 0.8		Strong electrolyte, slightly acidic	1	7.8 (as H_3PO_4)	0.02
Brewers dried yeast	Complex mixture				5	27	0.3
Soybean meal	Complex mixture				10	4.2	0.09

Products							
Lactic acid	90.08	—	122°C. at 15 mmHg	Weak electrolyte, pK = 3.9	130	38	10.9
Monosodium glutamate	169.13	ca. 0.6	d. 248°C., s. 200°C.	Weak electrolyte, pK_{a1} = 4.07	35	44	3.4
Sodium benzylpenicillin	356.38	—	d.	Electrolyte pK = 2.76	2	1600.	7.2
Streptomycin sulfate	581.58	—	d.	Weak electrolyte	4	1500.	13.6
Vitamin B_{12}	1357.44	—	d.	Weak electrolyte	0.023	363,000	1.8
α-Amylase	48,900	0.08	d.	Weak polyelectrolyte	20	—	—

[a] d. = Decomposes without boiling.

products range from low molecular weight compounds such as lactic acid and glutamic acid, through products of intermediate size, such as the antibiotics and steroids, to large molecules such as enzymes and dextran. Product concentrations may be as low as milligrams per liter (e.g., vitamin B_{12}), but values between 0.1 and 15% (w./v.) are more typical. The products may be acidic, basic, or neutral.

Steps in product recovery are outlined in Fig. 1. The first step is

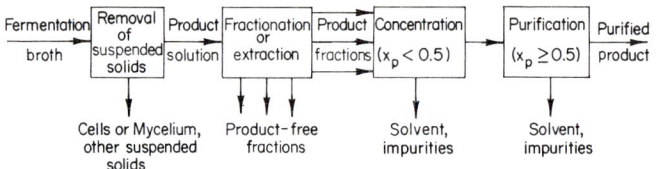

FIG. 1. Steps in the recovery of fermentation products.

usually the removal of suspended solids, which includes both cellular material and solids from the nutrients. Suspended solids may amount to 10% (v./v.) or even more in the case of dialysis cultures. If the product is the cells themselves or an intracellular product, further recovery operations proceed with the recovered cells. Even when the suspended solids are not the major product, they may be used for animal feed. In the more usual case, the desired product is dissolved in the broth. Further separation operations then focus on the clarified broth and the solution obtained from washing the recovered solids.

The next objective is reduction of the volume of solution being processed. Volume reduction may be accomplished by various techniques broadly classified as concentrations, fractionations, or extractions. Concentration operations achieve volume reduction through solvent removal by vacuum evaporation, reverse osmosis, or some other method. Fractionation techniques separate a mixture into different solution fractions. A product fraction contains the desired product with a reduced level of other solutes. Dialysis, electrodialysis, ultrafiltration, gel filtration, gas–liquid chromatography and ion-exchange chromatography, electrophoresis, thermal diffusion, and foam separation are examples of fractionation techniques. Extractions involve product recovery by transfer of the product to a different solvent phase. Solvent extraction, liquid ion exchange, and selective adsorption are typical extraction techniques. Not all separation techniques fit a single category. Ultrafiltration, for example, may achieve both concentration and fractionation. The choice of separation technique depends on the nature and concentration of the product and the components of the solution, and the order of the second and third steps shown in Fig. 1 may be reversed.

The final step or steps is purification of the product. Purifications are arbitrarily defined as operations starting with product streams in which the product weight fraction is greater than or equal to one-half. Besides some of the techniques mentioned previously, crystallization, drying, and zone refining are used in purification.

Clearly, a large number of separation techniques can be useful in the recovery of biochemicals. This paper will emphasize several methods that are particularly promising as means of processing fermentation products, although they have not yet been widely used for this purpose.

II. Removal of Suspended Solids

As indicated earlier, the removal of suspended solids is often the first step in processing a fermentation broth. Methods for solids removal have been discussed elsewhere and will therefore receive an abbreviated treatment here *(1–3)*. The methods used are classified in Fig. 2 according to the driving force producing the separation. Gravitational and mechanical methods find the widest industrial application. The first step may be a screening to remove large suspended solids. Then centrifugation, settling, or filtration follows, depending on whether mycelia, yeast, spores, bacteria, or viruses are to be recovered. In the manufacture of bakers' yeast and of fodder yeast, continuous centrifugation is employed *(3,4)*. The removal of mycelial organisms is usually accomplished by filtration. Actinomycetes offer

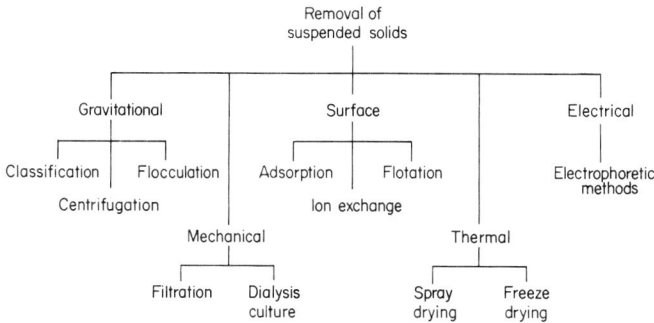

FIG. 2. Methods for removing suspended solids.

considerable resistance to filtration (2), and adjusting the broth to a pH of 2–3, followed by heating to near boiling and holding for up to an hour will greatly improve filtration characteristics by coagulating mycelial protein (3,5,6). Such vigorous treatment cannot be used with sensitive products like penicillin or enzymes. The filter is often precoated with a filtration aid such as diatomaceous earth to reduce filter blinding and to increase filtration rates, or the filtration aid may be added directly to the broth (7). Consumption of the filtration aid is typically about 2 to 3% (w./v.) in precoat applications and 5 to 10% (w./v.) in the case of direct addition (7). Filtration is usually conducted with rotary drum or continuous belt filters at rates ranging between 2 and 80 gal./ft.^{2}-hour (7).

Bacteria are considerably smaller than yeast or mycelial organisms and therefore cause more serious filter blinding. They are also more difficult to centrifuge because of their lower density and smaller diameter (2). For a sphere falling in a gravitational field in a Newtonian fluid, the terminal velocity is given by Stokes' law.

$$v = (gd^2_p/18\mu)(\rho_p - \rho_L) \qquad (1)$$

where

v = terminal settling velocity
d_p = diameter of the sphere
g = acceleration
μ = viscosity of the fluid
ρ_p = density of the sphere
ρ_L = density of the fluid

[Equation (1) applies only when the Reynold's number $(d_p v \rho_L/\mu)$ is less than 0.3.] Ambler has indicated that the allowable volumetric flow rate for a given continuous centrifuge is directly proportional to the terminal settling velocity of the particles (8). The permissible throughput for a clarifier is also roughly proportional to the terminal settling velocity. Thus bacteria, which have diameters approximately one-fifth those of yeast, are about twenty-five times more difficult to separate by centrifugation or sedimentation when their densities are equal. Johnson has pointed out that if inexpensive methods of flocculating bacteria could be developed, bacteria would offer economic advantages in protein production because of their versatility in substrate utilization (9). The beneficial effect of flocculation is evident from consideration of Eq. (1). If flocculation increases the average particle diameter only tenfold, the volumetric flow rate through a centrifuge or sedimentation device could, in theory, be increased 100-fold according to Eq. (1). However, flocs generally have a density lower than free cells, which would reduce the actual settling velocity. Also, high shear in a centrifugal field could partially disperse flocculated cells.

Autoflocculation commonly occurs with certain yeasts in brewing and it is essential to the success of the activated sludge process in the treatment of sewage and industrial wastes. Considerable study has been devoted to the flocculation of cells, particularly in connection with water and waste purification, but only a qualitative theory exists at present. Neutralization of the negative charge usually present on the surface of microorganisms and dehydration of the surface are necessary conditions for flocculation (10). Polyvalent cations are effective in flocculating yeast, bacteria, and algae (10). Typical concentrations of different flocculants are aluminum sulfate (1–2%), calcium chloride (0.1–0.5% at pH 8.0–9.5), and titanium tetrachloride (0.01–0.02%) (10). These flocs settle rapidly enough to permit separation by sedimentation. A variety of hydrophilic polymers flocculate bacteria (1,10,11). Examples include gelatin, carboxymethyl-cellulose, sodium alginate, polyacrylic acid, and cationic surface-active agents (10,11). The quantity of flocculant required is affected by the release of proteins, polysaccharides, and nucleic acids by the cell (12).

Recovery of the flocculant from separated cells is often not feasible and flocculants should therefore be inexpensive. If the cells are to be used as food or fodder, the flocculant should also be nontoxic and gelatin and other flocculants with positive nutritional value have an advantage. The economics of flocculation as an aid to centrifugation or to permit solids removal by classification or sedimentation depends on the process and flocculant under consideration, but it would be surprising if flocculation does not prove to offer economic advantages in at least some applications. For example, power requirements for centrifugal concentrating of unicellular algae are of the order of 1.7 to 3.1 kWh./lb. dry algae (13).

Another mechanical means of separating suspended solids is dialysis culture (14,15). With this technique cells are grown in a chamber separated from a nutrient solution by a dialysis membrane or membrane filter. Nutrients diffuse through the membrane into the growth chamber and products diffuse from the growth chamber into the nutrient reservoir (15). Cell concentrations ten times or more, higher than those obtained in normal batch culture, have been achieved (14).

Dialysis culture may offer economic advantages by improving rates and yields in product-limited fermentations and by reducing recovery costs since a major portion of the product will be recovered from the cell-free nutrient solution. Disadvantages of dialysis culture include the higher equipment cost, membrane fragility, incomplete utilization of nutrients, and dilution of the product.

Two types of flotation are effective in removing cells from fermentation broths. One process consists of bubbling a gas through a particle suspension containing a surfactant to promote foam formation. The particles become attached to the bubbles and thus concentrate in the foam. Separating and collapsing the foam yields a concentrated cell suspension. Larger bubbles favor dryer foams and therefore greater concentration. This process has been called both froth flotation *(1, 16)* and foam separation *(17–19)*, but under a recently established set of nomenclature, it should be called *microflotation (20)*. Results are best expressed in terms of the *enrichment ratio,* which is defined as the ratio of the particle concentration in the collapsed foam to the particle concentration in the residual suspension that remains after separation of the foam. The ratio of the particle concentration in the collapsed foam to the concentration in the original suspension defines the *concentration factor.* The enrichment ratio is more indicative of the effectiveness of a given separation and also more useful in the design of multistage operations. Boyles and Lincoln separated *Bacillus anthracis* spores from various media by microflotation and obtained concentration factors ranging from 1.4 to 16.8 *(21).* Enrichment ratios are often increased by the addition of collectors, which increase the attachment of particles to bubbles. Compounds that are effective collectors for bacteria include ionic starches *(1),* cetylpyridinium chloride *(1),* inorganic salts *(16),* and cationic surfactants *(17,18).* Secondary amines selectively collect spores, while fatty acid collectors concentrate vegetative cells and cell debris in preference to spores *(22).* Enrichment ratios as high as a million have been obtained using cationic surfactants at concentrations of 20 mg./liter *(18).*

The second type of bubble separation technique used to remove suspended cells employs an insoluble collector phase on the top of a cell suspension. Gas is bubbled through the suspension at a much lower rate than in microflotation, and cells attached to the bubbles rise to the surface where they transfer to the collector phase in a thin dry foam *(19,23).* The foam may then be collapsed to a stable surface phase in the foaming vessel and skimmed off *(19,23).* Although Rubin, et al. *(19)* originally proposed the name "microflotation" for this process, the correct name under the new nomenclature is *solvent sublation (20)*. In solvent sublation, long-chain fatty acids or amines form the collector phase and an alcohol may be added to the suspension as a *frother (19)*. The frother serves to reduce bubble size and enhance collection *(19,23)*. Two species of bacteria and two species of algae were efficiently removed by this method *(19,23)*. According to Rubin, the mechanism of solvent sublation is conversion of the cell to a hydrophobic particle through adsorption of the collector surfactant at the cell surface *(23)*. Flocculation of the cells with alum improves collection efficiency *(19,23)*.

In view of the problems associated with filtration or centrifugation of bacteria, microflotation or solvent sublation may offer advantages as methods of removal and preliminary concentration of suspended solids in industrial processes. Antifoam added during the course of fermentation must be overcome through the addition of a surfactant to promote foaming before microflotation can be successful. Capital costs for foam separation equipment are low, but not all bacteria can be successfully concentrated by adsorptive bubble separations and the degree of concentration is often not as high as that obtained by other methods. Higher concentrations might be achieved through staging and economies in reagent consumption by recycle of the collector. A process recently introduced in the French steel industry for the removal of iron dust and oil from water may find application in removal of microorganisms. Unlike other flotation processes, in electroflotation the bubbles are generated by electrolysis and are very small (less than 0.1 mm.) *(24)*. It is claimed that the smaller, more uniform bubble size results in improved separation efficiency through smoother flow around bubbles and reduced coalescence *(24)*.

Another means of cell removal that depends on surface phenomena is adsorption by ion-exchange resins *(1,13,25)*. Algae, bacteria, and viruses have been removed using either anion- or cation-exchange resins by contacting a particle suspension with resin particles. Separation of the particles is accomplished by separating the resin from the suspension and desorbing the particles from the resin by addition of a reagent, that either changes the pH or else exchanges ions with the resin, converting it to a form that no longer attracts the particles. Ion exchange suffers from several disadvantages in comparison with other techniques for concentrating suspended cells. First, recoveries per unit of resin are low because cells adsorb only on the surface of the resin. Thus for spherical cells 1 μ in diameter adsorbing as a monolayer on the surface of resin particles 0.1 mm. in diameter, only about 1.5% of the volume of a resin bed would be occupied by cells. This leads to a large capital investment in resin. For example, a cation-exchange column containing 2.78×10^{-3} gm. dry algae/ml., no longer collected additional algae efficiently *(13)*. Reagent requirements can also be considerable. One pound of sulfuric acid was required to elute and precipitate 0.07 pounds of algae *(13)*. A third disadvantage is that the resin may also exchange ions with the fermentation broth, perhaps partially removing product or converting to a form that no longer adsorbs cells.

Electrical methods of removing cells and viruses have been studied in the laboratory. The electrophoretic mobilities of bacteria are about 5 μ/second/V./cm. *(10)*, and the mobilities of viruses are even lower *(1,10)*. Thus electrophoresis is not likely to be an economic means of concentrating cells from industrial fermentation broths where high electrical conductivities lead to high energy requirements. Application to virus removal from water supplies, where electrolyte concentrations are low, may be attractive. A number of methods of removing viruses from water are described in Berg *(26)*.

III. Liquid Ion Exchange

Many commercial biochemicals are electrolytes that ionize to varying degrees in aqueous solution. Ion exchange is an efficient separation technique for ionic compounds. Ion exchangers are usually manufactured as small solid particles that contain ionogenic groups. Cations or anions are associated with each ionogenic group that are exchangeable for ions of the same sign in solutions contacted with the ion-exchange particles. Ion exchangers can distinguish between different ions, and tend to prefer ions of higher charge and smaller solvated volume *(27)*. Other factors, such as interaction with the ion exchanger matrix or complex formation, also influence selectivity *(27)*. Several authors have reviewed applications of ion exchangers in the fermentation industries *(27–30)*.

Although batch ion-exchange processes exist, it is usually more efficient to conduct ion-exchange processes otherwise. A column is filled with solid ion-exchange particles and a solution containing the ionic product is passed through the column. If the proper exchanger has been selected, the column initially removes almost all product ions from solution, replacing them with other ions of the same sign. Eventually, a point is reached where product ions are associated with most ionogenic groups and product concentrations in the column effluent become appreciable. Then flow of the product stream is terminated and the product ions are eluted from the column by passage of a concentrated solution of an appropriate acid, salt, or base through the column. Solid ion-exchange processes suffer from two disadvantages. First, operation is intermittent rather than continuous which increases the capital investment and labor costs. Continuous, countercurrent flow ion exchangers have been developed, but these are not entirely satisfactory (27). Second, conventional ion-exchange columns plug if suspended solids are not removed. Bartels et al. (30) developed a solid ion-exchange process for streptomycin recovery that required only screening of the fermentation broth to remove large particles. However, their method does not allow continuous operation.

Solvent extraction is a popular separation technique in the fermentation industries that offers two important advantages by the use of centrifugal extractors. One advantage is separation without removal of suspended solids, and the other is short separation times, which is important in the recovery of unstable products. Both advantages are important in the solvent extraction process for fermentation products such as penicillin (31).

By employing liquid exchange resins, the best features of solvent extraction and solid ion exchange can be combined to give a rapid, continuous, and selective separation technique (32). The basic aspects of the liquid ion-exchange process are illustrated by Fig. 3. The liquid ion exchanger, in this case a complex of an organic amine (R_3N) and an acid (HA), is dissolved in an organic diluent. When the organic phase contacts an aqueous solution containing anionic product (P^-), the product ions are extracted to the organic phase by anion exchange with the amine complex. Acid anions (A^-) replace product

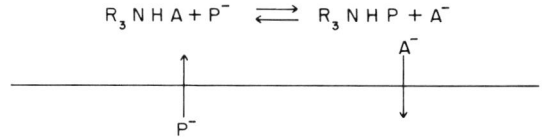

FIG. 3. Schematic of the liquid ion-exchange process.

anions in the aqueous phase. After removal of the product from the aqueous phase, the phases are separated. The product is stripped from the organic phase by contact with an aqueous reagent that either precipitates the product or transfers the product in high concentration to the aqueous phase by a second exchange reaction. A complete liquid anion exchange process is illustrated in Fig. 4. Concentration

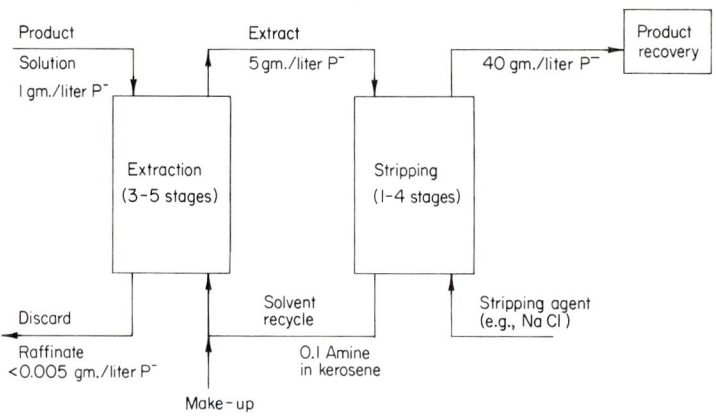

FIG. 4. Product separation by a typical liquid ion-exchange process.

and stage requirements in extraction and stripping are representative of the values obtained in commercial processes for uranium extraction. Normally no more than five stages are required for extraction or stripping operations in uranium recovery, but this depends on the extraction coefficients, the degree of recovery required, and the organic-to-broth ratio. Extraction and stripping may be conducted with centrifugal extractors, or alternatively in extraction columns when viscosities and product stability permit. Anderson and Lau obtained 2.21 theoretical stages in a centrifugal solvent extraction of Chloromycetin from a whole broth (31). Product recovery was 98.8% and the distribution coefficient was 16.1. [The distribution coefficient is defined as the ratio of the product concentration in the organic phase to the product concentration in the aqueous phase at equilibrium (31).] They achieved less than a single theoretical stage and 70% efficiency in centrifugal solvent extraction of penicillin, necessitating reextraction of the broth (31). However, as many as 12.5 theoretical stages have been attained with centrifugal extractors under favorable conditions (33). Liquid ion exchangers offer an added degree of freedom in process optimization. High distribution coefficients are obtained primarily by choice of an appropriate exchanger, although the diluent also influences equilibrium somewhat. An organic diluent that is cheap and that enhances extraction efficiency may then be selected. The pros and cons of liquid ion exchange are summarized in Table II. It should be noted that solvent extraction shares the disadvantages of liquid ion exchange.

TABLE II
FEATURES OF THE LIQUID ION-EXCHANGE PROCESS

Advantages
1. Continuous processing
2. Reduced resin requirements through higher mass transfer rates
3. Removal of suspended solids not necessary
4. Precipitation feasible
5. Concentration of ionogenic groups easily adjusted
6. Organic diluent may be selected to minimize cost and maximize extraction efficiency

Disadvantages
1. Loss of organic phase and resin
2. Taste and odor may be imparted to aqueous phase
3. Emulsion formation possible

Hundreds of compounds that may be classified as liquid ion exchangers have been studied, primarily in connection with the extraction of metals such as uranium, copper, thorium, zironium, and niobium. A number of reviews and articles have been written describing the metal extraction processes (34–37), the properties of liquid exchangers (38–40), and their analytical applications (41). Liquid ion-exchange compounds contain one or more ionogenic group plus hydrocarbon or other organic groups that give the molecule high solubility in organic solvents and low solubility in water. The most common liquid anion exchangers are organic amines with molecular weights between 250 and 600 (39). Liquid cation exchangers include alkyl phosphates, alkyl phosphoric acids and their esters, alkyl phosphites, phosphine oxides, alkyl phosphonates, and aromatic sulfonic acids (34,38,40). Table III lists some commercial liquid ion exchangers. The most popular diluents in metal extraction are kerosene or a heavy aromatic naphtha (34). A small amount of a long-chain alcohol such as nonanol or tridecanol is sometimes added to increase amine solu-

TABLE III
Commercial Sources of Liquid Ion Exchangers

Designation	Composition	Manufacturer[a]
Anion exchangers		
TIOA	Triisooctylamine	Carbide, Gulf
TLA	Trilaurylamine	ADM, Armour
RC-3749	Trifatty amine (mixed n-octyl and n-decyl alkyls)	General Mills
	Di(tridecyl)amine	Carbide
Amberlite LA-1	N-Dodecenyl (trialkylmethyl)amine (trialkylmethyl = homologous mixture, 12–15 carbon atoms)	Rohm and Haas
Amberlite LA-2	N-Lauryl(trialkylmethyl)amine (trialkylmethyl = homologous mixture, 12–15 carbon atoms)	Rohm and Haas
Primene JM-T	Trialkylmethylamine (homologous mixture, 18–24 carbon atoms)	Rohm and Haas
Amine S-24	Bis-(1-isobutyl-3,5-dimethylhexyl)amine	Union Carbide
Amine 21F81	1-(3-Ethylpentyl)-4-ethyloctylamine	Union Carbide
N-Benzylheptadecylamine	N-Benzyl-1-(3-ethylpentyl)-4-ethyloctylamine	Union Carbide
Alamine 336	Tricapryl amine	General Mills
Aliquat 336	Tricaprylmethylammonium chloride	General Mills
Cation exchangers		
D2EHPA	Di-(2-ethylhexyl)phosphoric acid	Union Carbide
HDPA	Heptadecyl phosphoric acid	Dow
DDPA	Dodecyl phosphoric acid	Dow
DBBP	Di-(n-butyl)-n-butylphosphonate	Monsanto, VC
DHHP	Di-(n-hexyl)-n-hexylphosphonate	Shea
TOPO	Tri-n-octylphosphine oxide	Eastman Kodak

[a] Archer-Daniels-Midland Co. (ADM), Minneapolis, Minnesota; Armour Chemical Division, Chicago, Illinois; Distillation Products Industries, Eastman Kodak Co., Rochester, New York; General Mills, Inc., Kankakee, Illinois; Gulf Oil Corp., Pittsburg, Pennsylvania; Monsanto Chemical Co., St. Louis, Missouri; Rohm and Haas Co., Philadelphia, Pennsylvania; Shea Chemical Corp., New York, New York; Union Carbide Chemicals Co., New York, New York; Virginia-Carolina Chemical Corp. (VC), Richmond, Virginia.

bility (34). Other potential diluents include solvents such as benzene, hexane, aliphatic alcohols, and chloroform (39).

Commercial processes for metal recovery by liquid ion exchange have been operating for about 15 years. Typical reagent costs for uranium recovery are presented in Table IV (35). According to Lewis, these constitute a major portion of the cost of uranium concentration (35). The cost of extracting copper from dilute solutions by liquid ion exchange has been estimated to be about three cents per pound of copper (35). If use of centrifugal extraction equipment is necessary, capital costs and power costs may also become large, but costs in this range make liquid ion exchange an attractive separation technique.

TABLE IV
REAGENT COSTS FOR URANIUM RECOVERY BY LIQUID ION EXCHANGE[a]

Reagent	Function	Consumption and cost (per 1000 gal. feed)[b]	Reagent cost (cents/lb. U_3O_8)[b]
Mixed solvent (amine, isodecanol, kerosene)	Extraction	0.25–0.50 gal. at 30 cents/gal.	0.8–1.8
Sodium chloride (1.5 M)	Stripping	16.5–24.8 lb. at 0.8 cents/lb.	1.6–2.4
Ammonia	Precipitation	1.65–2.48 lb. at 5 cents/lb.	1.0–1.5
		Total reagent cost	3.4–5.7

[a] See reference (35).
[b] Based on 1 gm./liter U_3O_8 in an aqueous feed.

Judging from the literature, liquid ion-exchange processes have found surprisingly few applications in the separation of fermentation products. Munden lists two patents, one for the recovery of basic antibiotics using alkyl acid phosphates and the other for the recovery of 6-aminopenicillanic acid (32). Kaplan and co-workers (42) have published details of a liquid ion-exchange process for the separation and purification of Terramycin. A solution of 5% cetylpyridinium bromide in isooctyl alcohol at pH 9.5–10.2 gave 95–97% extraction (42), corresponding to a distribution coefficient greater than twenty.

In view of the attractive features of the liquid ion-exchanger process, its use in the fermentation industry is likely to increase.

IV. Gel Filtration

Gel filtration is a comparatively new technique that separates molecules and other submicron particles, such as viruses, according to size. It is now widely used in analytical biochemistry and in the determination of molecular weight distributions of polydisperse synthetic polymers, but there have been few large-scale applications. Excellent coverage of gel filtration is available in the books by Determann *(43)* and by Morris and Morris *(44)*, and the reviews by Altgelt and Moore *(45)*, Flodin *(46)*, Porath and Flodin *(47)*, Tiselius *(48)*, and Gelotte *(49)*, plus the group of reviews in a recent issue of the journal *Laboratory Practice (50–55)*. For a series of recent results in the application of gel filtration to synthetic polymers, the reader is referred to the proceedings of the symposium edited by Johnson and Porter *(56)*. The present discussion has drawn heavily on these sources.

The gels employed in gel filtration consist of a dispersed substance (the gel-forming material) and a dispersing agent (the solvent). The dispersed substance is a three-dimensional network of linear macromolecules held together at junction points either by secondary valence forces (arising from hydrogen bonds, dipole-dipole interactions, and dispersion forces) or by primary valence forces (ionic and covalent bonds). The dispersing agent is distributed throughout the pores in the network.

Gels can be used to separate molecules of different sizes because small molecules can permeate the solvent in the gel phase, but larger molecules are partly or completely excluded from the gel phase by the macromolecular network, particularly in the neighborhood of junction points. Also, the diffusion rates of large molecules are reduced more in the gel than the diffusion rates of small molecules. The application of this principle is illustrated in Fig. 5. A column is filled with solvent and spherical or granular gel particles, which are represented by the large open circles (see Fig. 5). A mixture of large molecules (large filled circles) and small molecules (small filled circles) is dissolved in the solvent and applied to the top of the column in Fig. 5a. The small molecules distribute between interstitial solvent and the solvent in the gel phase, but the large molecules are excluded from the gel phase and must remain in the interstitial solvent. As additional solvent is added to the top of the column, the interstitial fluid is displaced, leaving behind those small molecules in the gel phase and contacting fresh gel (Fig. 5b). The fresh gel removes more small molecules from the solvent originally containing the mixture, while the fresh solvent following picks up the small molecules left behind in the gel. The large molecules are thus carried directly through the packed column by the eluting solvent, while the smaller molecules are retarded by distribution in the gel phase. The

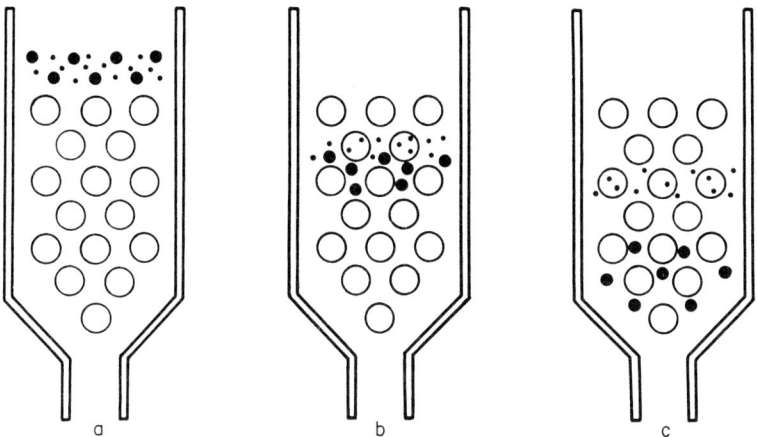

Fig. 5. Schematic representation of the gel filtration process.

larger molecules are the first to appear in the effluent and if the column is sufficiently long, they are completely separated from the small molecules (Fig. 5c). This type of partition chromatography has been called gel filtration (57), molecular sieve filtration (58), exclusion chromatography (59), molecular sieve chromatography (60), gel permeation chromatography (61), and gel chromatography (43). Gel filtration is the term most popular with workers in the biological field, while gel permeation chromatography is used by those applying the method to synthetic polymers. Gel filtration will be the name adopted here.

A. MEDIA FOR GEL FILTRATION

Steric exclusion of molecules from gels was first noted in the late 1940's in studies of ion-exchange gels. For example, Kunin and Meyers showed that as the degree of cross-linking of a polystyrene anion exchanger was increased, its capacity for penicillin decreased much more strongly than its capacity for hydroxyl ions (62). In 1954, Deuel and Neukom synthesized an uncharged gel by cross-linking galactomannan and used it to desalt colloids (63). Two years later, Lathe and Ruthven published an extensive study of the use of granulated starch gels for the separation of neutral molecules (64). They recognized the fundamental mechanism of gel filtration, but a number of disadvantages were associated with the use of starch gel, including ion-exchange effects and low flow rates. Similar problems were associated with the first attempt to employ agar gel (65).

A major advance was the discovery by Porath and Flodin of the separative power of gel produced by cross-linking soluble dextran with epichlorohydrin (57). The dextran gels so produced had very desirable properties. Also, varying the degree of cross-linking changed the porosity of the gel and thus the size range of molecules that could be fractionated by the gel. Commercial dextran gels in a variety of particle sizes and degrees of cross-linking soon became available (Sephadex, Pharmacia Fine Chemicals Inc., Uppsala, Sweden) and numerous separations have been achieved with them. Selected properties of some commercial dextran gels are presented in Table V.

TABLE V
PROPERTIES OF COMMERCIAL DEXTRAN GELS[a]

Sephadex type	Particle diameter (dry, μ)	Water regain (gm. H_2O/gm. dry Sephadex)	Bed volume (ml./gm. dry Sephadex)	Fractionation range for peptides and globular proteins
G-10	40–120	1.0 ± 0.1	2–3	700
G-25 Coarse	100–300	2.5 ± 0.2	4–6	1,000–5,000
Medium	50–150	2.5 ± 0.2	4–6	1,000–5,000
Fine	20–80	2.5 ± 0.2	4–6	1,000–5,000
Superfine	10–40	2.5 ± 0.2	4–6	1,000–5,000
G-50 Medium	50–100	5.0 ± 0.3	9–11	1,500–30,000
G-100	40–120	10.0 ± 1.0	15–20	4,000–150,000
G-200	40–120	20.0 ± 2.0	30–40	5,000–800,000

[a] Properties of Sephadex supplied by Pharmacia Fine Chemicals, Inc., Piscataway, New Jersey.

For the sake of brevity, G-15, G-75, and G-150 gels have been omitted. Low numbers in the G-series (e.g., G-10, G-15) designate gels with higher cross-linking that are suitable for separation of low molecular weight compounds, while higher numbers indicate less cross-linking and higher porosity. The more porous gels tend to be soft and compressible. This reduces acceptable column length and flow rates because if the pressure drop is too large, the gel bed compresses irreversibly. As is evident from Table V, swelling or water (solvent) regain increases with decreasing cross-linking. The solvent regain, S_r, is defined as the grams of solvent imbibed by one gram of dry gel-forming material during swelling (43). Solvent regain is practically independent of ionic strength for most dextran gels, which may be reversibly swollen and dried, but is strongly dependent upon the solvent. Gels with low cross-linking shrink in solutions of high ionic strength (3). Dextran gels are stable in basic, neutral, and weakly acidic solutions and may be sterilized in the wet state by autoclaving. Carbohydrate concentrations of 0.002 to 0.005% are liberated from column packings during continuous elution. Sephadex LH-20 is an alkylated form of Sephadex G-25 that swells in polar organic solvents and is useful in separating hydrophobic materials.

In 1961, Polson granulated agar and used it to separate high-molecular weight proteins (66). The agaropectin fraction of agar contains carboxyl and sulfate groups, which cause considerable adsorption and ion exchange effects. Commercial preparations of agarose, the neutral main component, are now available (Bio-Rad Laboratories, Richmond, California; Mann Research Laboratories, Inc., New York; Pharmacia Fine Chemicals, Inc., Piscataway, New Jersey). Agarose gels are softer than most dextran gels (Sephadex G-200 is the exception), and their use is therefore restricted to lower flow rates and shorter columns. These gels begin to dissolve at temperatures above 30°C. and are damaged by dehydration, freezing, organic solvents, hydrogen bond-breaking solutions (e.g., urea), and some concentrated salt solutions. The major advantage of agarose gels is their ability to fractionate large macromolecules, viruses, phages, subcellular particles, and bacteria (43).

Polyacrylamide was first introduced as a gel filtration medium by Hjerten (67). These gels offer several advantages. They are stable for pH values between 2 and 9, they do not shrink in solutions of high ionic strength, and they contain few ionic groups, thereby minimizing adsorption effects (45). Also, polyacrylamide gels are not attacked by microorganisms as are agarose and dextran gels. Polyacrylamide gels having a wide variety of fractionation ranges are available commercially (Bio-Rad Laboratories).

Haller recently introduced porous glass as a matrix for the separation of viruses (68). Several types of porous glass with pore sizes ranging from 200 to about 2500 Å are now available commercially in several particle sizes (Waters Associates, Inc., Framingham, Massachusetts; Bio-Rad Laboratories). Glass has the desirable advantages of chemical, thermal, and mechanical stability. Ion exchange and adsorption properties of porous glass beads may be greatly reduced through silanization by application of hexamethyldisilazane under a vacuum (69). Two other materials that show promise for gel filtration in aqueous solvents are cellulose and ethylene glycol methacrylate (70,71).

Gel filtration of materials dissolved in organic solvents have been accomplished most often using polystyrene gels and partially alkylated dextran gels, but expanded silica gel, porous glass, rubber, polyethylene, cellulose acetate, and polyvinyl alcohol also find application (45). Polystyrene gels are marketed under the trade name Styrogel (Waters Associates, Inc., Framingham, Massachusetts) with pore sizes ranging from 60 to 10^6 Å.

B. Theory

As explained earlier, separations result when smaller solute molecules are more retarded in their passage through a gel filtration column because of their ability to penetrate the stationary gel phase more effectively than larger molecules. Both steric exclusion from the gel phase and the slower diffusion rate of larger molecules can contribute to the separation. As solute zones pass through a gel filtration column, they become broader and less concentrated as a result of axial diffusion, axial mixing, and varying degrees of residence in the gel phase by molecules of a single type. Two dependent variables, elution volume and zone broadening, determine the success of a given separation. Their dependence on column, solvent, and solute parameters, and sample size is therefore of considerable interest.

1. Elution Parameters

The gel bed in a gel filtration column consists of the gel particles and the surrounding interstitial solvent. The gel particles may be regarded as having two components: the gel matrix and the imbibed solvent. Thus the total bed volume, V_t, is given by the following equation:

$$V_t = V_o + V_i + V_m \tag{2}$$

where

V_o = outer volume (volume of solvent between gel particles)
V_i = inner volume (volume of solvent imbibed in gel)
V_m = volume of gel matrix

The outer volume may be determined directly as the elution volume of a high molecular weight substance that is completely excluded from the gel particles. The inner volume cannot be measured directly, but for zerogels (gels from which all solvent has been removed), the inner volume can be calculated with Eq. (3).

$$V_i = G\, S_r / \rho_L \tag{3}$$

where

G = dry weight of gel in column
ρ_L = density of the solvent

Often the value of G is not known (for example, when the gel is supplied wet). In that case, the inner volume can be calculated using Eq. (4).

$$V_i = \frac{S_r \rho_G}{\rho_L(S_r/\rho_L + 1)}(V_t - V_o) \qquad (4)$$

where

ρ_G = density of the swollen (wet) gel

The elution volume of a solute may be correlated by means of a distribution coefficient, which is a measure of the degree of penetration of the gel by the solute. Equation (5) defines the distribution coefficient, K_d.

$$\begin{aligned}V_e &= V_o + K_d V_i \\ &= \text{elution volume}\end{aligned} \qquad (5)$$

Thus for large molecules, which are completely excluded from the gel phase, $K_d = 0$ and the elution volume is equal to the outer volume. For very small molecules, all the gel phase is accessible, $K_d = 1$, and the elution volume is the sum of the outer and the inner volumes. For molecules of intermediate size, which are partly excluded, K_d is between 0 and 1 and depends on the size of the molecule.

In real systems, two types of deviations from the ideal behavior occur. First, even small solutes are eluted with a K_d less than unity because some solvent solvates the gel matrix and is therefore not available to solutes. Second, adsorption sometimes occurs, resulting in K_d values greater than unity. For example, aromatic compounds in organic solvents often adsorb somewhat on styrene-divinylbenzene gels (72,73). Eaker and Porath took advantage of adsorption effects to separate amino acids on Sephadex G-10 (74).

In addition to application of Eq. (5) to column elution data, it is possible to determine K_d from batch equilibrium experiments. A solution having an initial solute concentration C_{orig} and volume V_o' is contacted with a swollen gel having an inner volume V_i for a length of time sufficient to reach equilibrium. The concentration of solute in the supernatant, C_{sup}, is then determined and K_d may be calculated from Eq. (6).

$$K_d = \frac{V_0'}{V_i} \frac{C_{orig} - C_{sup}}{C_{sup}} \quad (6)$$

Because it is not possible to directly measure the inner volume, which appears in the definition of K_d, Laurent and Killander have proposed another distribution coefficient, K_{av}, defined by Eq. (7) (75).

$$K_{av} = (V_e - V_o)/(V_t - V_o) \quad (7)$$

The distribution coefficients depend on solute properties, as well as the nature of the gel. A number of empirical or theoretical equations have been proposed to relate the distribution coefficient or elution volume to properties of the solute and column. Determann lists 16 such equations (43), but only a few will be presented here. Most theoretical models that have been proposed to represent the gel filtration process have assumed either a steric mechanism or a restricted diffusion mechanism, rather than a more realistic combination of both mechanisms. One of the first steric models was proposed by Porath (76). He assumed that the inner volume of the gel could be represented by conical pores of radius R' and that solutes could be represented by spheres of radius a. In this model, a greater volume of the conical pores is accessible to solutes with smaller molecular radii. From the above assumptions, Porath derived Eq. (8) (76).

$$K_d = K[1 - (2a/R')]^3 \quad (8)$$

where

$$K = \text{constant}$$

The volume of the conical pores is proportional to the swelling of the gel ($R'^3 = K'S_r$), and for flexible macromolecules with even segments the molecular weight is proportional to the square of the molecular radius ($M = K'' a^2$). Substituting for a and R' in Eq. (7) leads to an expression for the distribution coefficients in terms of experimentally measurable parameters.

$$K_d = K_1 [1 - K_2(M^{1/2}/S_r^{1/3})]^3 \quad (9)$$

Porath has shown that a graph of $K_d^{1/3}$ versus $M^{1/2}$ gives a straight line for dextran fractions, in agreement with Eq. (9) (76). Equation (9) has also been applied to date successfully by other authors (51).

Akers has avoided the assumption of any specific geometry in proposing that the regions of nonexcluded volume within the gel are distributed randomly with respect to the size of molecules they can accommodate (77). This assumption leads to the following result.

$$K_d = \text{erfc}\,[(a-a_0)/b_0]$$

$$= 1 - 2\pi^{-1/2} \int_0^{(a-a_0)/b_0} \exp(-a^2)\,da \qquad (10)$$

where

    erfc = error function complement
    a_0, b_0 = constants

Equation (10) also correlates elution data successfully (77).

Laurent and Killander proposed that the gel phase be viewed as a network of randomly oriented straight rods of infinite length (75). This leads to the following equation:

$$K_{av} = \exp\,[-\pi Z(a+r)^2] \qquad (11)$$

where

    Z = concentration of rods, length/volume of gel
    r = radius of rods

They found Eq. (11) correlated elution data of a large number of proteins and of dextran fractions (75).

The models discussed so far have assumed complete diffusional equilibrium between gel and interstitial solution and have depended on physical exclusion to explain the restricted permeation of the gel phase by large molecules. The slower diffusion rates of larger molecules may also play a part in gel filtration separations. Akers determined distribution coefficients from both column elution data and batch equilibration data for proteins and found that K_d (batch equilibration) $> K_d$ (column) for Sephadex G-200 and agar gels for large molecules (78). This effect was interpreted as restricted diffusion of the larger molecules due to interference by the pore walls. Akers obtained the following expression by equating the distribution coefficient with an expression for the ratio of rate of diffusion through cylindrical pores of radius R' to the rate of free diffusion (78).

$$K_d = [1-(a/R')]^2\,[1 - 2.104\,(a/R') + 2.09\,(a/R')^3 - 0.95\,(a/R')^5] \qquad (12)$$

This equation correlated data for Sephadex G-200 and agar gels (78).

Yau and Malone have developed a model which recognizes both the exclusion effect and the effect of diffusion (79). Their derivation is based on diffusion-limited penetration of a fraction of the inner volume of the column determined by the size of the solute. By further assuming that the solute diffusion coefficient is proportional to M^{-b}, they obtained Eq. (13).

$$V_e = V_o + V_g \left[\frac{k}{(\pi U M^m)^{1/2}} [1 - \exp(-UM^m/k^2)] + \mathrm{erfc}\left(\frac{(UM^m)^{1/2}}{k}\right) \right] \quad (13)$$

where

V_g = volume of gel accessible to solute
U = average velocity of mobile phase
k = constant
m = constant, usually between 1/2 and 2/3

Equation (13) is approximated by one empirical correlation at intermediate molecular weights and by a second empirical correlation at large molecular weights. Equation (13) correlated elution volumes of narrow polystyrene fractions as a function of molecular weight, but it gave only semiquantitative agreement at different flow rates because they treated V_g as independent of molecular weight (79).

There have been too few critical comparisons of the equations proposed to correlate elution data to permit choice of one expression as more correct than all others. Siegel and Monty determined the elution behavior of a number of proteins having different shapes and partial specific volumes on Sephadex G-200 (80). They found that Eqs. (8), (11), and (12) all correlated the data well, but that the data did not correlate with molecular weight. It was their contention that most proteins used as calibrating standards possess closely similar frictional ratios and partial specific volumes, and that a correlation with molecular weight is virtually indistinguishable from a correlation with molecular radius in that case (80). There are two reasons for the ability of a number of expressions to correlate gel filtration elution data. One is the error in determining the elution volume and the other is the fact that the elution volume varies by only about a factor of two. (For example, $V_i \simeq 0.5 V_t$ and $V_o \simeq 0.4 V_t$ for Sephadex G-25. Thus V_e varies from about 0.4 V_t to 0.9 V_t.) Thus an explicit model with two fittable parameters, such as Eq. (11), is adequate for most purposes.

With regard to the question of whether steric exclusion or restricted diffusion is more important in determining elution volume, most evidence favors a steric exclusion mechanism. A number of investigators have found elution volume to be independent of flow rate (43, 46,78), which is strong evidence for the steric exclusion mechanism. Flow rate did influence elution volumes in the results of Yau and Malone and as commercial applications of gel filtration develop, it is likely that diffusion effects will be encountered as the higher flow rates dictated by economy are employed. Distribution coefficients determined from column performance agree with distribution coefficients determined by batch equilibrium for some gels, offering further support for the steric exclusion mechanism (78). Finally, Obrink et al. (81) showed that elution behavior on Sephadex G-200 was nearly identical at temperatures of 9° and 60°C., the difference resulting from a structural change in the gel. The weak temperature dependence they observed is evidence for a steric exclusion mechanism (81).

2. Zone Broadening

In addition to elution volume, the degree of broadening of a band of solute as it passes through a gel filtration column has an important influence on a separation. Zone broadening is caused by molecular diffusion and fluid mixing in the axial direction, the statistical distribution of solute molecules between the gel and mobile phases, nonequilibrium between the two phases, and irregular flow patterns in the column. In small columns, the last factor can be made negligible by careful packing of spherical gel particles of uniform size, but channeling is difficult to control in large-diameter columns.

The theoretical plate concept is often used to analyze zone broadening in chromatography (82). The gel filtration column is conceptually divided into a number of sections, each having a height equivalent to a theoretical plate (H). The number of theoretical plates may be calculated from elution data with the formula

$$N = (V_e/\sigma)^2 = (4V_e/w)^2 \qquad (14)$$

where

N = number of theoretical plates
σ = standard deviation of elution curve
w = base line width between lines drawn tangent to inflection points in elution curve

For Gaussian elution curves, the two expressions for N in Eq. (14) are identical, but for non-Gaussian curves, only the first expression should be used. It should also be noted that Eq. (14) may be used only if the sample is introduced into the column as a narrow pulse. The height equivalent to a theoretical plate may be determined with Eq. (15).

$$H = L/N \tag{15}$$

where

L = length of the column

The height equivalent to a theoretical plate is a measure of zone broadening, increased values indicating broader elution curves and less efficient separations. Theoretical plate heights determined by Flodin on Sephadex G-25 are presented in Table VI (83). In the case of uridylic acid increasing the flow rates increased H by decreasing the time for equilibrium between gel and mobile phases. The decreased plate height obtained using the smaller gel particles confirms this conclusion. In the case of hydrochloric acid, the increased value of H at the lower flow rate was caused by molecular diffusion of the solute in the axial direction. Axial diffusion was more important in the case of hydrochloric acid because its diffusion coefficient is much higher than that of uridylic acid.

By analogy with gas–liquid chromatography, Giddings and Mallik (84) have proposed Eq. (16) to correlate theoretical plate heights in gel filtration.

$$H = \frac{4}{3}\frac{D_s V_o}{V_e U} + \frac{1}{20}\frac{V_e}{V_o}\left(1 - \frac{V_e}{V_o}\right)\frac{d_p^2 U}{D_s}$$
$$+ \sum_{j=1}^{4} (1/2\, \lambda_j d_p + D_s/W_j d_p^2 U)^{-1} \tag{16}$$

where

d_p = diameter of gel particles
D_s = diffusivity of solute
U = mobile phase velocity
λ_j, W_j = geometric factors of order unity

The first term in Eq. (16) accounts for molecular diffusion in the axial direction, the second term represents the effects of nonequilibrium in the gel phase, and the last term accounts for nonequilibrium and mixing in the mobile phase (84). Giddings and Mallik (84) compared

TABLE VI
EFFECT OF FLOW RATE AND GEL PARTICLE SIZE ON THEORETICAL
PLATE HEIGHTS IN GEL CHROMATOGRAPHY[a]

Compound	Flow rate (ml./cm.^{2} minute)	H(180–300 μ) (mm.)	H(35–50 μ) (mm.)
Uridylic acid	0.053	0.39	0.15
	0.48	2.62	
	1.01	5.49	
Hydrochloric acid	0.074	1.05	0.23
	0.51	0.51	

[a] Data of Flodin (83).

the predictions of Eq. (16) with twelve published experimental values for theoretical plate height and found little agreement, experimental values tending to be considerably larger than calculated values in a number of cases. Among the reasons given is the possibility of axial mixing in the sample injection and solute detection sections of the equipment used to determine elution behavior. A model has been proposed by Billmeyer et al. (85), which does not assume that the elution curve can be represented by a normal distribution. They wrote simultaneous partial differential equations for the gel and mobile phases and solved them to obtain Eq. (17).

$$H = \frac{2D_l}{U} + \frac{2(V_t - V_o)KU}{V_t k_G (1 + K)^2} \qquad (17)$$

where

$$K = \frac{(V_i + V_m)^2}{V_i V_o} K_d$$

k_G = mass transfer coefficient between gel and mobile phase
D_l = longitudinal dispersion coefficient

The longitudinal dispersion coefficient was given by Eq. (18).

$$D_l = \phi D_s + \lambda U d_p + \frac{hR_c^2 U^2}{\phi D_s + \lambda U d_p} \qquad (18)$$

where

R_c = column radius
d_p = particle diameter
ϕ, λ, h = constants

In a subsequent work, Billmeyer and Kelly determined H for a number of flow rates for a small solute and also for a large solute that did not penetrate the gel (86). For the latter case, the second term in Eq. (17) is zero. A commercial apparatus was used in these experiments and it was found that peak broadening occurred in the injection and detection sections in addition to the column. The equipment was modified to reduce these effects and methods were developed to correct for the remaining dispersion. Theoretical plate heights were greater for the large solute than for the small solute. This was attributed to the longer residence time of the smaller solute and the greater axial dispersion of the large solute caused by the increased solution viscosity and decreased radial diffusion (86). For the large solute, the observed values of D_l/Ud_p (inverse of the Peclet number) were about three times higher than those measured in classical studies of dispersion of small solutes in packed beds (86). Viscous solutions give rise to considerable zone broadening by causing viscous fingering, as shown by Flodin, who recommended that solution viscosities be kept below 5 cP. (83).

Zone broadening limits the number of components that may be separated by gel filtration. Giddings (87) showed that the maximum number of components resolvable by gel filtration can be estimated with Eq. (19).

$$M_c = \text{maximum number of components separable by a gel filtration column}$$
$$= 1 + 0.2\,N^{1/2} \tag{19}$$

Thus for a column with 400 plates, a maximum of five components could be resolved. Increasing the ratio of the volume of the mixture to be separated to the volume of the column further decreases the number of resolvable components, although gel filtration columns can accommodate surprisingly large mixture volumes. For example, Fig. 6 shows the effect of mixture volume on a desalting operation conducted by Flodin (83). In the lower graph, the mixture volume was nearly 40% of the total column volume (83).

3. Pressure Drop

Many of the gels employed in gel filtration are compressible, particularly those gels that are not highly cross-linked. At flow rates above a certain threshold, irreversible deformation of the gel particles occurs, reducing the flow rate. Figure 7 shows the effect of column pressure drop on flow rate for three dextran gels, illustrating deforma-

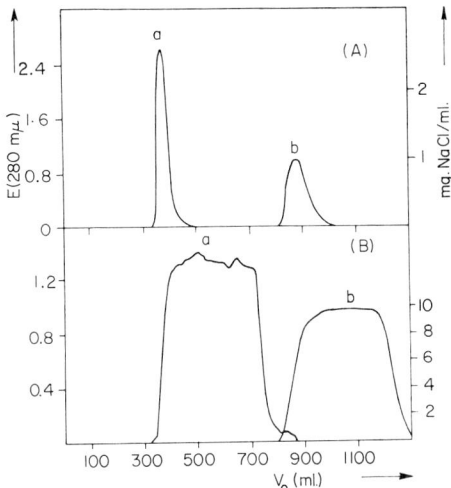

FIG. 6. Gel filtration of sodium chloride (b) and hemoglobin (a) on Sephadex G-25; Column volume, 1070 ml. (A) Sample volume 10 ml. (100 mg. hemoglobin and 100 mg sodium chloride). (B) Sample volume 400 ml. (400 mg. hemoglobin and 4 g. sodium chloride. [From Determann (43).]

tion in the G-200 gel (43). In addition to flow rate, the weight of gel packing contributes to deformation and as a rule of thumb, bed depths of greater than one meter should be avoided. Greater lengths can be obtained by connecting several columns in series. Upward flow instead of downward flow should be used with soft gels to reduce the chance of damage to the bed.

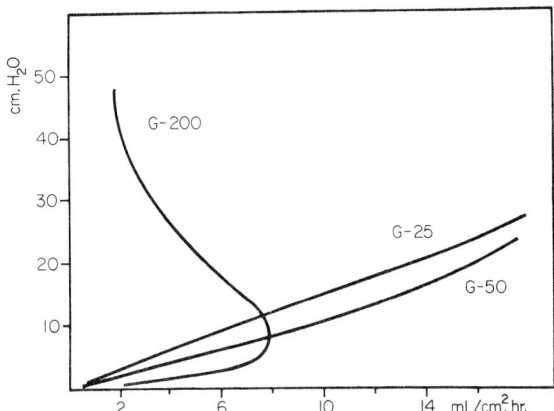

FIG. 7. Effect of flow rate on column pressure drop. [From Determann (43).]

Below the flow rate at which deformation occurs, the pressure drop per unit length of bed is directly proportional to the flow rate *(88)*. Thus for a given flow rate, the total pressure drop in the bed is directly proportional to bed length. The total pressure drop across the bed determines the point at which the bed deforms and the admissible flow rate for a column is therefore inversely proportional to the length of the bed *(88)*.

C. APPLICATIONS

The different modes of conducting gel filtration are shown in Fig. 8. Gel filtration is usually conducted in a column by application of a mixture to the top of the column followed by elution with solvent *(elution chromatography)*. When separation is difficult to achieve, the column effluent may be returned to the column entrance and the mobile phase cycled through the column a number of times, increasing the separation of components with each succeeding pass. This mode of operation is called *recycling chromatography* and was first used by Porath and Bennich *(89)*.

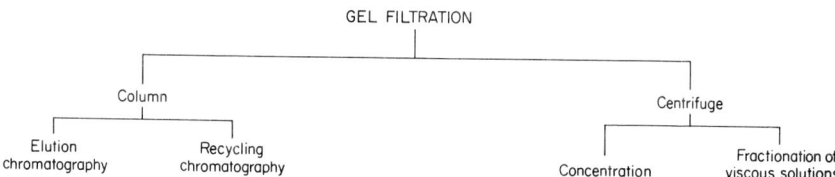

FIG. 8. Modes of conducting gel filtration separations.

Gel filtration may also be carried out in a basket centrifuge to achieve concentration of a solution or to separate components of a viscous solution *(90, 91)*. Solutions of high molecular weight compounds may be concentrated by equilibrating dry gel with the solution and then separating the interstitial liquid from the gel particles using a basket centrifuge. During the equilibration the gel swells, imbibing solvent, but excluding the large molecules which remain in the interstitial fluid. Gel filtration of viscous solutions is accomplished by swelling a gel, removing the interstitial solvent by centrifugation, adding the viscous mixture to the gel, and centrifuging to remove the interstitial solution. The small solutes permeate the gel phase, thereby reducing their concentration in the interstitial fluid. As much as 90% of the small solutes can be removed in a single step by this procedure *(90)*. The gel may then be regenerated by washing with solvent *(90, 91)*.

Gel filtration has found wide application in biological and biochemical research as a separation technique and several thousand articles have been published reporting uses. Viruses, subcellular particles, enzymes, proteins, amino acids, nucleic acids, nucleotides, hormones, steroids, vitamins, polysaccharides, saccharides, antibiotics, and even salts have been separated by gel filtration. Desalting of the above compounds is also a common operation. Many of these compounds are commercial products, and gel filtration may provide a cheaper alternative to the present separation techniques.

Several preparative scale applications have been reported. One case of commercial operation is the removal of an allergy-causing penicilloylated protein impurity from penicillins. The method was developed at Beecham Research Laboratories, Ltd. and is described in two articles *(92, 93)* and three patents *(94, 95)*. The proteinaceous impurity was discovered by gel filtration of the sodium salt of 6-aminopenicillanic acid *(92)*. The impurity has a molecular weight in excess of 10^6 and produces anaphylactic reactions in guinea pigs. A similar impurity was discovered in commercial benzylpenicillin *(92)*, and in empicillin, propicillin, methicillin, cloxacillin, and phenethicillin *(95)*. Recycling chromatography was employed in the original purification process *(92)*. Pharmacia has calculated the conditions necessary for a large-scale separation and they are presented in Table VII *(96)*. A complete and automated system capable of processing 27 liters of

TABLE VII
PURIFICATION OF BENZYLPENICILLIN BY GEL FILTRATION[a]

Gel filter type:	Sephamatic GF08-10
	(Vol. 500 liters)
Gel type:	Sephadex G-25 medium
Charge solution:	50 liters of 40% penicillin solution per cycle
Eluant:	Physiological saline, 500 liters/cycle
Product fraction:	375 liters of purified benzylpenicillin in 98% yield
Cycle time:	110 minutes
Capacity:	27 liters of contaminated solution/hour
Capital investment:	
Filter (complete, automated)	$52,000
Gel	7,800
	$59,800

[a] From references *(96)* and *(97)*.

40% benzylpenicillin per hour would cost $52,000, and the gel would cost an additional $7800 *(97)*. The gel life is 5000–10,000 running hours (1–2 years) *(97)*. Assuming 5000 operating hours per year, a 2-year life for the gel, a 10-year life for the gel filter, and 10% linear depreciation costs alone amount to 7 cents per liter. Additional costs include labor, utilities, the eluant, reconcentration of the penicillin solution, and interests costs on the capital investment.

Protein-enriched milk has been prepared by gel filtration *(98)*. Skimmed milk is separated into a high molecular weight fraction containing the milk proteins and a low molecular weight fraction containing lactose, salts, and vitamin B_2 *(98)*. The high protein fraction is then combined with skim milk and whole milk and the mixture concentrated by evaporation, resulting in a product with a protein content 60% higher than ordinary milk and with comparable taste properties *(98)*. Sixty liters of milk was fractionated per hour using three programmed columns each having a capacity of 16 liters *(98)*.

A third large-scale application is the purification of diphtheria and tetanus toxoids using two 13-liter columns in series *(99)*. The first column filled with Sephadex G-100, separated low molecular weight material that was discarded. The second column was filled with Sephadex G-200 and served in the fractionation of the high molecular weight components, resulting in a 1.68-fold increase in toxoid purity. The existing system is used to process 100,000 human doses of diphtheria toxoid per week *(99)*.

Based on the limited amount of experience obtained with large-scale gel filtration, cost estimations are difficult. Cost also depends strongly on the operating conditions, such as total volume processed, volume of the gel bed, the degree of dilution of the product, and the difference between K_d of the product and the K'_d of the impurities. For example, the approximate cost for one desalting operation handling 140 liters per hour would be about 3 cents per liter *(97)*. Approximate costs of three different gel filters are given in Table VIII *(97)*. With costs in this range, plus mild operating conditions and the ability to separate concentrated solutions in a short time, gel filtration should find many applications in commercial biochemical processes.

V. Membrane Separations

A membrane may be defined as an imperfect barrier between two fluids *(100)*. The barrier is physically distinct from the two fluid phases by virtue of its different chemical and molecular structure. The

TABLE VIII
APPROXIMATE COST OF COMPLETELY AUTOMATED GEL FILTERS
(NOT INCLUDING GEL)

Gel filter type	Diameter (cm.)	Height (cm.)	Volume (liter)	Cost
GF-04-10	40	100	125	$ 18,000
GF-08-10	80	100	500	$ 52,000
GF-18-10	180	100	2500	$130,000

[a] F.O.B. Pharmacia Fine Chemicals, Piscataway, New Jersey.

membrane is usually a solid, but it may also be a liquid or a vapor gap. Two important properties of a membrane are the *permeability*, which is the rate at which a given component is transferred through a membrane under defined conditions *(101)*, and the *permselectivity*, which is the ratio of the permeation fluxes of two components through a membrane under conditions of identical driving forces *(102)*. The ideal membrane would permit rapid transfer of a desired product, but it would act as a perfect barrier to all other components of a solution, thus giving high product permeability and infinite permselectivity. Conversely, a second type of ideal membrane would retain the desired product while passing all other components in a fluid mixture at a high rate. Recent developments in membrane technology have produced membranes approaching both ideals in selectivity for certain products, but still higher permeabilities would be desirable.

Permselectivity results from the influence of membrane structure and composition on the mechanism of transport of molecules, ions, and particles through the membrane. A *microporous* membrane is a solid barrier containing many fine pores of submicroscopic dimensions. Transport occurs by both diffusion and viscous flow through the pores, and selectivity results from a sieving mechanism based primarily on molecular size and shape. Membranes are available with pores as small as 3 Å *(103)*. Membranes with small pores may be used to fractionate solutions or to separate solutes from solvent. For pore sizes in the submicron and micron range, microporous membranes are known as membrane filters, which find application in the filtration of bacteria, viruses, and other fine particles *(104)*.

A *diffusive* membrane does not have pores. Transport occurs when a molecule in one fluid dissolves in the membrane, diffuses across it, and is transferred into the fluid on the other side of the membrane. Both the diffusivity and the solubility of a molecule in the membrane determine permeability for the molecule. Molecular size still has an important influence, but solvent-membrane intermolecular forces play a more important role in diffusive membranes than in microporous membranes. For example, if the membrane contains a high concentration of fixed positive charges and exchangeable negative ions, it will exhibit anion-exchange properties. It will be highly permeable to anions but will offer very low permeability to cations because coulombic repulsion between the cations and the fixed charges will prevent many cations from entering the membrane.

Membranes may be *homogeneous* or *heterogeneous*. A homogeneous membrane consists of a single phase on the microscopic level and may be either crystalline or amorphous *(100)*. Membranes that are heterogeneous on the submicroscopic level, such as those formed from gels, are classified as homogeneous. Heterogeneous membranes consist of more than one phase. One example is colloidal particles of ion exchange resin dispersed in a plastic. Membranes formed by *in situ* precipitation of a compound in a porous media are another example. Membranes may also be reinforced by impregnating a material such as fine nylon cloth or paper with a membrane-forming material, leading to a heterogeneous structure. Properties of *asymmetric* membranes vary across the membrane.

At least thirteen types of transport phenomena result from gradients in concentration, pressure, temperature, or electromotive force across a membrane *(100)*. The membrane separation techniques of greatest commercial interest that result from combination of these phenomena are classified according to the primary driving force in Fig. 9. Separations achieved primarily by a concentration driving force include dialysis, osmosis, and pervaporation. The fluids on

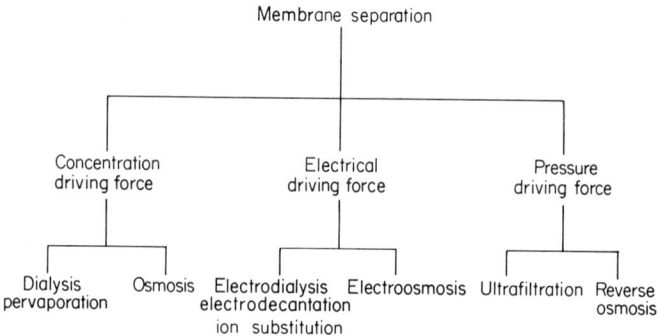

FIG. 9. Classification of common membrane separations.

either side of the membrane are both liquids in a dialysis and osmosis. *Dialysis* is the transfer of solute molecules across a membrane by diffusion from a concentrated solution to a more dilute solution. Countercurrent flux of solvent may also result from the concentration gradient in solvent concentration. *Osmosis* involves transport of solvent from a dilute solution to a concentrated solution across a membrane impermeable to solute. *Pervaporation* or *liquid permeation* processes involve a liquid on one side of a membrane and a vapor on the other side *(102, 105)*. One or more volatile components of the liquid are transferred across the membrane and are evaporated into the vapor phase. Heat must be supplied to the liquid phase for the vaporization and either a sweep gas or a vacuum is applied to the vapor phase to reduce the concentration of the diffusing components *(102)*.

The processes of *electrodialysis, electrodecantation,* or *ion substitution* are based on solute transfer due to an applied electric field. *Electroosmosis* or *endoosmosis* is the solvent flow across a membrane that results from an applied electrical potential. High pressures on a solution will force components of a solution to flow through a membrane at varying rates. Names for this process include *reverse osmosis, ultrafiltration,* and *hyperfiltration. Reverse osmosis* usually implies that transport is primarily limited to solvent, while *ultrafiltration* is a broader term that also encompasses separations in which both small solutes and solvent pass through the membrane.

Some highlights in the history of membrane phenomena are listed in Table IX. Osmosis was first observed over 200 years ago by Abbe Nollet *(106)*. The discovery of dialysis, the synthesis of inorganic membranes, and studies on osmosis were noteworthy events during the nineteenth century. The theory of membrane phenomena attracted scientists such as van't Hoff, Gibbs, Fick, Graham, and Van Laar. Early work used naturally occurring materials for membranes, such as apple skin, gelatin, natural rubber, pig and sheep gut, and parchment *(100)*. In the early twentieth century, use of regenerated cellulose and cellulose esters marked a step forward, and some of these membrane materials are still in widespread use today. During this period, considerable attention was devoted to desalting and purification of large biological molecules by dialysis and ultrafiltration. During the last thirty years, membranes formed from purely synthetic organic polymers, certain types of highly permeable heterogeneous membranes, and microporous membranes and membrane filters having very fine controlled pore sizes have become available com-

TABLE IX
LANDMARKS IN THE HISTORY OF MEMBRANE SEPARATIONS

Discovery	Year	Investigator
Osmosis through animal membranes	1748	Abbé Nollet (106)
Solute transport through membranes	1828	H. Dutrochet (107)
Dialysis separation of colloids and crystalloids	1854	Thomas Graham (108)
Artificial membranes	1867	Traube (106)
Quantitative measurements of osmotic pressure	1877	Pfeffer (106)
Electrodialysis	1900	Schwein (101)
	1903	H. N. Morse and J. A. Pierce (109)
Pore size control in nitrocellulose membranes for dialysis and ultrafiltration	1906	H. Bechhold (110)
Theory of unequal distribution of electrolytes separated by membranes	1911	F. G. Donnan (111)
Observation of ion selective properties of membranes	1927	L. Michaelis (101)
Development of ion-exchange membranes	late 1940's early 1950's	Juda, Bodamer, Wyllie, and others (29)
Desalination by ultrafiltration	1953	Charles E. Reid (112)

mercially (100). A number of useful books, reviews, and symposia dealing with membrane separations have appeared recently (100–106, 112–120). A large amount of recent work has centered on the use of membrane techniques in desalination of sea water and brackish water (106, 112, 113). With the availability of new membranes, membrane techniques are receiving renewed consideration as means of purification of sewage and industrial wastes, food processing, and separation in the chemical, petroleum, and drug industries.

In the analysis of membrane separations, it is convenient to use the stage concept. A simple membrane stage such as those found in gas permeation, ultrafiltration, and some pervaporation operations is shown in Fig. 10. Feed enters a compartment, transfers material across a membrane, and the remaining solution leaves the stage as the retentate stream (102). The material passing through the membrane forms the permeate stream. The permeate is richer in component A and the retentate is poorer in A than the feed. In the cases of dialysis, electrodialysis, and sometimes pervaporation, a sweep stream is employed downstream of the membrane to increase the driving force for transfer.

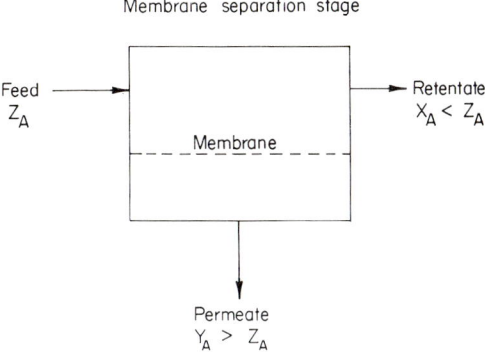

FIG. 10. Idealized membrane separation stage.

If permselectivity is not adequate to effect the desired separation in a single stage, multistage membrane operations may be used. Design methods for multistage membrane separations are discussed by Kammermeyer *(114, 117)* and by Michaels and Bixler *(102)*.

The physical design of a single membrane stage has an important influence on the cost and performance of a membrane separation process. A membrane separation unit should possess the following features *(102, 119)*:

1. Efficient contact of liquid mixture and sweep stream with membrane and efficient permeate removal.
2. Efficient application of driving force for transfer.
3. Mechanical support for membrane and limited undesirable effects from pinholes and other membrane defects.
4. Efficient mixing to minimize concentration polarization and to make effective use of all membrane area.
5. Low pressure drop through unit.
6. Reversible operation to reduce membrane fouling.
7. Simple replacement of membrane by either easy disassembly of unit or use of an inexpensive disposable unit.
8. Large membrane area-to-equipment volume ratio.
9. Minimum cost.

Membrane separation units have one of three different geometries: tubular, planar, or spiral. Dialysis and electrodialysis are usually conducted in equipment with membranes planar and parallel to one another. The plate-and-frame dialyzer is an example of such a membrane stack. Efficient application of the electrical driving force makes a membrane stack the most attractive geometry for electrodialysis. Disadvantages of the plate-and-frame design include multiple membrane handling and high labor costs, uneven flow patterns, expensive membrane support media, small area-volume ratios, and expensive pressure containers in the case of ultrafiltration *(119)*. Some of these disadvantages of the membrane stack have been eliminated in more recent stack units that utilize close membrane spacing, new support media, and turbulence promoters.

Tubular and spiral geometries have found application in some ultrafiltration devices. Havens Industries and American Standard market tubular ultrafiltration units. Porous fiberglass tubes 1/2 inch in diameter are lined with a replaceable membrane. The high pressure feed is to the inside of the tube and permeate collects outside the tubes. The design permits use of turbulence promoters. Dupont and Dow employ very fine tubes in their ultrafiltration units. The Dupont fibers are only 50 μ in diameter. Here permeation is from outside the tube to inside. Advantages of the fibers are high membrane areas per unit volume (about 10,000 ft.^{2} membrane/ft.^{3}) and elimination of the need for membrane support.

Membrane-support double sandwiches may be wrapped in a spiral to give a third type of membrane separation device with several hundred square feet of membrane per cubic foot *(119)*.

A. DIALYSIS

Dialysis was the first membrane separation to be widely used on a commercial scale. The first major application, the recovery of sodium hydroxide from rayon steep liquor, was introduced in the 1930's *(105)*. It has also been widely used in the biochemical laboratory for separating small molecules and salts from proteins and other large molecules. Recent applications include dialysis culture, which was discussed earlier, and the artificial kidney *(119)*.

Dialysis separates solutes by concentration diffusion across a membrane. A concentrated solution flows past one side of a membrane and a solvent sweep stream flows past the other side of the membrane, usually countercurrently. The feed solution on the feed side is often called the *dialysate* and the solution on the other side is called the *diffusate*. The transfer of solute from dialysate to diffusate involves diffusion through the liquids on both sides of the membrane, as well as through the membrane itself. The mass transfer rate in a countercurrent flow plate-and-frame dialyzer may be calculated with Eq. (19) *(101, 114)*.

$$W = U_o A\, C_{lm} \tag{19}$$

where

W = solute transferred (moles/time)
U_o = overall mass transfer coefficient
A = membrane area
C_{lm} = log mean concentration difference of diffusing solute

$$= \frac{(C_f - C_d) - (C_o - C_i)}{\ln\left[(C_f - C_d)/(C_o - C_i)\right]}$$

C_f = solute concentration in feed
C_d = concentration in outgoing diffusate
C_o = concentration in outgoing dialysate
C_i = concentration in incoming diffusate

The overall mass transfer coefficient includes the mass transfer resistances of both liquid films and of the membrane, as shown by Eq. (20).

$$U_o^{-1} = K_m^{-1} + k_{L1}^{-1} + k_{L2}^{-1} \tag{20}$$

where

K_m = membrane permeability to solute
k_{L1} = liquid film mass transfer coefficient on dialysate side of membrane
k_{L2} = liquid film mass transfer coefficient on diffusate side of membrane

In a thorough study of dialysis Lane and Riggle *(114)* present a method of calculating membrane permeability. Newman and Wilke *(121)* have recently developed a correlation for liquid phase mass transfer coefficients in dialysis that accounts for the effects of both free and forced convection. Their correlation is given by Eq. (21).

$$k_L = (D_s/x)\, 0.66\, (\text{Sc Gr})^{1/4} + b^{-1}\left[1.45\, U_{av}\, (\text{Sc Gr})^{-1/2}\, x^2\right] \\ - b^{-2}\left[2.64\, U_{av}\, (\text{Sc Gr})^{-3/4}\, x^3\right] \tag{21}$$

where

k_L = average liquid phase mass transfer coefficient (cm./second)
x = height of liquid-membrane interface (cm.)
D_s = solute diffusivity (cm.^{2}/second)

U_{av} = average linear bulk velocity (cm./second)
b = distance between adjacent membranes in stack (cm.)
Sc = Schmidt number (dimensionless)
 = $\mu_L/\rho_L D_s$
Gr = Grashof number (dimensionless)
 = $g\,(\rho_L{}^2 - \rho_{LM}{}^2)\,(x^3/2\,\mu_L{}^2)$
ρ_L = density of bulk solution
ρ_{LM} = density of solution at membrane

A significant amount of solvent may also transfer across the membrane, diluting the feed stream. The solvent flux may be calculated using Eqs. (19) through (21) by substituting solvent concentration diffusivity, and permeability for corresponding solute properties.

Dialysis may be competitive for some commercial biochemical separations, such as desalting high molecular weight compounds. Advantages include continuous operation, mild conditions, membrane lives approaching ten years, and no need for removal of suspended solids. Significant disadvantages are the dilution of the transferred solute and the low transfer rates, which lead to large membrane areas and high capital costs. Even with a large concentration driving force, transfer rates may only be in the neighborhood of 0.2 lb./ft.² hour *(120)*. Dialysis culture partly overcomes the low rates by providing long contact times.

B. Ultrafiltration

Ultrafiltration or reverse osmosis separates a solute from a solution by forcing the solvent to flow through a membrane by application of a hydraulic-pressure gradient. Figure 11 is a schematic representation

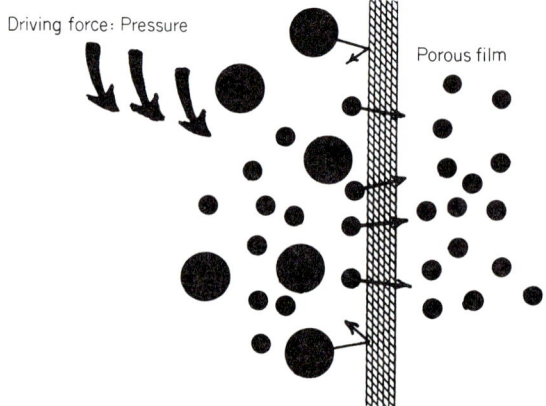

Fig. 11. Schematic representation of ultrafiltration (courtesy of Ionics, Inc.).

of the process. The permselectivity of the membrane depends strongly on molecular size; small molecules pass through the membrane, while large molecules are retained. Michaels proposes that the term "reverse osmosis" be used to describe processes in which the solute dimensions are within an order of magnitude of the solvent dimensions *(116)*. Ultrafiltrations are defined as pressure separations involving solutes whose molecular dimensions are ten or more times larger than those of the solvent and are below $1/2\ \mu$ *(116)*. Ultrafiltration received considerable attention as a laboratory technique for separating colloids and crystalloids in the first part of this century. Early ultrafiltration work is reviewed by Ferry *(122)*. The new generation of membranes that have been developed, along with studies on desalination by reverse osmosis has caused a revival of research on the technique, and several recent reviews have been published *(106, 112, 116)*.

Desirable features in ultrafiltration membranes include high permeability to solvent, high retention efficiencies for solutes above a characteristic size, durability, chemical and thermal stability, minimum effect of solute type or concentration on solvent permeability, resistance to plugging, reproducible properties, and low cost *(116)*. Ultrafiltration may function by either a sieving (microporous) or diffusive mechanism. Some general features of these two basic types of ultrafiltration are summarized in Tables X and XI. Because micro-

TABLE X
FEATURES OF DIFFUSIVE ULTRAFILTRATION

1. Solute and solvent transport by molecular diffusion
2. Does not plug
3. Solute retention efficiency increases with increasing flux
4. Rate of transport more strongly temperature dependent than for microporous membranes
5. More selective
6. Preferred for small solutes
7. Anisotropic membranes common

porous membranes transport occurs through pores, solute molecules whose dimensions are close to the pore size will block some of the pores, thereby causing plugging and reducing permeability. Permselectivity will be altered because the plugging will give a more narrow pore size distribution. Solutes whose dimensions are near the bottom of the pore size distribution are the most effective because they can enter more pores, selectively blocking smaller pores *(116)*.

TABLE XI
FEATURES OF MICROPOROUS ULTRAFILTRATION

1. Solute and solvent transport primarily by viscous flow through very small pores.
2. Susceptible to plugging by solute molecules whose dimensions approximate membrane pore size
3. Retention efficiency independent of flow rate
4. Temperature dependence of solvent permeability and solvent fluidity are the same.
5. Best used with large solutes
6. Anisotropic membranes new (and promising) development

The partially plugged membrane that results has a larger mean pore size and lower permselectivity to smaller solutes. To avoid plugging of microporous membranes, it is thus best to select a membrane with pore sizes well below solute dimensions. Recently developed asymmetric microporous membranes exhibit superior resistance to plugging. They are manufactured with conical pores whose diameters increase with distance through the membrane. This also leads to higher solvent permeability for a given pore entrance diameter.

A diffusive membrane has no pores and thus cannot plug (116). However, permeability of both types of membranes can be reduced if the retained solute accumulates at the membrane to an extent that would form a cake. Cake formation may occur with concentrated highly hydrophilic materials such as certain proteins and is often difficult to reverse by shear. Membrane fragility and high pressure driving forces unfortunately prevent periodic reversal of flow direction to reduce plugging and fouling.

Because transport through a diffusive membrane involves dissolution in the membrane and molecular diffusion through it, solute rejection efficiency is more strongly affected by solute dimensions and polarity. A commonly cited example is the xylenes. *Para-* and *ortho-*xylene differ in molecular diameter by less than 10%, but their permeabilities through certain diffusive membranes differ by a factor of two (102). Because molecular diffusion through solids is very slow, permeabilities per unit membrane thickness are much lower for diffusive membranes. To overcome this, anisotropic diffusive membranes are widely used that have a very thin (ca. 0.1–10 μ) diffusive layer attached to a coarse thicker (20–1000 μ) microporous support. Properties of some commercial membranes are given in Table XII. It should be noted that the permeabilities are not corrected for membrane thickness. Also, permeabilities may be significantly lower for solutions.

TABLE XII
PROPERTIES OF SOME COMMERCIAL ULTRAFILTRATION MEMBRANES[a]

Name	Type	Permeability to water at 100 psi (gal./ft.^{2}-day)	Molecular weight cutoff	Manufacturer
Diffusive				
Gel cellophane	Homogeneous cellulosic	1.5	ca. 5000	Dupont, Union Carbide
P Membrane	Homogeneous cellulosic	35	60,000	Schleicher and Schull
Loeb-type				
CA-type A	Anisotropic, cellulose acetate	0.4	ca. 300	General Atomics
CA-type B	Anisotropic, cellulose acetate	2.7	ca. 700	General Atomics
CA-type C	Anisotropic, cellulose acetate	1.4	ca. 500	General Atomics
Diaflo				
UM-1	Anisotropic, polyelectrolyte	175	10,000	Amicon
UM-2	Anisotropic, polyelectrolyte	60	1,000	Amicon
UM-3	Anisotropic, polyelectrolyte	25	350	Amicon
Microporous				
PEM membrane	Isotropic, cellulose triacetate	8.5	50,000	Gelman
Diaflo XM-50	Anisotropic	350	50,000	Amicon
VF	Isotropic	1,000	ca. 200,000	Millipore

[a] See reference (116).

1. Theory of Diffusive Ultrafiltration

Some important equations in diffusive ultrafiltration will be presented here. More complete coverage of the theory of ultrafiltration and membrane transport is given elsewhere *(102, 105, 106, 112, 116, 118)*. For a homogeneous, isotropic membrane, the steady-state solvent flux is given by Eq. (22).

$$J_w = K_w \left[(\Delta P - \Delta \Pi)/t \right] \tag{22}$$

where

J_w = steady-state solvent flux through membrane (moles/time-area)
ΔP = change in total pressure across membrane
$\Delta \Pi$ = change in osmotic pressure across membrane
   = $(RT/\bar{V}_w) \ln (a_{w1}/a_{w2})$
$\bar{V}_w$ = molar volume of solvent
a_w = solvent activity
t = thickness of membrane
K_w = specific permeability
   = $\bar{C}'_w \bar{D}_w \bar{V}_w / RT$
$\bar{C}'_w$ = mean solvent concentration in membrane
$\bar{D}_w$ = mean diffusivity of solvent in membrane
R = ideal gas constant
T = absolute temperature

Equation (22) shows that the solvent flux is inversely proportional to the membrane thickness. Note that the driving force for transfer is the total pressure drop *less* the osmotic pressure difference. For high pressures, dilute solutions or high molecular weight solutes, the osmotic correction is not as important. All three variables in the numerator of the equation for the specific permeability are temperature dependent. However, for convenience, specific permeability may be successfully correlated using the Arrhenius equation *(123)*.

The solute flux at a single point in the membrane is given by Eq. (23).

$$J_s = - D'_s \, (dC'_s/dt) + k'_s \, (C'_s/\bar{C}_w) \, J_w \tag{23}$$

where

J_s = solute flux through the membrane
C'_s = local solute concentration in the membrane
D'_s = local solute diffusion coefficient
k'_s = coupling coefficient (between zero and unity)

The first term on the right hand side of Eq. (23) accounts for solute transport by molecular diffusion, while the second term represents enhanced solute transport by convective solvent drag. The *solute rejection efficiency* is defined by Eq. (24).

$$R_s = 1 - C_{s2}/C_{s1} \qquad (24)$$

where

R_s = solute rejection efficiency
C_{s2} = downstream solute concentration in solution
C_{s1} = upstream solute concentration in solution

By assuming that the diffusivity and distribution coefficient of the solute are independent of concentration, it is possible to derive Eq. (25) for small solvent fluxes.

$$R_s = \alpha' t J_w/(1 + \alpha' t J_w) \qquad (25)$$

where

$\alpha' = (1/k_s D'_s)(1 - k_s k'_s/\bar{C}_w)$
k_s = distribution coefficient
   $= C'_s/C_s$

Examination of Eq. (25) predicts that the rejection efficiency increases with increasing flux. This has been observed in experiments with sulfite pulping liquors *(124)*. At high pressure differences, a limiting rejection efficiency given by Eq. (26) is attained *(116)*.

$$R_s = 1 - k_s k'_s/\bar{C}_w \qquad (26)$$

2. Theory of Microporous Ultrafiltration

Transport of pure solvent across an isotropic microporous membrane may be calculated with Eq. (27).

$$J_{wo} = K'_w \Delta P/\mu_w t \qquad (27)$$

where

K_w = hydraulic permeability
   $= \epsilon \bar{R}^2/20$
J_{wo} = solvent flux in absence of solute
ϵ = porosity of membrane (void fraction)
$\bar{R}$ = hydraulic mean pore radius

Temperature dependence thus arises solely from the variation of solvent viscosity, because the porosity and mean pore size of the membrane are temperature independent. The presence of a solute that is partly retained reduces solvent flux by increasing the resistance to flow through larger pores. Equation (28) describes this effect.

$$J_w/J_{wo} = \alpha + \beta (1 - \alpha) \qquad (28)$$

where

J_w = presence of solute
α = fraction of solvent flux which passes through pores smaller than the solute
β = resistance factor (less than unity)

Solute transport is assumed to be purely convective and is given by Eq. (29).

$$J_s = [K'_w (1 - \alpha) \Delta P/t] \, C_{s1} \qquad (29)$$

The solute rejection efficiency may thus be obtained from Eq. (30).

$$R_s = [1 + \beta (1 - \alpha)/\alpha]^{-1} \qquad (30)$$

Equation (30) predicts that the solute rejection efficiency is independent of the flux for microporous ultrafiltration. However, at sufficiently low fluxes, molecular diffusion reduces rejection efficiency.

Actual fluxes may be lower than those predicted with the preceding equations because of *concentration polarization*. This phenomenon is illustrated in Fig. 12. Solute is carried to the membrane by the flow of solvent through membrane, causing solute to accumulate to a concentration C_s^m higher than the concentration in the bulk solution, C_s *(125)*. The solute is carried away from the membrane by diffusive and molecular transport back into the upstream solution and by transport through the membrane. The increased solute concentration at the membrane increases the osmotic pressure opposing ultrafiltration and lends a greater driving force for solute transport through the membrane. Calculation of membrane fluxes should thus be based on the effective solute concentration at the membrane rather than the concentration in the bulk solution. Concentration polarization also contributes to cake formation. Use of mixers or turbulence promoters is thus desirable to reduce concentration polarization by increasing convective transport of solute away from the membrane.

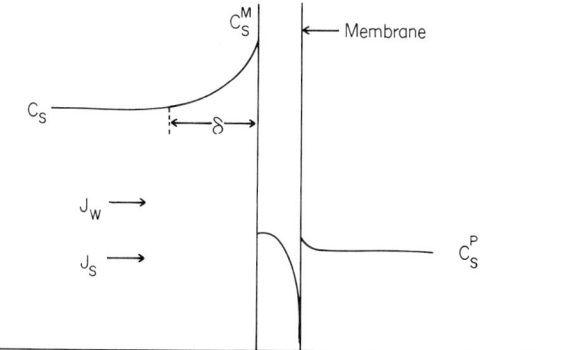

FIG. 12. Concentration profiles in ultrafiltration.

3. Applications

Ultrafiltration is most often used for concentration, but it can also be used to fractionate mixtures of large and small solutes. Fairly high concentrations of suspended solids can be tolerated if sufficient agitation is used to prevent cake formation on the membrane, and it may be practical to ultrafilter fermentation broths. It offers the advantage of mild operating conditions and greater speed than dialysis, but loss of enzyme activity due to interaction with the membrane has still been observed (123). One major operating expense in ultrafiltration is frequent replacement of membranes, which presently have lifetimes between 6 months and 2 years in most applications. Equipment and energy costs are also significant because of the high pressures required, which vary from 100 psi for very large solutes to a 1000 psi for low molecular weight solutes. Only limited experience has been obtained with large-scale operation, making costs difficult to estimate, but an equipment cost of $10 to $50 per square foot of membrane area may be anticipated (119). For example, Dorr-Oliver, Inc. manufactures an ultrafiltration pilot plant with 240 square feet of membrane that costs $10,000 (126). The cost of desalting brackish water containing 5000 ppm dissolved solids has been estimated as 52.5 cents per thousand gallons of product water for a plant producing 10^6 gallons per day (119). Costs for the much smaller capacities and more complicated separations of the pharmaceutical industry would be considerably higher. Other potential applications of ultrafiltration include waste treatment (127), processing of spent liquors from the pulp and paper industry (124), concentration and desalting of foods and juices (127), concentration of latices (127), and the artificial kidney (128).

TABLE XIII
Comparison of Separation Techniques

Technique	Primary variables	Conditions	Restrictions	Applications
Liquid ion exchange	Affinity for resin	Liquid phase	Ionic or complex-forming compounds	Extraction, concentration
Gel filtration	Size and shape	Liquid phase, mild	None	Fractionation, concentration
Dialysis, ultrafiltration	Diffusivity, membrane affinity	Liquid phase, mild	None	Concentration, fractionation
Electrodialysis	Diffusivity, membrane affinity	Liquid phase, mild	Ionic compounds	Fractionation, concentration
Pervaporation	Volatility, membrane affinity, diffusivity	Liquid and vapor phase, warm	Some components volatile	Concentration, fractionation, extraction
Gas-liquid chromatography	Volatility, solubility in liquid phase, diffusivity	Vapor phase, warm	All components volatile	Fractionation, purification, concentration
Foam fractionation	Surface activity or affinity for surfactant	Liquid, mild	None	Fractionation, concentration

VI. Conclusions

The recovery of biochemicals from fermentation broths is complicated by low product concentration, product sensitivity, and the presence of other materials, and requires ingenuity in the design of commercial processes. Liquid ion exchange, gel filtration, and ultrafiltration are three separation techniques that may find wide application in the recovery and purification of biochemicals. Salient features of these and several other promising separation techniques are summarized in Table XIII.

References

1. R. R. Freeman, *Biotechnol. Bioeng.* **6**, 87–125 (1964).
2. S. Aiba, A. E. Humphrey, and N. F. Millis, "Biochemical Engineering." Academic Press, New York, 1965.
3. F. C. Webb, "Biochemical Engineering." Van Nostrand, Princeton, New Jersey, 1964.
4. J. M. Holderby and W. A. Moggio, *J. Water Pollution Control Federation* **32**, 171–181 (1960).
5. Elmer Gaden, Jr., *Chem. Eng.* **64** [No. 5], 237–252 (1957).
6. S. Shirato and S. Esumi, *J. Ferment. Technol.* **41**, 87 (1963).
7. P. E. Dlouhy and D. A. Dahlstrom, *Chem. Eng. Progr.* **64** [No. 4], 116–121 (1968).
8. C. M. Ambler, *J. Biochem. Microbiol. Technol. Eng.* **1**, 185–205 (1959).
9. M. J. Johnson, *Science* **155**, 1515–1519 (1967).
10. H. Nakamura, *J. Biochem. Microbiol. Technol. Eng.* **3**, 395–403 (1961).
11. C. Lamanna and M. F. Mallette, "Basic Bacteriology," 3rd Ed. Williams & Wilkins, Baltimore, Maryland, 1965.
12. W. C. McGregor and R. K. Finn, *3rd Intern. Fermentation Symp., Rutgers, New Jersey, September, 1968*.
13. C. G. Golueke and W. J. Oswald, *J. Water Pollution Control Federation* **34**, 471–498 (1965).
14. P. Gerhardt and D. M. Gallup, *J. Bacteriol.* **86**, 919–929 (1963).
15. P. Gerhardt and D. M. Gallup, *Appl. Microbiol.* **11**, 506–512 (1963).
16. A. M. Gaudin, N. S. Davis, and S. E. Bangs, *Biotechnol. Bioeng.* **4**, 211–222 (1962); *ibid.* pp. 223–230.
17. H. W. Bretz, S. L. Wang, and R. B. Grieves, *Appl. Microbiol.* **14**, 778–783 (1966).
18. R. B. Grieves and Shing-Ling Wang, *Biotechnol. Bioeng.* **8**, 323–336 (1966).
19. A. J. Rubin, E. A. Cassel, O. Henderson, J. D. Johnson, and J. C. Lamb, III, *Biotechnol. Bioeng.* **8**, 135–151 (1966).
20. B. L. Karger, R. B. Grieves, R. Lemlich, A. J. Rubin, and F. Sebba, *Separation Sci.* **2**, 401–404 (1967).
21. W. A. Boyles and R. E. Lincoln, *Appl. Microbiol.* **6**, 327–334 (1958).
22. A. M. Gaudin, A. L. Mular, and R. F. O'Connor, *Appl. Microbiol.* **8**, 91–97 (1960).
23. A. J. Rubin, *Biotechnol. Bioeng.* **10**, 89–98 (1968).
24. P. Ellwood, *Chem. Eng.* **75**, [No. 16], 82–84 (1968).
25. S. L. Daniels and L. L. Kempe, *Chem. Eng. Progr. Symp. Ser.* **69** [No. 62] 142–148 (1966).

26. Gerald Berg (ed.), "Transmission of Viruses by the Water Route." Wiley (Interscience), New York, 1966.
27. F. Helfferich, "Ion Exchange." McGraw-Hill, New York, 1962.
28. R. I. Mateles, F. H. Deindoerfer, and A. E. Humphrey, *Biotechnol. Bioeng.* **6**, 5–84 (1964).
29. R. Kunin, "Ion Exchange Resins." Wiley New York, 1958.
30. C. R. Bartels, G. Kleiman, J. N. Korzun, and D. B. Irish, *Chem. Eng. Progr.* **54** [No. 8], 49–51 (1958).
31. D. W. Anderson and E. F. Lau, *Chem. Eng. Progr.* **51** [No. 11], 507–512 (1955).
32. J. E. Munden, *Process Biochem.* **1**, 431–434 (1966).
33. R. E. Treybal, "Liquid Extraction," 2nd ed. McGraw-Hill, New York, 1963.
34. W. D. Jamrack, "Rare Metal Extraction by Chemical Engineering Techniques." Macmillan, New York. 1963.
35. C. J. Lewis, *Chem. Eng.* **72**, 101–106 (1965).
36. D. J. Crouse, Jr., and K. B. Brown, *Ind. Eng. Chem.* **51**, 1461–1464 (1959).
37. E. G. Joe, G. M. Ritcey, and A. W. Ashbrook, *J. Metals* **18**, 18–21 (1966).
38. E. Hogfeldt, *In* "Ion Exchange" (Jacob A. Marinsky, ed.), Vol. 1, pp. 139–171. Dekker, New York, 1966.
39. C. F. Coleman, K. B. Brown, J. G. Moore, and D. Crouse, *Ind. Eng. Chem.* **50**, 1756–1762 (1958).
40. C. A. Blake, Jr., C. F. Baes, Jr., and K. B. Brown, *Ind. Eng. Chem.* **50**, 1763–1767 (1958).
41 E. Cerai, *Chromatogr. Rev.* **6**, 129–153 (1964).
42. S. I. Kaplan, N. L. Isaeva, and I. N. Trubnikova, *Med. Prom. SSSR* **16**, 25–31 (1961).
43. Helmut Determann, "Gel Chromatography: Gel Filtration, Gel Permeation, Molecular Sieves, A Laboratory Handbook." Springer, Berlin, 1968.
44. C. J. O. R. Morris and P. Morris, "Separation Methods in Biochemistry." Pitman, London, 1964.
45. K. H. Altgelt and J. C. Moore, *In* "Polymer Fractionation" (Manfred J. R. Cantow, ed.), pp. 123–179. Academic Press, New York, 1967.
46. P. Flodin, Dissertation, University of Uppsala, Uppsala, Sweden, 1962.
47. J. Porath and P. Flodin, *Proc. Colloq. Protides Biol. Fluids* **10**, 290–297 (1963).
48. A. Tiselius, J. Porath, and P. A. Albertsson, *Science* **141**, 13–20 (1963).
49. B. Gelotte, *In* "New Biochemical Separations" (A. T. James and L. J. Morris, eds.), p. 93. Van Nostrand, Princeton, New Jersey, 1963.
50. J. Porath, *Lab. Pract.* **16**, 838–840 (1967).
51. D. M. W. Anderson and J. F. Stoddart, *Lab. Pract.* **16**, 841–846 (1967).
52. C. C. Maitland, *Lab. Pract.* **16**, 847–850, 862 (1967).
53. P. Andrews, *Lab. Pract.* **16**, 851–856 (1967).
54. G. L. Kellett, *Lab. Pract.* **16**, 857–862 (1967).
55. Celia A. Male, *Lab. Pract.* **16**, 863–870 (1967).
56. J. F. Johnson and R. S. Porter (eds.), "Analytical Gel Permeation Chromatography," *J. Polymer Sci. C. Polymer Symposia* No. 21, Interscience, New York (1968).
57. J. Porath and P. Flodin, *Nature* **183**, 1657–9 (1959).
58. H. Fasold, G. Gundlach, and F. Turba, *In* "Chromatography" (Erich Heftmann, ed.), 2nd ed. Reinhold, New York, 1961.
59. K. O. Pedersen, *Arch. Biochem. Biophys. Suppl.* **1**, 157–168 (1962).
60. S. Hjerten and R. Mosbach, *Anal. Biochem.* **3**, 109–118 (1962).
61. J. C. Moore, *J. Polymer Sci.* **A2**, 835–843 (1964).

62. R. Kunin and R. J. Meyers, *Discussions Faraday Soc.* **7**, 114–118 (1949).
63. H. Deuel and H. Neukom, *Advan. Chem. Ser.* **11**, 51–61 (1954).
64. G. H. Lathe and C. R. J. Ruthven, *Biochem. J.* **62**, 665–674 (1956).
65. A. Polson, *Biochim. Biophys. Acta* **19**, 53–57 (1956).
66. A. Polson, *Biochim. Biophys. Acta* **50**, 565–567 (1961).
67. S. Hjerten, *Arch. Biochem. Biophys. Suppl.* **1**, 147–151 (1962).
68. W. Haller, *Nature* **206**, 693–697 (1965).
69. "Materials for Ion Exchange, Gel Filtration, and Asorption," p. 46. BioRad Lab., Richmond, California, 1967.
70. L. F. Martin and S. P. Rowland, *J. Polymer Sci.* **A1**, 5, 2563–2578 (1967).
71. Z. Adamcova, *Collection Czech. Chem. Commun.* **33**, 1, 336–340 (1968).
72. B. Cortis-Jones, *Nature* **191**, 272–3 (1961).
73. B. Gelotte, *J. Chromatog.* **3**, 330–342 (1960).
74. D. Eaker and J. Porath, *Separation Sci.* **2**, 507–550 (1967).
75. T. C. Laurent and J. Killander, *J. Chromatog.* **14**, 317–330 (1964).
76. J. Porath, *Pure Appl. Chem.* **6**, 233–244 (1963).
77. G. K. Akers, *J. Biol. Chem.* **242**, 3237–3238 (1967).
78. G. K. Akers, *Biochemistry* **3**, 723–730 (1964).
79. W. W. Yau and C. P. Malone, *J. Polymer Sci.* **B5**, 663–9 (1967).
80. L. M. Siegel and K. J. Monty, *Biochim. Biophys. Acta* **112**, 346–362 (1966).
81. B. Obrink, T. C. Laurent, and R. Rigler, *J. Chromatog.* **31**, 1, 48–55 (1967).
82. J. C. Giddings, "Dynamics of Chromatography." Dekker, New York, 1965.
83. P. Flodin, *J. Chromatog.* **5**, 103–115 (1961).
84. J. C. Giddings and K. L. Mallik, *Anal. Chem.* **38**, 997–1000 (1966).
85. F. W. Billmeyer, Jr., G. W. Johnson, and R. N. Kelley, *J. Chromatog.* **34**, 316–21 (1968).
86. F. W. Billmeyer, Jr., and R. N. Kelley, *J. Chromatog.* **34**, 322–331 (1968).
87. J. C. Giddings, *Anal. Chem.* **39**, 1027–1028 (1967).
88. M. K. Joustra, A. Emneus, and P. Tibbling, *Proc. Colloq. Protides Biol. Fluids* **15**, 575–579 (1967).
89. J. Porath and H. Bennich, *Arch. Biochem. Biophys. Suppl.* **1**, 152–156 (1962).
90. B. Gelotte and A. Emneus, *Chem. Eng. Technol.* **38**, 445–451 (1966).
91. A. Emneus, *J. Chromatog.* **32**, 243–257 (1968).
92. F. R. Batchelor, J. M. Dewdney, J. G. Feinberg, and R. D. Weston, *Lancet* i, 1175–1183 (1967).
93. E. F. Knudsen, O. P. W. Robinson, E. A. P. Croydon, and E. C. Tees, *Lancet* i, 1184–1188 (1967).
94. J. G. Feinburg, British Patent 1,078,847 assigned to Beecham Group Ltd. (1967).
95. Beecham Group Ltd., Australian Patents 29,477 and 29,478 (1966).
96. "Industrial Gel Filtration with the Sephamatic System." Pharmacia Fine Chem. Co.
97. Vel Cubrilovic, private communication (1968).
98. E.-G. Samuelsson, P. Tibbling, and S. Holm, *Food Technol.* **21** [No. 11], 121–124 (1967).
99. W. C. Latham, C. B. Michelsen, and G. Edsall, *Appl. Microbiol.* **15**, 616–621 (1967).
100. H. Z. Friedlander and R. N. Rickles. *Anal. Chem.* **37**, 27A–68A (1965).
101. S. B. Tuwiner, "Diffusion and Membrane Technology." Reinhold, New York, 1962.
102. A. S. Michaels and H. J. Bixler, *In* "Progress in Separation and Purification"

(Edmund S. Perry, ed.), Vol. 1, pp. 143-186. Wiley (Interscience), New York, 1968.
103. A. Michaels, *Ind. Res.* **10** [No. 4], 48-54 (1968).
104. C. Gelman, *Anal. Chem.* **37** [No. 6], 29A-37A (1965).
105. N. N. Li, R. B. Long, and E. J. Henley, *Ind. Eng. Chem.* **57** [No. 3], 18-29 (1965).
106. Ulrich Merten (ed.), "Desalination by Reverse Osmosis." M.I.T. Press, Cambridge, Massachusetts, 1966.
107. H. Dutrochet, "Nouvelles Resherches sur l'Endosmose et l'Exosmose." Paris, 1828.
108. T. Graham, *Phil. Trans. Roy. Soc. London* **144**, 177 (1854); *ibid.* **151**, 183 (1861).
109. H. N. Morse and J. A. Pierce, *Z. Phys. Chem.* **45**, 589 (1903).
110. H. Bechhold, *Z. Phys. Chem.* **60**, 257 (1907); *ibid.* **64**, 328 (1908).
111. F. G. Donnan, *Z. Elektrochem.* **17**, 572 (1911).
112. J. S. Johnson, Jr., L. Dresner, and K. A. Kraus, *In* "Principles of Desalination" (K. S. Spiegler, ed.), pp. 345-439. Academic Press, New York, 1966.
113. L. H. Schaffer and M. S. Mintz, ref. 112, pp. 199-289.
114. G. P. Monet (ed.), *Chem. Eng. Progr. Symp. Ser.* **55**, 24 (1959).
115. J. M. Pirie (ed.), "The Less Common Means of Separation." Inst. Chem. Eng., London, 1964.
116. Alan S. Michaels, *In* "Progress in Separation and Purification" (Edmond S. Perry, ed.), Vol. 1, pp. 297-334. Wiley (Interscience), New York, 1968.
117. Karl Kammermeyer, ref. 116, pp. 335-372.
118. R. N. Rickles, *Ind. Eng. Chem.* **58**, 18-35 (1966).
119. H. Z. Friedlander and R. N. Rickles, *Chem. Eng.* **73** [No. 4], 111-115 (1966); *ibid.* [No. 6], pp. 121-124; *ibid.* [No. 8], pp. 163-168; *ibid.* [No. 10], pp. 153-156; *ibid.* [No. 11], pp. 145-148; *ibid.* [No. 12], pp. 217-224.
120. D. A. Pattison, *Chem. Eng.* **75** [No. 12], 38-42 (1968).
121. R. D. Newman and C. R. Wilke, "Fluid Phase Mass Transfer in Dialysis," Univ. California Radiation Lab. Report UCRL-17117 (1966).
122. J. D. Ferry, *Chem. Rev.* **18**, 373-456 (1936).
123. D. I. C. Wang, T. Sonoyama, and R. I. Mateles, *17th Canad. Chem. Eng. Conf., Niagara Falls, Canada, 1967.*
124. A. J. Wiley, A. C. F. Ammerlaan, and G. A. Dubey, *Res. Conf. Reverse Osmosis, San Diego, California, 1967.*
125. T. K. Sherwood, P. L. F. Brian, and R. E. Fisher, and L. Dresner, *Ind. Eng. Chem. Fundamentals* **4**, 113-118 (1965).
126. R. A. Fiedler, private communication (1968).
127. G. G. Havens and D. B. Guy, *59th Natl. Am. Inst. Chem. Eng. Meeting, Salt Lake City*, 1967.
128. W. Dorson and Meyer Markowitz, *Chem. Eng. Progr. Symp. Ser.* **64** [No. 84], 85-89 (1968).

Ergot Alkaloid Fermentations

WILLIAM J. KELLEHER

*Division of Pharmacognosy,
Pharmacy Research Institute,
University of Connecticut, Storrs, Connecticut*

I. Introduction	211
II. Mycology	212
III. Chemistry	214
IV. Pharmacology	218
V. Occurrence	218
VI. Extraction and Analysis	219
VII. Clavine Alkaloid Fermentations	221
VIII. Lysergic Acid Alkaloid Fermentations	222
A. Simple Lysergic Acid Derivatives	222
B. Peptide Alkaloids	233
C. Culture Maintenance	235
IX. Biosynthesis	236
X. Summary	241
References	242

I. Introduction

The drug, ergot, consists of the dried sclerotium of *Claviceps purpurea* (Fries) Tulasne (Fam. Hypocreaceae) developed on rye plants. Few drugs have had a history so rich as this one. Barger (1931) assembled an exhaustive chronicle of cases of mass poisoning resulting from ingestion of ergotized grain as well as an account of the early medicinal use of this drug. Records of ergot poisoning, also known as "holy fire," "Saint Anthony's fire," and more recently as "ergotism," date back to the early middle ages. The major symptoms of ergotism were found to depend on the geographical region and on the identity of the plant that was parasitized, an observation that is consistent with the now-known existence of "chemical races" of the ergot fungus. The convulsive and gangrenous types were the most commonly observed, and both of these, in their severest forms, could terminate in death. Epidemics involving tens of thousands of people have been attributed by some writers to ergot poisoning. Other more skeptical writers feel that such maladies as bubonic plague, syphilis, and malnutrition could not be adequately excluded on the basis of recorded descriptions of the symptoms. However, scores of examples, each involving thousands of people, stand as unchallenged cases of ergotism. In one region in

Russia in 1926–27, some 11,000 cases of ergotism were brought to the attention of the authorities. Sporadic outbreaks, on a much more limited scale, continue to occur to the present day.

The first record of the use of ergot as a drug—to induce labor—occurred in a German herbal published in 1582. Recognition of ergot as the cause of ergotism came in the 17th century. In the 18th century, ergot was recognized as a fungus and evidence of its use (in midwifery) appeared. It was first introduced into official medicine in the United States shortly after the turn of the 19th century. In the years that followed, it was introduced into pharmacopeias around the world. In time, the crude drug or its extracts was replaced by purified alkaloids obtained from them and the major medical application turned to uses other than stimulation of uterine contraction.

II. Mycology

The ergot fungus has a complex life cycle; this is depicted in Fig. 1. As normally encountered, it occurs as a sclerotium which, when fully developed, takes the approximate form of the seed that it replaces on the host plant. At the end of the growing season, the sclerotia fall to the ground and remain dormant until the following spring. At that time, fruiting occurs. Ascospores are discharged from the stroma and are carried by air currents to plants of the new spring crop where infection of the ovary takes place. During the early stages of infection, numerous conidia are formed. A syrupy liquid, somewhat turbid because of its high content of conidia, exudes from the infected flowering heads. This liquid is called honeydew. Insects attracted to the honeydew transport the conidia to neighboring plants and thus effect a widespread secondary infection. After the production of honeydew ceases, the mycelia continue to grow in place of the developing seed and eventually undergo numerous cell divisions and compaction to form the dense, brittle sclerotium.

In an extensive review on *Claviceps* parasitism, Brady (1962) lists a few hundred host plants and 28 species of *Claviceps*. Fuenties *et al.* (1964) and Loveless (1967a,b) have since described three new species. Host specificity varies considerably and therefore must usually be supplemented by other characteristics in order to distinguish a species of *Claviceps*. Size, shape, and color of the sclerotium are of minor importance in classification; the germinated sclerotium (the perfect stage), on the other hand, provides the main specific characters of a species. The size and shape of the conidia of the honeydew

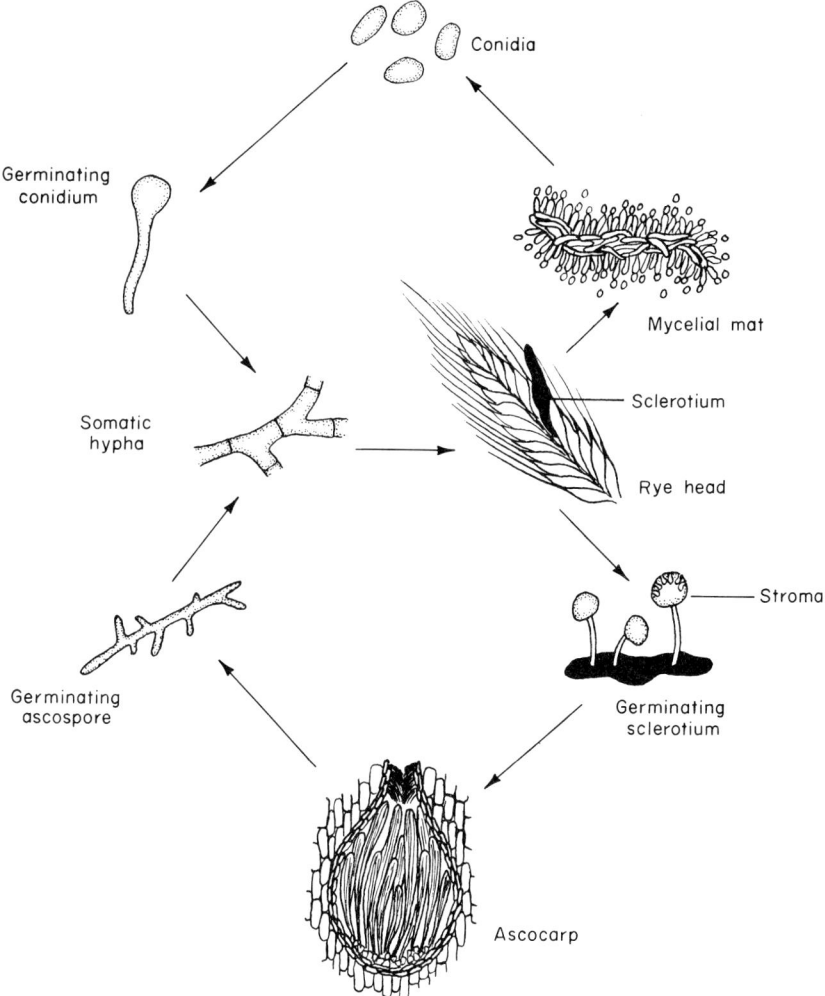

FIG. 1. Life cycle of the ergot fungus.

state have also been found to be of some value in identification (Loveless, 1964).

Some authors have further subdivided *Claviceps* species into "physiological" or "biological races" (Barger, 1931; Brady, 1962). These are grouped according to the spectrum of plants which serve as hosts for the given species. In addition, the type or types of alkaloids produced by a *Claviceps* species has given rise to the so-called "chemical races" of *Claviceps* (Hofmann, 1964).

III. Chemistry

The alkaloids of ergot are classified as indole alkaloids. With one exception, chanoclavine, they all possess the basic tetracyclic ring structure which is known as ergoline (Fig. 2). Among the several known types of indole alkaloids, ergolines are unique in that the "C"-ring closure is on the 4 position of the indole nucleus.

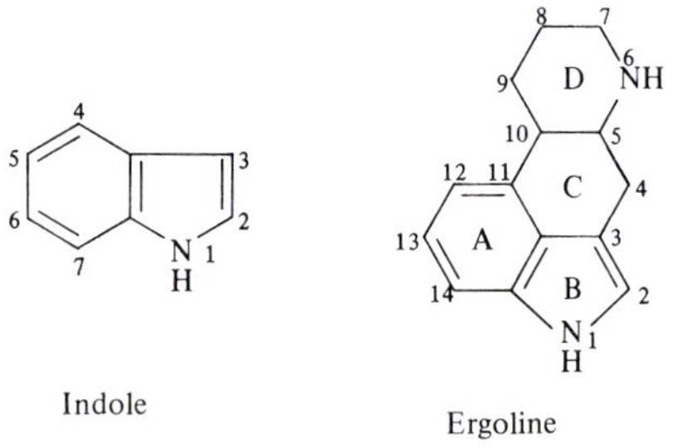

FIG. 2. The indole and ergoline ring structures.

The naturally occurring ergot alkaloids can be grouped into two categories on the basis of their chemical structure. These are the lysergic acid derivatives whose general structure is given in Fig. 3 and the clavine alkaloids whose general structure is shown in Fig. 4.

Among the lysergic acid derivatives, only one $\Delta^{8,9}$ derivative has thus far been found in nature; this is the acid itself (Kobel *et al.*, 1964). All of the others are $\Delta^{9,10}$ derivatives and these fall into two categories: the simple amides (Fig. 3) and the so-called peptide alkaloids. The latter consist of lysergic acid joined to a cyclic peptide *via* an amide linkage. Some representative peptide alkaloids are shown in Fig. 5. All of the naturally occurring lysergic acid alkaloids are derivatives of *d*-lysergic acid; these have the configuration about C-5 shown in Fig. 3. The opportunity for the existence of two isomers of *d*-lysergic acid (or *l*-lysergic acid) exists by virtue of the asymmetry of C-8. These are called *d*-lysergic acid and *d*-isolysergic acid and the alkaloids derived from these isomers usually have names ending in *-ine* or *-inine*, respectively. Alkaloids of both of these types have been found in nature. However, these are readily and reversibly interconverted in dilute alkali (Stoll *et al.*, 1954) and the presence of both isomers in an

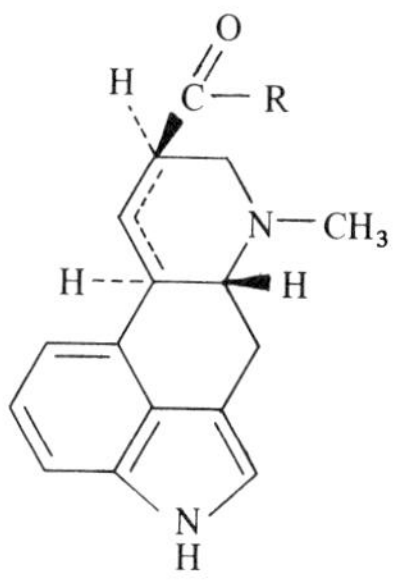

FIG. 3. Structures of some simple lysergic acid derivatives (all have been reported to be produced in saprophytic culture).

	R
$\Delta^{8,9}$ Derivatives	
$\Delta^{8,9}$ d-Lysergic acid (6-methyl-$\Delta^{8,9}$-ergolene-8-carboxylic acid)	—OH
$\Delta^{9,10}$ Derivatives	
Ergine	—NH_2
d-Lysergic acid methylcarbinolamide	—NH-CH-CH$_3$ | OH
Ergonovine (ergometrine, ergobasine)	—NH-CH-CH$_3$ | CH$_2$OH

extract might, in many instances, be attributable to isomerization during the extraction procedure.

The peptide side chains of the peptide alkaloids contain additional asymmetric carbon atoms. Many of the peptide alkaloids undergo an acid-catalyzed reversible isomerization to the so-called *aci*-alkaloids; these vary from the parent alkaloid in the configuration of C-2 in the side chain (McPhail *et al.*, 1966). Ultraviolet irradiation of aqueous acidic solutions of lysergic acid derivatives results in the addition of water across the 9,10-double bond, thereby giving rise to another category of derivatives called *lumi*-alkaloids (Stoll and Schlientz, 1955).

All of the clavine alkaloids thus far discovered possess the configuration about C-5 shown in Fig. 4. (The stereochemistry of molliclavine has not yet been determined.) In addition, the clavine alkaloids may

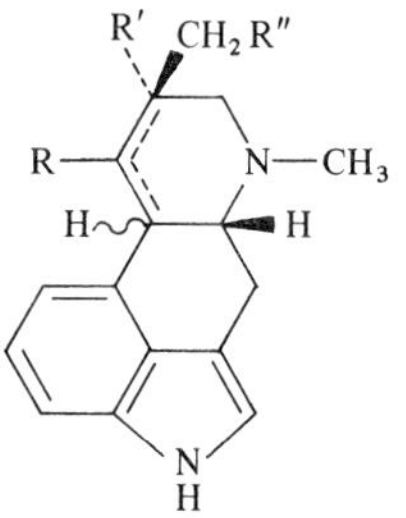

FIG. 4. Structures of some clavine alkaloids.

	D-Ring Δ	C-10	R	R'	R''
Agroclavine	8-9	(10R)	H	none	H
Elymoclavine	8-9	(10R)	H	none	OH
Molliclavine	8-9	?	OH	none	OH
Lysergene	8-17 and 9-10	—	H	none	none
Lysergine	9-10	—	H	H	H
Isolysergine[a]	9-10	—	H	H	H
Lysergol	9-10	—	H	H	OH
Isolysergol[a]	9-10	—	H	H	OH
Setoclavine	9-10	—	H	OH	H
Isosetoclavine[a]	9-10	—	H	OH	H
Penniclavine	9-10	—	H	OH	OH
Isopenniclavine[a]	9-10	—	H	OH	OH
Festuclavine	none	(10R)	H_2	H	H
Pyroclavine[a]	none	(10R)	H_2	H	H
Costaclavine	none	(10S)	H_2	H	H
Fumigaclavine A	none	?	CH_3COO	H	H
Fumigaclavine B	none	?	OH	H	CH_3

[a] C-17 projects to the rear as in isolysergic acid.

possess asymmetric carbon atoms at positions 8 and/or 10. There are several pairs of clavine alkaloids which are isomeric about C-8. In most instances, the isomer with C-17 in the same relative position as in isolysergic acid is named with the prefix, *iso*.

Chanoclavine is unique among the alkaloids of ergot in that it does not possess the complete ergoline ring structure (Fig. 6). Several years after its isolation and characterization by Hofmann et al. (1957), the natural occurrence of four isomers was discovered (Stauffacher and Tscherter, 1964). Structural elucidation of these isomers showed that the alkaloid originally known as chanoclavine is identical to the newly named chanoclavine-I.

Fig. 5. Structures of some peptide-type alkaloids.

	R	R'	R''
Ergotamine	H	H	$CH_2\text{-}C_6H_5$
Ergotaminine[a]	H	H	$CH_2\text{-}C_6H_5$
Aci-ergotamine (2S)	H	H	$CH_2\text{-}C_6H_5$
Aci-ergotaminine[a] (2S)	H	H	$CH_2\text{-}C_6H_5$
Ergosine	H	H	$CH_2\text{-}CH(CH_3)_2$
Ergosinine[a]	H	H	$CH_2\text{-}CH(CH_3)_2$
Ergocristine	CH_3	CH_3	$CH_2\text{-}C_6H_5$
Ergocristinine[a]	CH_3	CH_3	$CH_2\text{-}C_6H_5$
Ergokryptine	CH_3	CH_3	$CH_2\text{-}CH(CH_3)_2$
Ergokryptinine[a]	CH_3	CH_3	$CH_2\text{-}CH(CH_3)_2$
Ergocornine	CH_3	CH_3	$CH\text{-}(CH_3)_2$
Ergocorninine[a]	CH_3	CH_3	$CH\text{-}(CH_3)_2$
Ergostine	H	CH_3	$CH_2\text{-}C_6H_5$
Ergostinine[a]	H	CH_3	$CH_2\text{-}C_6H_5$

[a] Configuration as in isolysergic acid.

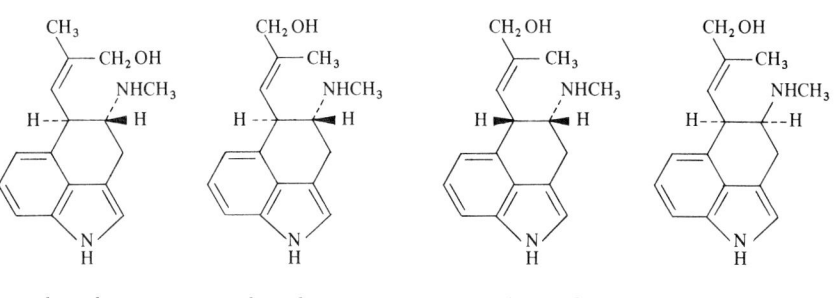

Isochanoclavine-I   Chanoclavine-I   (+)− and (−)−Chanoclavine-II

FIG. 6. The naturally occurring chanoclavines.

IV. Pharmacology

Virtually all of the lysergic acid derivatives have some type and degree of pharmacological activity (Goodman and Gilman, 1965). The nature of this activity depends very markedly on the identity of the amide substituent. The diethylamide of lysergic acid, better known as LSD, is an extremely potent hallucinogenic agent. Ergonovine, the 2-aminopropanolamide, stimulates the sympathetic nervous system; its principle use is to check postpartum bleeding. Ergotamine, on the other hand, blocks the sympathetic nervous system and finds use today primarily in the treatment of migraine headache. Most of the clavine alkaloids are without pronounced pharmacological activity but a few of the dihydro derivatives have been reported to be potent oxytocics or uterine contractants.

V. Occurrence

Up until 1960, the only known occurrence of ergot alkaloids was in a few species of *Claviceps* growing either parasitically or saprophytically. The first, and perhaps the most significant extension of the occurrence came when Hofmann and Tscherter (1960) reported the isolation of lysergic acid amide, isolysergic acid amide, and chanoclavine from the seeds of *Rivea corymbosa* (L.) Hallier filius and *Ipomoea tricolor* Cav., both members of the morning-glory family, the Convolvulaceae. This discovery triggered the investigation of many other members of the Convolvulaceae by a number of other groups; the results of these studies are summarized in a review by Der Marderosian (1967). In general, the alkaloid-containing genera were largely restricted to *Ipomoea, Rivea,* and *Argyreia,* and the alkaloids, to the simple lysergic acid derivatives and clavine alkaloids. One study was especially significant because it resulted in the isolation of the peptide alkaloids, ergosine and ergosinine, from the seeds of one of these species, viz., *Ipomoea argyrophylla* Vatke (Stauffacher *et al.,* 1965).

Somewhat less dramatic than the discovery of ergot alkaloids in higher plants was the notable discovery of their production by fungi other than *Claviceps* species. Spilsbury and Wilkinson (1961) screened 50 selected fungi for alkaloid production. They isolated festuclavine and two new alkaloids, fumigaclavine A and fumigaclavine B, from *Aspergillus fumigatus* and obtained chromatographic evidence for the presence of the last-named alkaloid in *Rhizopus arrhizus* Fischer, a Phycomycete. Yamano *et al.* (1962) tested 8 strains of *A. fumigatus* and found that all of these produced various clavine alkaloids. The Sandoz pharmaceutical company (1963) obtained a patent on a process for the production of lysergic acid derivatives by several species of *Aspergillus*. Paper chromatographic evidence for the production of a clavine alkaloid by a strain of *Penicillium roqueforti* (Taber and Vining, 1958) received substantial strengthening several years later by the isolation of costaclavine from *Penicillium chermesinum* Biourge (Agurell, 1964).

VI. Extraction and Analysis

The earliest procedures used for the extraction and purification of ergot alkaloids were designed to obtain the peptide alkaloids from sclerotia. A typical procedure would consist of alkalinizing the powdered, defatted sclerotia, extracting with chloroform, and shaking the chloroformic extract with an aqueous tartaric acid solution; or, if further purification was desired, the tartaric acid solution could be made alkaline and extracted with chloroform. The chloroform layer would then contain all of the alkaloidal bases. Estimation of the alkaloids in the appropriate solution would follow by a biological or colorimetric method. Relatively mild acidic and basic conditions were used because of the instability of the alkaloids. With the discovery of ergonovine in 1935 (Dudley and Moir, 1935; Kharasch and Legault, 1935; Stoll and Burckhardt, 1935; Thompson, 1935), methods such as this one became obsolete. For this alkaloid, like other simple lysergic acid derivatives, does not partition strongly into organic solvents from mildly alkaline aqueous solutions; the free base has a much greater water solubility and much poorer chloroform solubility than the peptide alkaloid bases. This distinction has led to the classification of ergot alkaloids into the so-called water-soluble and water-insoluble types. The former category now includes the simple lysergic acid derivatives and most of the clavine alkaloids; the latter includes the peptide-type alkaloids.

Some of the methods devised for the analysis of both the water-soluble and the water-insoluble alkaloids are described in a review by Foster (1955). The method adopted for inclusion in the official pharmaceutical compendia of the United States consisted of extracting the powdered ergot with chloroform to which a small amount of methanolic ammonia had been added. The chloroformic extract is concentrated to a small volume, diluted with ether, and the ethereal solution extracted with dilute aqueous sulfuric acid. The sulfuric acid solution is made alkaline to phenolphthalein with ammonia and the water-insoluble alkaloids contained therein removed by extraction with carbon tetrachloride. The water-soluble alkaloids are removed from the aqueous layer by extracting with ether after saturating the aqueous phase with sodium chloride.

In many of the early studies on the production of ergot alkaloids in saprophytic culture, the yields were often very low and there was a great deal of uncertainty as to whether the alkaloids occurred primarily in the mycelia or in the beer. In order to best accommodate both of these possibilities, the entire culture was freeze-dried and the freeze-dried material extracted and analyzed for alkaloids (Vining and Taber, 1959). The number of variations in extraction techniques are almost as numerous as the number of groups of investigators which have studied the ergot alkaloid fermentation; it would serve no useful purpose to describe all of these.

In most of the high-yielding clavine and lysergic acid alkaloid-producing fermentations, virtually all of the alkaloids appear in the culture filtrate and direct colorimetric analysis of the filtrate has been widely used. One noteworthy exception is the recently described fermentation for the production of peptide-type alkaloids; in this case, approximately 90% of the alkaloids occur in the mycelia and extraction of the mycelia must be undertaken to obtain the alkaloids in a form suitable for colorimetric analysis (Amici *et al.*, 1967a).

In recent times, the determination of alkaloids has been accomplished almost exclusively by some modification of a colorimetric method employing *p*-dimethylaminobenzaldehyde (PDAB) or the so-called van Urk reagent (van Urk, 1929). This reaction has been studied in detail by Dibbern and Rochelmeyer (1963). Michelon and Kelleher (1963) described a modified analytical procedure which possessed several advantages over those previously employed. Regardless of the procedure used, the reaction with PDAB gives an estimation of the total amount of alkaloid in a sample. Normally, a single alkaloid, representative of the major type of alkaloid present in the sample, is used as a standard and the results of the analysis are expressed in terms of concentration of this reference alkaloid.

The Keller reaction for indoles has been modified to make it suitable for quantitative analysis of ergot alkaloids (Rieder and Böhmer, 1958). Hofmann (1964) has reported qualitative differences in the results given by the Keller reaction with various ergot alkaloids; the results suggest that the reaction can be especially valuable in distinguishing between a number of the peptide-type alkaloids. Fluorimetric methods have been utilized successfully for the determination of several ergot alkaloids (Doepfner, 1962).

Certain types of alkaloids, e.g., those with a 9,10-double bond can be distinguished from alkaloids which do not contain this double bond (in conjugation with the indole nucleus) by their ultraviolet absorption and their fluorescence spectra. Others may be distinguished by the quality of the color given with PDAB; most alkaloids give a blue color, but several of the clavine alkaloids give a green color. Comprehensive tabulations of a number of chemical and physical properties on the individual ergot alkaloids have been presented in recent years (Agurell and Ramstad, 1962a; Hofmann, 1964; Voigt, 1959).

In most instances, it is necessary to separate an individual alkaloid from all others before it can be determined. Chromatographic procedures of all types — paper, thin-layer, and ion-exchange — have been described for this purpose. Most of the methods are published as portions of studies in which they constitute a means to an end rather than an end in themselves. Consequently, chromatographic methods for the separation of the various ergot alkaloids can be found in many of the references cited in this essay on such topics as the isolation, production, and biosynthesis of ergot alkaloids. Yamatodani (1960) has presented some valuable data on the distribution coefficients of various alkaloids in binary systems composed of chloroform, benzene, or ethyl acetate and aqueous buffers. The distribution coefficients of each of seven alkaloids was determined over the pH range of 2.5 to 8.5 at intervals of one-half pH unit. The results indicated that liquid-liquid extraction under carefully controlled conditions of pH possesses great potential as means of separating certain alkaloids on a preparative scale.

VII. Clavine Alkaloid Fermentations

Since the comprehensive review by Abe and Yamatodani (1964), there has been very little additional work on clavine alkaloid-producing fermentations. This has presumably been the result of the publication of a report on a highly successful lysergic acid alkaloid-producing fermentation by Arcamone et al. (1960, 1961). The well-known pharmacological activities of the lysergic acid alkaloids make them much more attractive candidates for study than the pharmacologically uninteresting clavine alkaloids. Because of these considerations, the clavine alkaloid fermentations will not be treated in detail in this chapter.

When the clavine alkaloid fermentations are viewed from the historical perspective, it is readily seen that they provided several notable milestones: (1) they were the first successful type of alkaloid-producing fermentation; (2) they focused attention on *Claviceps* strains growing parasitically on grasses other than rye; (3) they permitted the formulation of basal media and the definition of cultural conditions favorable to alkaloid production; and (4) they provided the biological systems with which most of the biosynthetic studies were performed.

VIII. Lysergic Acid Alkaloid Fermentations

A. SIMPLE LYSERGIC ACID DERIVATIVES

Efforts of scores of research workers since the initial publication of Bonns (1922) have been directed toward obtaining a fermentation process which would give consistently high yields of lysergic acid alkaloids. This long-sought goal was finally achieved in 1960 when a group at the Istituto Superiore di Sanita in Rome published a preliminary report of their findings (Arcamone et al., 1960). The subsequently published detailed results of this study will long remain a landmark in the history of studies on the ergot alkaloid fermentation (Arcamone et al., 1961). It was not long after this publication that other groups of investigators described *C. paspali* fermentations which produced a similar spectrum of lysergic acid derivatives (Pacifici et al., 1962; Gröger and Tyler, 1963). Kobel et al. (1964) then described an isolate of *C. paspali* which produced high yields of $\Delta^{8,9}$ lysergic acid in submerged culture. This product is readily converted to $\Delta^{9,10}$ lysergic acid by heating with dilute aqueous alkali. Two years later, Castagnoli and Mantle (1966) published a brief note on the production of this same acid by submerged cultures of a strain of *C. purpurea*.

1. Organism

The organism used by Arcamone et al. (1961) was a strain of *C. paspali* isolated from a sclerotium found on an infected plant of *Paspalum distichum* L. in the neighborhood of Rome. The early selections were made on the basis of the ability of the isolates to infect rye embryos germinated and grown on an agar medium. A highly infective isolate, found to produce 10-20 μg./ml. of alkaloids (as ergonovine) in submerged culture, was further improved by successive selection and screening of colonies arising from monohyphal elements. Briefly, the procedure consisted of homogenizing a 7-day-old submerged culture in a Waring blendor, plating out a 0.1-ml. aliquot of the homogenate, and screening the resultant colonies individually for alkaloid production in submerged culture. A second selection was conducted in the same manner using mycelia from the highest producing isolate of the first selection. Successive selections were carried out by repetition of this process. The average alkaloid production after 9 days for isolates of the first, second, and third selections, respectively, were 37, 62.5, and 370 μg./ml. (The last value was calculated with the assumption that the authors meant 0.9% instead of 9% in their Table 3.) All of the isolates possessed identical macroscopic features on potato-dextrose agar. In submerged culture, however, the low-producing isolates were white, the medium-producing isolates brown, and the high-producing isolates violet in color. The pigment was present in both the mycelium and in the culture filtrate.

Pacifici et al. (1963) carried out similar selections with an isolate of C. paspali found on initial screening to produce 50 μg./ml. total alkaloids (as ergonovine maleate). In three successive selections, the average alkaloid production was 23, 209, and 408 μg./ml., respectively. On the fourth selection, however, the average alkaloid production decreased to 152 μg./ml.; some evidence indicated that this decline might be attributable to changes in the trace element composition of the tap water used to prepare the media. Selections similar to these, but employing a defined medium for screening in submerged culture, were found to be even more successful (Mary et al., 1965).

Gröger and Tyler (1963) prepared isolates from two different batches of C. paspali sclerotia found on *Paspalum dilatatum* of Australian origin. From one batch, 250 sclerotia were sterilized and placed on nutrient agar; 96 were viable and, of these, 62 were able to produce detectable quantities of alkaloids in submerged culture. Reported values for alkaloid levels after 7 days of incubation ranged up to 560 μg./ml. The second batch of sclerotia yielded only two alkaloid-producing isolates. Experience in our own laboratory with sclerotia collected from *Paspalum* species in various parts of the world showed that it is necessary to use these within a few months after collection to avoid loss of viability (unpublished results).

Kobel et al. (1964) were able to define conditions which supported the production of conidia by their isolate of C. paspali. Alkaloid production was improved by ultraviolet irradiation of conidia and screening of monoconidial isolates obtained from the survivors (Kobel and Schreier, 1964). Other investigators have not observed conidia in their alkaloid-producing isolates of C. paspali.

2. General Procedure

All of the four laboratories that have reported high alkaloid-producing *C. paspali* fermentations use a specially prescribed procedure in going from the stock culture to the final alkaloid-producing stage. Arcamone *et al.* (1961) described a procedure wherein a mycelial fragment was transferred from a 5- to 10-day-old slant to a liquid medium; this was incubated on a rotary shaker until heavy growth developed, then homogenized and used to inoculate a second shaken culture, which after 48 hours of incubation, was used to inoculate the third and alkaloid-producing stage of the process. Pacifici *et al.* (1963) employed essentially the same procedure as that described above; but in later studies, investigators from the same laboratory modified it to include homogenization of the mycelial fragment taken from the slant (Mary *et al.*, 1965). Both laboratories found that homogenization led to a significant improvement in the uniformity among the cultures. Arcamone *et al.* (1961), in addition, studied the effect of age of inoculum and showed that optimum alkaloid production was given by the 48-hour-old seed culture.

Gröger and Tyler (1963) separated the mycelium from the seed culture and washed the mycelial pad with sterile distilled water before using it as inoculum for the alkaloid-producing stage. They found that homogenization of the inoculum resulted in lower alkaloid yields. Kobel *et al.* (1964) inoculated the seed culture with a suspension of conidia and, after abundant growth appeared, used the seed culture directly as inoculum for the alkaloid-producing stage.

The media used for the inoculum were different from the media used for the alkaloid-producing stage of the fermentations in all of the laboratories mentioned above. Arcamone *et al.* (1961) grew their inoculum on a medium consisting of 4% mannitol, 1% succinic acid, 0.1% chick pea meal, 0.1% KH_2PO_4, 0.03% $MgSO_4 \cdot 7\ H_2O$, ammonia to give pH 5.2, and tap water. For the alkaloid-producing stage, they used a medium consisting of 5% mannitol, 3% succinic acid, 0.1% KH_2PO_4, 0.03% $MgSO_4 \cdot 7\ H_2O$, ammonia to give pH 5.2, and tap water. The Connecticut group used the same media in their initial selection and screening experiments (Pacifici *et al.*, 1963), but later devised the defined media shown in Table I for both the inoculum and the alkaloid-producing stages. Gröger and Tyler (1963) examined several seed-culture media and found that corn steep solids had a decisive influence on the alkaloid-producing capacity of the mycelia even though the mycelia were washed free of spent medium before being used as inoculum for the alkaloid-producing stage. Kobel *et al.*

TABLE I
DEFINED MEDIA FOR THE PRODUCTION OF LYSERGIC ACID DERIVATIVES[a]

	Seed medium (g./liter)	Production medium (g./liter)
Mannitol	40.0	50.0
Succinic Acid	10.0	30.0
KH_2PO_4	1.0	1.0
$MgSO_4 \cdot 7\ H_2O$	0.800	0.800
$Ca(NO_3)_2 \cdot 4\ H_2O$	0.300	0.300
$NaNO_3$	0.100	0.100
$FeSO_4 \cdot 7\ H_2O$	0.050	0.050
$MnSO_4 \cdot 4\ H_2O$	0.010	0.010
$ZnSO_4 \cdot 5\ H_2O$	0.005	0.005
$CuSO_4 \cdot 5\ H_2O$	0.005	0.005
NH_4OH	to pH 5.2	to pH 5.2

[a] From Mary et al. (1965).

(1964) selected a medium consisting of 4.5% malt extract for their seed culture; this was chosen on the basis of its ability to support germination of the conidia. In every instance when alkaloid production was determined, it was found that no production occurred in the seed cultures.

All of the processes described above involved submerged cultures. Rotary shaking machines operating at 200–250 r.p.m. were used by all of the groups; Kobel et al. (1964) also used a reciprocating shaker for certain experiments. Arcamone et al. (1961) studied the effect of temperature on alkaloid production and found 23°C. to be optimum. The other investigators employed temperatures of 24° or 25°C.

3. Media

A mannitol-succinate-salts medium was used for the alkaloid-producing stage by all of the groups except Kobel et al. (1964); these investigators used a sorbitol-succinate-salts medium. Arcamone et al. (1961) provided evidence which established pH 5.2 as the optimum for alkaloid production. The other groups used values of 5.1, 5.2, or 5.4.

Various sugars and organic acids were substituted for mannitol and succinic acid, respectively, and the results showed that filter-sterilized glucose and sucrose could effectively replace mannitol and that malic or tartaric acid could replace succinic acid without causing a loss of alkaloid production (Arcamone et al., 1961). In this study, the substitution was made only in the final (third) stage and the alkaloid analyses were performed only after 9 days of incubation. Kelleher (unpublished results) studied the effect of each of the 30 combinations of 6 sugars and 5 organic acids when these were used in all three stages of the submerged culture process. The results, shown in Table II, indicate that sorbitol and fumaric acid are the preferred carbon

TABLE II
EFFECT OF VARIOUS COMBINATIONS OF SUGARS AND ORGANIC ACIDS ON ALKALOID PRODUCTION BY SUBMERGED CULTURES OF C. paspali[a,b]

	Succinic	Fumaric	Malic	Tartaric	Citric
Mannitol	359	376	256	50	170
Sorbitol	855	1632	476	188	86
Maltose	23	228	24	84	None
Sucrose	None	None	11	39	24
Glucose	None	None	None	None	52
Lactose[c]	None	None	None	None	None

[a] Each sugar and organic acid were substituted, on a gram for gram basis, for mannitol and succinic acid, respectively, in both the seed medium and the production medium shown in Table I. Each fermentation was conducted in duplicate.

[b] All of the values are peak alkaloid levels and are expressed as μg./ml. total alkaloids (as ergonovine maleate).

[c] There was no growth on the lactose-containing media.

sources for high alkaloid production. The time taken to reach peak alkaloid levels in media containing sorbitol and/or fumarate was somewhat lengthened and failure to continue the fermentations until peak levels were established might explain the discrepancy between these results and those reported by Arcamone et al. (1961). In our studies, a great deal of difficulty was encountered in initiating first-stage cultures on the sorbitol-fumarate medium. Also, the generally slower growth rate in the first two stages often caused the total fermentation time (slant to peak alkaloid production) to approach 30 days. These disadvantages could be overcome and the advantages of the sorbitol-fumarate medium could be retained by conducting the first two stages on the mannitol-succinate medium and the final stage on the sorbitol-fumarate medium. In this way, the first-stage culture almost always had a duration of 7 days, the second stage 2 days, and the third stage 10–12 days to peak alkaloid production. In one experiment in which the common first and second stages were grown on the mannitol-succinate medium, the mannitol-succinate third-stage (control) culture reached a peak alkaloid level of 462 μg./ml. in 8 days and the sorbitol-fumarate third-stage culture reached a peak level of 1754 μg./ml. in 11 days (both values are the averages of triplicate fermentations). It is worth noting that sorbitol and fumaric acid are substantially cheaper than mannitol and succinic acid, respectively.

Rosazza *et al.* (1967) performed a definitive study of the inorganic requirements for both growth and alkaloid production. These investigators used the eight-salt mannitol-succinate medium shown in Table I as their control medium. Their results showed that the requirement for peak growth was essentially the same as the requirement for maximal alkaloid production in the case of potassium and magnesium. With phosphorus, however, maximum growth was reached at lower levels than those required for maximal alkaloid production. Sulfur was required in smaller amounts for peak alkaloid production than for maximal growth. In the case of iron and zinc, the requirement for maximal alkaloid production was greater than for growth. Manganese and copper could be sufficiently depleted to demonstrate their essentiality for alkaloid production, but their requirements for growth could not be demonstrated.

4. Fermentation Kinetics

Periodic analyses of the major culture components were conducted by Arcamone *et al.* (1961) and by Mary *et al.* (1965). The results are shown in Figs. 7A and 8, respectively. Figure 7B shows the pattern of change in a replacement culture prepared by filtering the mycelia from a 9-day-old culture and suspending it in fresh medium. In general, the pattern given by the replacement culture is very similar to that given by the primary culture except that the time dimension is compressed. Mary *et al.* (1965) compared the results of analysis of four different fermentations whose peak alkaloid production ranged from 75 to 700 μg./ml. The only parameter which showed a consistent correlation with high alkaloid production was the rapidity of uptake of phosphorus.

5. Media Supplements

The demonstration of the origin of the ergoline ring system from tryptophan and mevalonic acid suggested that precursor feeding might be effective in increasing alkaloid yields. Arcamone *et al.* (1961) tested several amino acids and found that only DL-tryptophan caused

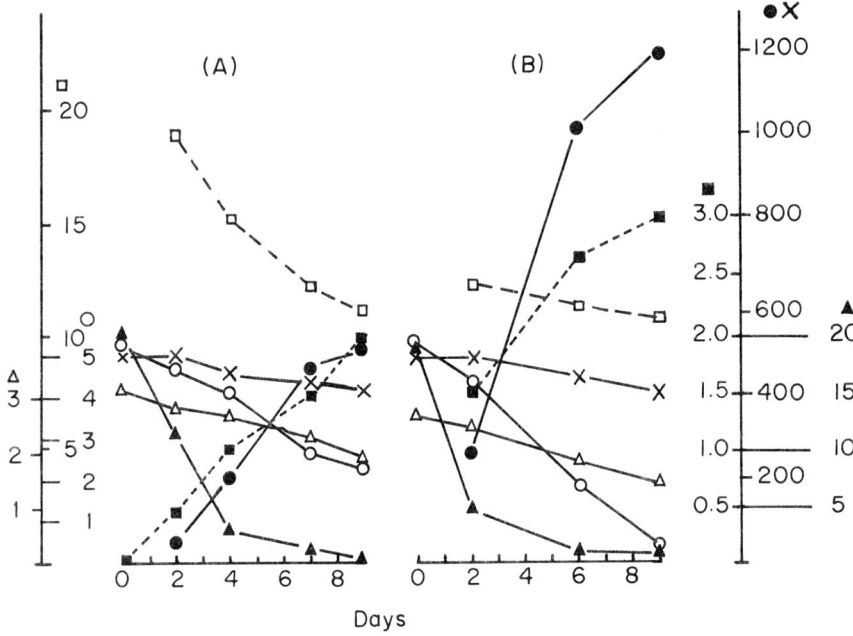

Fig. 7. Course of a typical replacement fermentation of *Claviceps paspali*, Strain F-550, in shake flasks in medium B. (A) First fermentation; (B) replacement fermentation. △, Succinic Acid, g %; ○, mannitol, g %; □, Q_{O_2}; ■, dry weight, g %; ▲, inorganic phosphate, mg. P %; ●, lysergic acid derivatives, μg./ml.; X, ammonia, mg. N %. [After Arcamone et al. (1961).]

a substantial increase in alkaloid production when added to the culture medium at the beginning of the fermentation (Table III). Pacifici et al. (1963) also obtained stimulatory effects from L-tryptophan and found that acetamide and glycine enhanced alkaloid production as well (Table III). These latter two supplements were added as potential sources of acetate; this is known to supply the carbon atoms of mevalonic acid, but is highly inhibitory to growth when added to the cultures even in very low concentrations. Gröger and Tyler (1963) not only failed to obtain an increase in alkaloid production upon adding L-tryptophan, L-tyrosine, L-phenylalanine or mevalonic acid to mycelia suspended in buffer or to growing cultures, but they observed a consistent decrease in alkaloid production as a result of these additions. Kelleher (unpublished results) obtained similar results with a high alkaloid-producing isolate obtained after five selections on the eight-salt mannitol-succinate medium (Mary et al.,

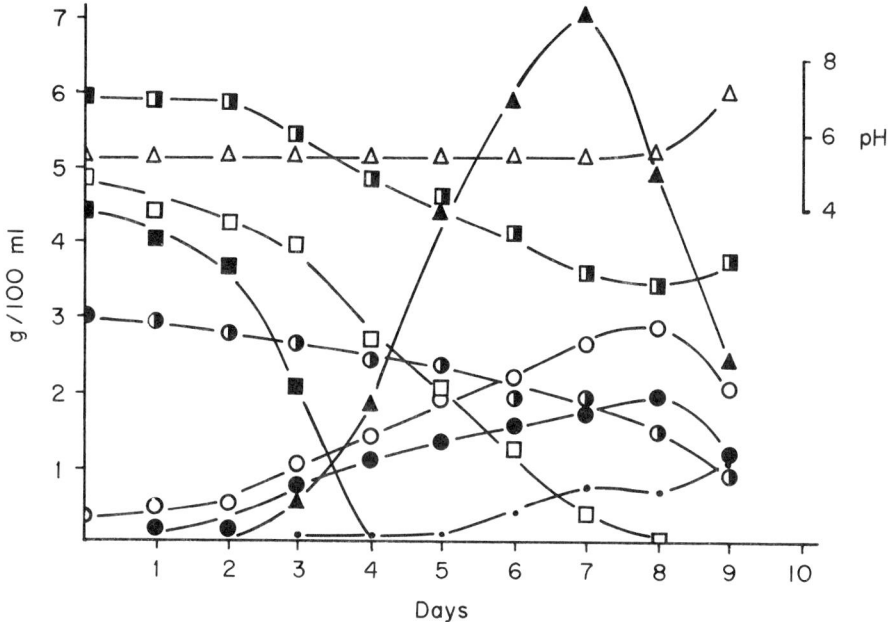

FIG. 8. Changes in the medium components during growth of submerged cultures of *C. paspali*. ○, Mycelial dry weight; ◑, succinic acid; ●, mycelial nitrogen × 10; □, mannitol; ▣, ammonia nitrogen × 10; ■, phosphorus × 200; △, pH; ▲, alkaloid × 100; -··-, organic nitrogen × 10. [After Mary *et al.* (1965).]

1965). Analyses of the culture filtrates during the first 3 or 4 days of incubation revealed, however, that alkaloid levels in the tryptophan-supplemented cultures were higher than those in unsupplemented cultures.

The correlations between high alkaloid production and rapid rate of phosphate depletion prompted Kim *et al.* (1968a) to undertake a series of experiments with agents that interfere with phosphate metabolism. The inorganic nutrition studies of Rosazza *et al.* (1967) had shown that reduction of the phosphate level in the basal eight-salt medium caused a decrease in alkaloid production. Therefore, the direct imposition of a phosphate deficiency was not effective. The well-known role of arsenate in interfering with various aspects of phosphate metabolism made it a promising candidate for study. A preliminary study in which various concentrations of arsenate were added to the culture medium, revealed that concentrations of arsenate between one-fiftieth and one-twentieth the molar concentration of phosphate present in the medium (7.35 mM) gave approximately a 100% increase in alkaloid pro-

TABLE III
Effect of Various Precursors on Lysergic Acid Alkaloid Production

Expt. No.	Precursor	Alkaloid production (μg./ml.)	Reference
I.	None	504	Arcamone et al. (1961)
	DL-Tryptophan, 400 μg./ml.	1430	
	DL-Tryptophan, 800 μg./ml.	1800	
II.	None	224	Pacifici et al. (1963)
	L-Tryptophan, 50 μg./ml.	332	
	L-Tryptophan, 100 μg./ml.	402	
	L-Tryptophan, 500 μg./ml.	446	
	L-Tryptophan, 1000 μg./ml.	274	
III.	None	203	Pacifici et al. (1963)
	Acetamide, 1 mg./ml.	227	
	Acetamide, 5 mg./ml.	430	
	Acetamide, 10 mg./ml.	632	
	Acetamide, 20 mg./ml.	443	
IV.	None	205	Pacifici et al. (1963)
	Glycine, 100 μg./ml.	341	
	Glycine, 250 μg./ml.	475	
	Glycine, 500 μg./ml.	631	
	L-Tryptophan, 500 μg./ml., plus acetamide, 10 mg./ml.	1172	

TABLE IV
Effect of Arsenate on Growth and Alkaloid Production in Low- and High-Phosphate Media[a,b]

Medium		Mycelial dry weight (g.)	Peak alkaloid concentration[c] (μg./ml.)
KH_2PO_4 (%)	Na_2HAsO_4 (mM)		
0.1	None	2.49	622
0.1	0.184	2.23	1442
0.1	1.84	None	None
1.0	None	2.70	930
1.0	0.184	2.61	1146
1.0	1.84	2.16	1225

[a] From Kim et al. (1968a).

[b] A lysergic acid alkaloid-producing isolate of *C. paspali* was used; experimental media were variations of the production medium shown in Table I.

[c] Each value is the average of triplicate fermentations; alkaloid concentrations are expressed as ergonovine maleate.

duction without causing a significant decrease in mycelial dry weight. The results of another experiment indicated that the ratio of arsenate to phosphate rather than the absolute concentration of arsenate was the important factor in determining the final effect (Table IV).

Daily analyses of the orthophosphate content of the culture filtrate from the fermentations shown in Table IV revealed that arsenate decreased the rate of phosphate uptake in all of the fermentations to which it was added and decreased the extent of phosphate uptake only in the high-phosphate medium.

In an effort to better define the mode of action of arsenate, several respiratory inhibitors and one uncoupler of oxidative phosphorylation were tested. Only the uncoupler, dinitrophenol, gave a significant (49%) increase in alkaloid production.

The observation that high alkaloid-producing cultures often showed an increased tendency to foam when samples were filtered with suction, suggested that these cultures might produce a surface active agent which, in turn, would affect alkaloid production (Kim et al., 1968b). This prompted a study of the effects of a large number of surfactants representing the anionic, cationic, and nonionic types. Among the surfactants found to have the greatest stimulatory effect on alkaloid production were some hydrophilic nonionic types. Table V presents the results obtained with some Tween-type surfactants (polyoxyethylene sorbitan esters of fatty acids) as well as with blends of Tween 80 and Arlacel 80 (a sorbitan monooleate).

These experiments also showed that certain lipophilic surfactants (e.g., Arlacel 80 and oleic acid) virtually eliminated alkaloid production while causing only moderate decreases in mycelial growth when added to the culture medium in concentrations as low as 10 μl./100 ml. This observation emphasizes the necessity of removing traces of soaps and other cleansing agents from experimental glassware and it offers a possible explanation for much of the variation that has been encountered with this and with similar alkaloid-producing fermentations.

Gröger and Tyler (1963) obtained nearly a threefold increase in alkaloid production upon adding 1,2-propanediol to their medium at a concentration of 3%.

6. Effect of Aeration

Arcamone et al. (1961) measured the concentration of dissolved oxygen in the culture medium contained in the shaken flasks and found that it was near the saturation level during the entire course of the fermentation. Reduction of the dissolved oxygen concentration by

TABLE V
ENHANCEMENT OF ALKALOID PRODUCTION BY SURFACTANTS[a,b]

Surfactant[c]	HLB[d]	Peak alkaloid[e] concentration (μg./ml.)
None	—	263
Tween 80 + Arlacel 80	7	604
Tween 80 + Arlacel 80	9	657
Tween 80 + Arlacel 80	11	919
Tween 80 + Arlacel 80	13	831
Tween 80	15	744
Tween 81	10	461
Tween 61	9.6	464

[a] From Kim et al. (1968b).
[b] A lysergic acid alkaloid-producing isolate of *C. paspali* was used; basal media were those shown in Table I.
[c] Additions were made to give a concentration of 0.1% for each single surfactant or combination of surfactants.
[d] HLB, hydrophile-lipophile balance (Atlas Chemical Industries, Inc.).
[e] Each value is the average of duplicate fermentations; alkaloid concentrations are expressed as ergonovine maleate.

the employment of tight cotton plugs or by reducing the oxygen content of the atmospheric gas stream caused a decline in both growth and alkaloid production. Gröger and Tyler (1963) obtained a small increase in alkaloid production by employing indented culture flasks.

7. Production in Stirred Fermentors

The course of the fermentation in 500-liter stirred fermentors containing 300 liters of medium was found by Arcamone et al. (1961) to be very similar to the course of the fermentation in shaken flasks (Fig. 7). Aeration was supplied solely by sparging air through a ring sparger for the first 6 or 7 days; stirring was not initiated until the oxygen electrode indicated that the dissolved oxygen level had fallen below 20–30% of saturation. Agitation in the early stages of the fermentation caused far-reaching mechanical damage to the hyphae. Kobel et al. (1964) described a relatively straightforward procedure for conducting their fermentation in a stirred fermentor containing 170 liters of medium.

8. Replacement Culture

Arcamone et al. (1961) had shown that repeated transfer of the culture after it had reached the alkaloid-producing stage did not impair its capacity for growth but it did result in a stepwise decline in alkaloid-producing capacity. However, when the alkaloid-producing mycelia was first separated from the spent medium (e.g., by filtration) it retained its full alkaloid-producing capacity upon transfer to fresh medium. The course of the fermentation in the replacement culture is shown in Fig. 7B.

B. Peptide Alkaloids

The production of peptide alkaloids in low yields by surface cultures of *C. purpurea* was achieved several years ago in a number of laboratories (Abe *et al.*, 1958; Rassbach *et al.*, 1956; Stoll *et al.*, 1957; Taber and Vining, 1957). None of these processes, however, showed a great deal of promise for future development.

In 1966, two reports appeared describing the production of peptide alkaloids in yields of over 1 g./liter of culture (Tonolo, 1966; Amici *et al.*, 1966). Both processes used isolates of *C. purpurea* that did not produce conidia and both employed media of high osmotic pressure. Ergotamine was the major alkaloid produced in both cases. Amici *et al.* (1967a) reported that approximately 90% of the alkaloid is found in the mycelium.

In a detailed presentation of their results, Amici *et al.* (1967a) described some metabolic differences between their high alkaloid-producing strain 275 FI and three strains, V (low alkaloid-producing), C (nonalkaloid-producing), and W (nonalkaloid-producing) obtained from sectors of a giant colony of strain 275FI. The course of the fermentation for each of these strains in submerged culture was determined by periodic measurement of ten parameters over a 14-day duration. The status of these fermentations, in terms of these parameters, after 14 days of incubation is shown in Table VI.

Notable differences between these strains reside in their total lipid and sterol contents and in their ability to utilize sucrose and citric acid. High alkaloid production appears to be associated with the capacity to accumulate high concentrations of lipids and sterols and with the capacity to utilize large amounts of both sucrose and citric acid. Strain 275 FI consumes large amounts of sucrose and citric acid; strain V utilizes large amounts of sucrose but small amounts of citric acid; strain C uses large amounts of citric acid but small amounts of sucrose; and strain W utilizes only limited quantities of both substances.

The initial distinction between the strains derived from strain 275 FI was based on their color on the agar plates: strain V was deep purple-violet; strain C was cream-white; and strain W, white. The

TABLE VI
PRINCIPAL METABOLIC CHARACTERISTICS OF STRAINS 275 FI, V, C, AND W
AS OBSERVED IN SUBMERGED CULTURE IN MEDIUM T-25 AFTER 14 DAYS[a]

Characteristic	Strain			
	275 FI	V	C	W
Alkaloids (μg./ml.)	1150	80	–	–
Dry weight (mg./ml.)	38.5	28	13.7	22.5
Lipids (mg./ml.)	15	4.5	1.5	3.8
Dry weight without lipids (mg./ml.)	23.5	23.5	12.2	18.7
Protein N (mg./100 ml.)	92	95	98	97
Sterols (μg./ml.)	316	136	51	145
Sucrose utilized (g./liter)	160	120	75	50
Citric acid utilized (g./liter)	14	7.5	15	8
Ammonia N utilized (mg./100 ml.)	246	250	228	212
Total N of filtrate (mg./100 ml.)	87	80	82	95

[a] From Amici et al. (1967a). Medium T-25 contained the following (in grams/liter): sucrose, 300; citric acid, 15; KH_2PO_4, 0.5; $MgSO_4 \cdot 7 H_2O$, 0.25; yeast extract, 0.1; KCl, 0.12; $FeSO_4 \cdot 7 H_2O$, 0.007; and $ZnSO_4 \cdot 7 H_2O$, 0.006. The pH was brought to 5.2 with aqueous ammonia and the solution diluted to volume with tap water.

parent stain 275 FI was pinkish-white on the agar plates. Considerable evidence was presented in an additional publication by Amici et al. (1967b) to establish that strain 275 FI was indeed a heterokaryon and not simply a mixture of strains V, C, and W. Among the observations offered in support of this conclusion were the following: (1) the sectors also appear on giant colonies initiated with single fragments of hyphae; (2) the sectoring process appears late in the growth of the colony and the origin of each sector is therefore at a point distant from the center of the colony; (3) various mixtures of the three strains failed to produce alkaloids in submerged culture; (4) giant colonies prepared from strains V, C, or W did not form sectors; (5) anastomoses were rarely observed on slides of the single strains but were abundant in mixed cultures; (6) microscopic examination of strain 275 FI showed that each cell usually contains five nuclei; and (7) strain 275 FI has not been shown to produce conidia.

These same investigators attempted to "synthesize" strain 275 FI by seeding petri plates with a mixture composed of equal amounts of small hyphal fragments from strains V, C, and W. An area of overgrowth was observed only on the contact line between colonies of strains V and C. Mycelial growth, resulting from a thick suspension of hyphal fragments from strains V and C only, was homogenized and the homogenate was filtered through a silk cloth, appropriately diluted, plated out, and incubated. Three types of colonies appeared: one similar to V (neo-V); one similar to C (neo-C); and one similar to 275 FI (neo-275 FI). Ten of each of these colonies were screened in submerged culture for alkaloid production and the results shown in Table VII were obtained. These data provide additional evidence

TABLE VII
PRODUCTION OF ALKALOIDS IN SUBMERGED CULTURE[a,b]

Colonies from strain 275 FI	Colonies from mated cultures of strains V and C		
	Neo-V	Neo-C	Neo-275 FI
1100	35	0	70
1200	60	0	160
1150	105	0	265
1300	90	0	360
1420	95	0	205
980	0	0	310
1300	85	0	275
1060	105	0	380
1320	100	0	150
1040	105	0	265

[a] From Amici et al. (1967b).
[b] Ten colonies of 275 FI and of each of the derived strains were tested. Results are expressed as micrograms per milliliter.

that strain neo-275 FI is a heterokaryon and that the heterokaryotic condition is favorable to the production of alkaloids. Additional studies with numerous sclerotia of C. purpurea showed that heterokaryosis is frequent in sclerotia but is not common in cultures which produce conidia. Since alkaloid production is most commonly associated with the sclerotia, further support is given to the suggestion that alkaloid production is associated with the heterokaryotic state.

C. CULTURE MAINTENANCE

Most investigators working with lysergic acid alkaloid-producing isolates of *Claviceps* species have for some time admitted to experiencing considerable difficulty in maintaining the alkaloid-producing capacity of their isolates. If the recent findings of Amici et al. (1967b) on the requirement for the heterokaryotic condition apply to the isolates of others, there is a sound basis for this difficulty. In all except one group (Kobel *et al.*, 1964), conidia have not been obtained from these isolates despite numerous attempts to achieve their production. The vegetative cells must, therefore, be maintained. Despite the central importance of the organism itself, very little has been published on maintenance and storage. Mary *et al.* (1965) stored their isolates of *C. paspali* on potato-dextrose agar in tightly closed screw-cap tubes at 2°–3°C. Our experience has been that cultures stored in this manner will retain their viability and alkaloid-producing capacity for up to 6 months (unpublished results). Thereafter, the culture must be transferred to a fresh slant or be used as the source of new isolates for selection and screening. Alkaloid-producing capacity usually declines with successive transfer to fresh slants but the selection and screening procedure, while very time consuming, will usually provide a new isolate with an alkaloid-producing capacity at least equal to that of the parent culture.

Hwang (1966) reported on the storage of strains of *C. purpurea* and *C. paspali* under liquid nitrogen for periods up to 38 months; alkaloid production, however, was not tested. Mizrahi and Miller (1968) were able to preserve both the viability and alkaloid-producing capacity of a *C. paspali* isolate on various media at storage temperatures of −70° or 4°C. for periods of 6 months to 2 years. No data were given in this brief report.

Kelleher (unpublished results) conducted a study of liquid nitrogen storage to determine optimum conditions for retention of both viability and alkaloid-producing capacity. Suspensions of mycelia in various media were sealed in glass vials, frozen at controlled rates, placed under liquid nitrogen, then thawed at controlled rates and screened for alkaloid production in submerged culture. The results are given in Table VIII. Long-term storage studies using the conditions found to be optimum in the preliminary study (1° C./minute freezing rate, 5° C./minute thawing rate, 20% glycerol suspending medium) are still in progress and thus far have shown essentially full retention of viability and alkaloid-producing capacity after 2 years of storage.

IX. Biosynthesis

The story of the biosynthesis of the ergot alkaloids has been one of the most fascinating chronicles in the study of alkaloid biosynthesis. Approximately 100 scientific publications have been devoted to one or

TABLE VIII
SUMMARY OF ALKALOID PRODUCTION BY CULTURES
STORED UNDER LIQUID NITROGEN[a,b,c]

Suspending medium	Slow freeze[d]			Medium freeze[d]			Fast freeze[d]		
	Slow thaw	Medium thaw	Fast thaw	Slow thaw	Medium thaw	Fast thaw	Slow thaw	Medium thaw	Fast thaw
5% DMSO	444	522	295	128	575	105	113	509	135
10% DMSO	287	212	54	32	618	254	36	524	255
20% DMSO	271	564	114	245	318	241	0	381	87
5% Glycerol	594	296	262	184	199	127	371	562	350
10% Glycerol	494	367	7.5	589	304	75	426	398	376
20% Glycerol	581	738	333	398	314	61	289	545	357

[a] A lysergic acid alkaloid-producing isolate of *C. paspali* was used; screening in submerged culture was conducted in the media shown in Table I.

[b] Except in the instances where there was a loss of viability, each value represents the average of five fermentations, each originated from a separate storage ampule.

[c] Values expressed in μg./ml. alkaloids (as ergonovine maleate).

[d] The rates that correspond to "slow" and "medium" are 1°C./minute and 5°C./minute; the fast freeze was obtained by immersing the unfrozen ampules directly in a dry ice–ethanol bath and the fast thaw, by immersing the frozen ampules in a 25°C. water bath.

another aspect of this problem. Obviously, adequate consideration of this subject would easily consume the entire space alloted to this review. A number of reviews dealing primarily with these biosynthetic studies have appeared in recent years and interested readers are referred to these and to the original research papers cited therein (Weygand and Floss, 1963; Agurell, 1966a; Floss, 1968; Ramstad, 1968).

Studies on the biosynthesis of ergot alkaloids proceeded independently along a number of parallel avenues: (1) identification of the precursors of the ergoline ring structure; (2) elucidation of the mechanism by which these precursors unite to form alkaloids; (3) demonstration of interconversions among the alkaloids; and (4) clarification of the origin of the amide substituents of the lysergic acid derivatives. Almost all of these studies, with the exception of a few dealing with the lysergic acid derivatives, were conducted with growing cultures or replacement cultures of clavine alkaloid-producing strains of *Claviceps*. However, it is interesting that when studies have been performed with parasitically-grown cultures or with alkaloid-producing higher plants, the results have been consistent with those obtained with saprophytically grown *Claviceps* cultures (Mothes et al., 1958; Gröger et al., 1963).

The first evidence for the role of tryptophan as a precursor of the ergot alkaloids was provided by the isolation of alkaloids, labeled in the lysergic acid moiety, from sclerotia taken from rye plants which had been injected with DL-tryptophan-β-^{14}C (Mothes et al., 1958). Subsequent studies with saprophytic cultures of *Claviceps* species by these and by other investigators documented this finding (Gröger et al., 1959; Plieninger et al., 1959; Taber and Vining, 1959) and further revealed that the carboxyl carbon of tryptophan was lost during incorporation (Gröger et al., 1959). Floss et al. (1964) established that L-tryptophan was the immediate precursor and that it underwent an inversion of configuration at the α-carbon upon incorporation into alkaloids. D-Tryptophan was first converted to the L-isomer *via* the α-keto acid prior to its incorporation into alkaloids. Tryptamine, N_ω-methyltryptophan, N_α-methyltryptophan, and 4-hydroxytryptophan did not serve as precursors (Baxter et al., 1961; Floss and Gröger, 1963; Floss et al., 1965). The origin of the N-methyl group of the alkaloids was shown by Baxter et al. (1964) to be methionine.

The nontryptophan-derived portion of the lysergine structure is an isoprene unit and, indeed, the incorporation of radioactivity from ^{14}C- and ^{3}H-labeled mevalonic acid was clearly demonstrated (Gröger et al., 1960; Taylor and Ramstad, 1960). Other investigators took this work a step further and established by degradation that virtually all of the radioactivity from 2-^{14}C-mevalonic acid resided in C-17 of the alkaloids (Baxter et al., 1962; Bhattacharji et al., 1962). Plieninger et al. (1961, 1967) also demonstrated the incorporation of labeled isopentenylpyrophosphate and dimethylallylpyrophosphate. The origin of the atoms of agroclavine is illustrated in Fig. 9.

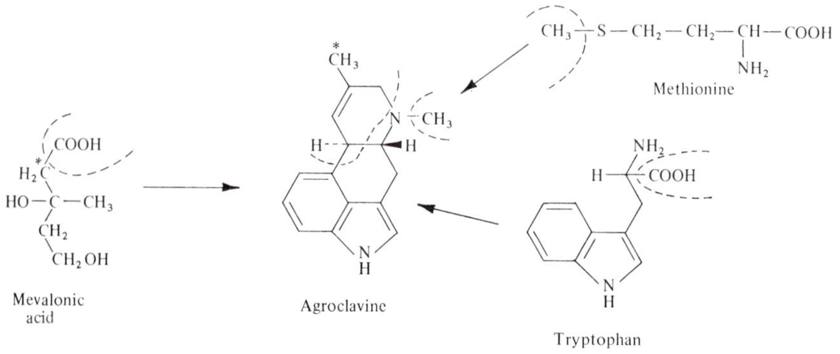

FIG. 9. Origin of the atoms of ergoline alkaloids.

The mechanism of the initial condensation of the isoprene unit with tryptophan is still obscure. Results on the incorporation of possible condensation products prepared by synthesis, however, have supported the assignment of a precursor role to 4-dimethylallyltryptophan and 4-dimethylallyltryptamine (Plieninger et al., 1967; Weygand et al., 1964). The occurrence of one of these compounds, 4-dimethylallyltryptophan, in *Claviceps* cultures has recently been

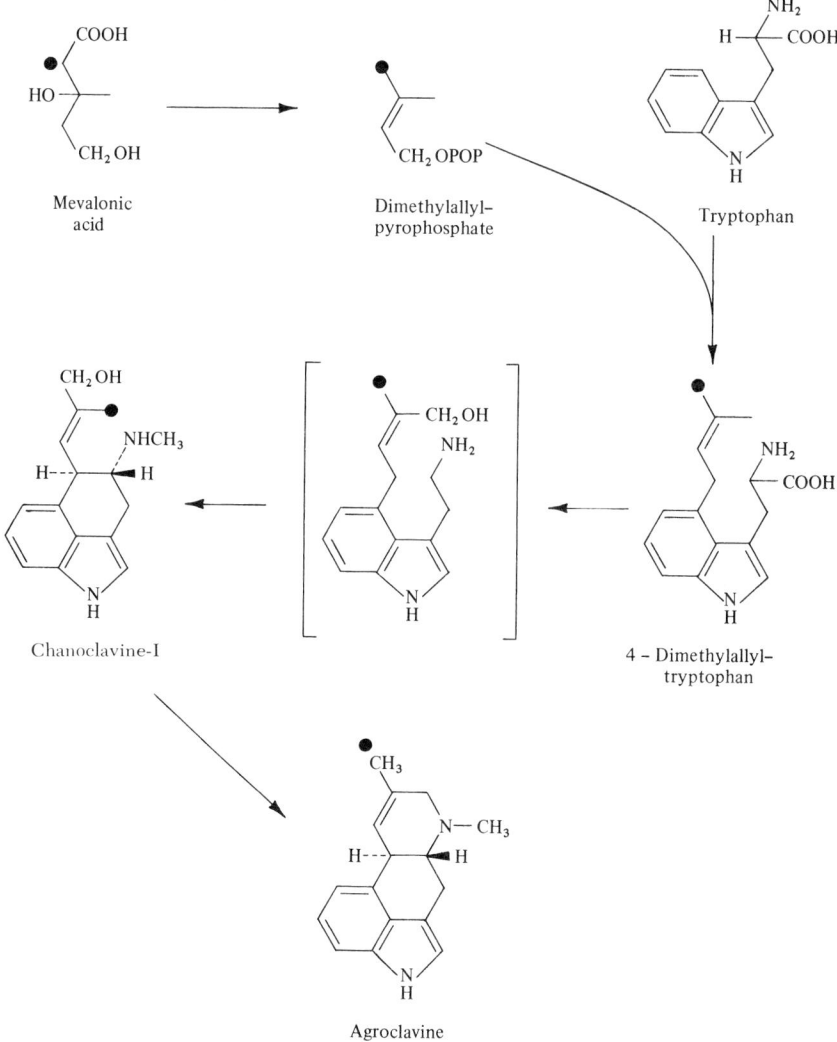

Fig. 10. Abbreviated scheme for ergoline alkaloid biosynthesis.

reported by Robbers and Floss (1968) and Agurell and Lindgren (1968).

The next demonstrated precursor in the sequence was chanoclavine-I (Fehr et al., 1966). Isochanoclavine-I was not incorporated into alkaloids to a significant extent but all of the chanoclavine isomers were labeled to a small degree from 2-^{14}C-mevalonic acid. In the cases of all of these isomers, approximately 90% of the radioactivity was located in the C-methyl group (Fig. 10). Taking into account this finding together with the previous demonstration of the incorporation of the C-2 of mevalonic acid almost exclusively into the C-17 of the alkaloids, it becomes apparent that at least one isomerization is required during the incorporation of chanoclavine-I into alkaloids. The demonstration by Floss (1967) that the 4S-hydrogen of mevalonic acid is specifically eliminated during incorporation of this precursor into alkaloids suggests that the label from C-2 of mevalonic acid should be located in the *trans*-methyl group of the supposed intermediate, dimethylallylpyrophosphate. In chanoclavine-I, however, the methyl group which carries the label is in the *cis*-position. An isomerization of the allylic double bond, therefore, must also occur at this early stage of the biosynthesis.

Additional documentation for the role of chanoclavine-I in the biosynthetic pathway to alkaloids was given by experiments (Gröger et al., 1966) in which the origin of chanoclavine-I from tryptophan as well as the incorporation of chanoclavine-I into agroclavine, elymoclavine, and lysergic acid methylcarbinolamide was demonstrated.

The failure to obtain incorporation of desoxychanoclavine-I as well as its N-demethyl derivative into alkaloids suggests that hydroxylation and possibly N-methylation occur before C-ring closure (Fehr, 1967). Positive evidence for the origin of the oxygen atoms of chanoclavine-I and elymoclavine has not been obtained but Floss et al. (1967) have shown that water is not the source.

Agroclavine appears to be the entry point into the tetracyclic ergoline alkaloids. A thorough discussion of the interrelationships between the various clavine alkaloids has been given by Abe and Yamatodani (1964) and by Agurell (1966a). In regard to their conversion to lysergic acid alkaloids, it has been demonstrated that agroclavine is first oxidized to elymoclavine (Agurell and Ramstad, 1962b) and then to lysergic acid alkaloids (Mothes et al., 1962; Agurell and Johansson, 1964; Floss et al., 1966). Lysergic acid serves as a precursor to lysergic acid amide and methylcarbinolamide (Agurell, 1966b). Lysergic acid amide, however, is not a precursor of lysergic acid methylcarbinolamide (Agurell, 1966c). Castagnoli and Tonolo

(1966) found that of the total radioactivity introduced into lysergic acid methylcarbinolamide from alanine-UL-^{14}C, approximately 40% resided in the side chain. Gröger et al. (1968) subsequently found that both alanine-2-^{14}C and alanine-^{15}N were incorporated with high efficiency into the carbinolamide side chain of lysergic acid methylcarbinolamide.

The enzymatic production of chanoclavine-I and chanoclavine-II was recently reported (Cavender and Anderson, 1968). The incubation mixture consisted of tryptophan-^{14}C, isopentenylpyrophosphate, methionine, adenosine triphosphate, and an enzyme preparation from a clavine alkaloid-producing strain of C. purpurea. Insufficient detail was given in the abstract to allow further comment, but the availability of a cell-free alkaloid-synthesizing system certainly constitutes a major milestone in the study of ergot alkaloid biosynthesis.

The very important subject of regulation of alkaloid biosynthesis has been virtually untouched. Many investigators have reported stimulatory or inhibitory effects of various compounds, many of which could be implicated in alkaloid biosynthesis. Others have sought and discovered correlations between certain metabolic activities and alkaloid accumulation. No attempts have been made, however, to relate these to the activities of various enzymes involved with the formation of the primary precursors of the alkaloids or with the biosynthesis of the alkaloids, themselves. Only in one instance has a detailed and comprehensive enzymatic study been performed; this dealt with the regulation of the biosynthesis of the aromatic amino acids in C. paspali (Lingens et al., 1967). Thus far, these results have not been related to any particular state or capacity for alkaloid production.

X. Summary

After some 40 years of efforts, the production of high yields of lysergic acid alkaloids by submerged cultures of *Claviceps* species has been achieved. The first successful processes involved the production of simple lysergic acid derivatives and these are now well established. The products of these fermentations may be readily hydrolyzed to lysergic acid which, in turn, can be chemically converted to various pharmacologically active amides. The extreme complexity of the total synthesis of lysergic acid — a 15-step process involving the preparation of some 140 intermediates (Kornfeld et al., 1956) — makes it very unlikely that a synthetic process will replace the biosynthetic process in the near future.

The production of ergotamine by submerged cultures of *C. purpurea* has paved the way for the production of peptide alkaloids in general. The association of the heterokaryotic condition with alkaloid-producing capacity appears to have provided the long-awaited key to the successful production of these alkaloids. These developments have been exploited commercially and they will, no doubt, secure the position of the fermentation process as the primary source of ergot alkaloids.

The wealth of knowledge that has been accumulated on the mechanism of biosynthesis of the ergot alkaloids together with the embryonic studies on the regulation of enzymatic processes related to alkaloid production constitute a sound basis for continued work in these areas. The next decade can be expected to bring an understanding of the process of ergot alkaloid biosynthesis that will be unsurpassed in the field of biosynthesis of secondary metabolities.

REFERENCES

Abe, M., and Yamatodani, S. (1964). *Progr. Ind. Microbiol.* **5**, 203-229.
Abe, M., Yamano, T., Kozu, Y., and Kusumoto, M. (1958). U.S. Patent 2,835,675 (May 21).
Agurell, S. L. (1964). *Experentia* **20**, 25.
Agurell, S. (1966a). *Acta Pharm. Suecica Suppl.* **3**, 71-100.
Agurell, S. (1966b). *Acta Pharm. Suecica Suppl.* **3**, 23-32.
Agurell, S. (1966c). *Acta Pharm. Suecica Suppl.* **3**, 33-36.
Agurell, S., and Johansson, M. (1964). *Acta Chem. Scand.* **18**, 2285-2293.
Agurell, S., and Lindgren, J. E. (1968). *Tetrahedron Letters*, pp. 5127-5128.
Agurell, S., and Ramstad, E. (1962a). *Lloydia* **25**, 67-77.
Agurell, S., and Ramstad, E. (1962b). *Arch. Biochem. Biophys.* **98**, 457-470.
Amici, A. M., Minghetti, A., Scotti, T., Spalla, C., and Tognoli, L. (1966). *Experentia* **22**, 415-416.
Amici, A. M., Minghetti, A., Scotti, T., Spalla, C., and Tognoli, L. (1967a). *Appl. Microbiol.* **15**, 597-602.
Amici, A. M., Scotti, T., Spalla, C., and Tognoli, L. (1967b). *Appl. Microbiol.* **15**, 611-615.
Arcamone, F., Bonino, C., Chain, E. B., Ferretti, A., Pennella, P., Tonolo, A., and Vero, L. (1960). *Nature* **187**, 238-239.
Arcamone, F., Chain, E. B., Ferretti, A., Minghetti, A., Pennella, P., Tonolo, A., and Vero, L. (1961). *Proc. Roy Soc. (London)* **B155**, 26-54.
Barger, G. (1931). "Ergot and Ergotism." Gurney and Jackson, London.
Baxter, R. M., Kandel, S. I., and Okany, A. (1961). *Chem. & Ind.*, pp. 1453-1454.
Baxter, R. M., Kandel, S. I., and Okany, A. (1962). *J. Am. Chem. Soc.* **84**, 2997-2999.
Baxter, R. M., Kandel, S. I., Okany, A., and Pyke, R. G. (1964). *Can. J. Chem.* **42**, 2936-2938.
Bhattacharji, S., Birch, A. J., Brack, A., Hofmann, A., Kobel, H., Smith, D. C. C., Smith, H., and Winter, J. (1962). *J. Chem. Soc.*, pp. 421-425.
Bonns, W. W. (1922). *Am. J. Botany* **9**, 339-353.
Brady, L. R. (1962). *Lloydia* **25**, 1-36.
Castagnoli, N., and Mantle, P. G. (1966). *Nature* **211**, 859-860.

Castagnoli, N., and Tonolo, A. (1966). *Proc. 9th Intern. Congr. Microbiol. Moscow*, pp. 31-40.
Cavender, F., and Anderson, J. A. (1968). *Abstr. 24th Southwest Regional Meeting, Am. Chem. Soc., December 4-6, Austin, Texas*, p. 55A.
Der Marderosran, A. (1967). *Lloydia* **30**, 23-38.
Dibbern, H. W., and Rochelmeyer, H. (1963). *Arzneimittelforsch.* **13**, 7-16.
Doepfner, W. (1962). *Experentia* **18**, 256-257.
Dudley, H. W., and Moir, J. C. (1935). *Science* **81**, 559-560.
Fehr, T. (1967). Untersuchungen über die Biosynthese der Ergotalkaloide. Dissertation. Eidgenössische Technische Hochschule, Zürich. p. 39.
Fehr, T., Acklin, W., and Arigoni, D. (1966). *Chem. Commun.*, pp. 801-802.
Floss, H. G. (1967). *Chem. Commun.*, pp. 804-805.
Floss, H. G. (1968). *Ber. Deut. Botan. Ges.* **80**, 705-711.
Floss, H. G., and Gröger, D. (1963). *Z. Naturforsch.* **18b**, 519-522.
Floss, H. G., Mothes, U., and Günther, H. (1964). *Z. Naturforsch.* **19b**, 784-788.
Floss, H. G., Mothes, U., Onderka, D., and Hornemann, U. (1965). *Z. Naturforsch.* **20b**, 133-136.
Floss, H. G., Günther, H., Gröger, D., and Erge, D. (1966). *Z. Naturforsch.* **21b**, 128-131.
Floss, H. G., Günther, H., Gröger, D., and Erge, D. (1967). *J. Pharm. Sci.* **56**, 1675-1677.
Foster, G. E. (1955). *J. Pharm. Pharmacol.* **7**, 1-15.
Fuentes, S. F., de la Isla, M. de L., Ullstrup, A. J., and Rodriquez, A. E. (1964). *Phytopathology* **54**, 379-381.
Goodman, L. S., and Gilman, A. (1965). "The Pharmacological Basis of Therapeutics." Macmillan, New York.
Gröger, D., and Tyler, V. E. (1963). *Lloydia* **26**, 174-191.
Gröger, D., Wendt, H. J., Mothes, K., and Weygand, F. (1959). *Z. Naturforsch.* **14b**, 355-358.
Gröger, D., Mothes, K., Simon, H., Floss, H. G., and Weygand, F. (1960). *Z. Naturforsch.* **15b**, 141-143.
Gröger, D., Mothes, K., Floss, H. G., and Weygand, F. (1963). *Z. Naturforsch.* **18b**, 1123-1124.
Gröger, D., Erge, D., and Floss, H. G. (1966). *Z. Naturforsch.* **21b**, 827-832.
Gröger, D., Erge, D., and Floss, H. G. (1968). *Z. Naturforsch.* **23b**, 177-180.
Hofmann, A. (1964). "Die Mutterkornalkaloide." Ferdinand Enke Verlag, Stuttgart.
Hofmann, A., and Tscherter, H. (1960). *Experentia* **16**, 414.
Hofmann, A., Brunner, R., Kobel, H., and Brack, A. (1957). *Helv. Chim. Acta* **40**, 1358-1373.
Hwang, S-W. (1966). *Appl. Microbiol.* **14**, 784-788.
Kharasch, M. S., and Legault, R. R. (1935). *Science* **81**, 388.
Kim, B. K., Kelleher, W. J., and Schwarting, A. E. (1968a). *Lloydia* **31**, 422.
Kim, B. K., Kelleher, W. J., and Schwarting, A. E. (1968b). *Lloydia* **31**, 422.
Kobel, H., and Schreier, E. (1964). *Abstr. Papers 148th Meeting Am. Chem. Soc.*, 8Q, *Aug.-Sept., Chicago, Illinois*, p. 8Q.
Kobel, H., Schreier, E., and Rutschmann, J. (1964). *Helv. Chim. Acta* **47**, 1052-1064.
Kornfeld, E. C., Fornefeld, E. J., Kline, G. B., Mann, M. J., Morrison, D. E., Jones, R. G., and Woodward, R. B. (1956). *J. Am. Chem. Soc.* **78**, 3087-3114.
Lingens, F., Goebel, W., and Uesseler, H. (1967). *European J. Biochem.* **2**, 442-447.
Loveless, A. R. (1964). *Trans. Brit. Mycol. Soc.* **47**, 205-213.

Loveless, A. R. (1967a). *Trans. Brit. Mycol. Soc.* **50**, 15–18.
Loveless, A. R. (1967b). *Trans. Brit. Mycol. Soc.* **50**, 19–22.
McPhail, A. T., Sim, G. A., Frey, A. J., and Ott, H. (1966). *J. Chem. Soc.* **B**, pp. 377–395.
Mary, N. Y., Kelleher, W. J., and Schwarting, A. E. (1965). *Lloydia* **28**, 218–229.
Michelon, L. E., and Kelleher, W. J. (1963). *Lloydia* **26**, 192–201.
Mizrahi, A., and Miller, G. (1968). *Appl. Microbiol.* **16**, 1100–1101.
Mothes, K., Weygand, F., Gröger, D., and Grisebach, H. (1958). *Z. Naturforsch.* **13b**, 41–44.
Mothes, K., Winkler, G., Gröger, D., Floss, H. G., Mothes, U., and Weygand, F. (1962). *Tetrahedron Letters*, pp. 933–937.
Pacifici, L. R., Kelleher, W. J., and Schwarting, A. E. (1962). *Lloydia* **25**, 37–45.
Pacifici, L. R., Kelleher, W. J., and Schwarting, A. E. (1963). *Lloydia* **26**, 161–173.
Plieninger, H., Fischer, R., Lwowski, W., Brack, A., Kobel, H., and Hofmann, A. (1959). *Angew. Chem.* **71**, 383.
Plieninger, H., Fischer, R., Keilich, G., and Orth, H. D. (1961). *Ann.* **642**, 214–224.
Plieninger, H., Immel, H., and Völkl, A. (1967). *Ann.* **706**, 223–229.
Ramstad, E. (1968). *Lloydia* **31**, 327–341.
Rassbach, H., Büchel, K. G., and Rochelmeyer, H. (1956). *Arzneimittel-Forsch.* **6**, 690–691.
Rieder, H. P., and Böhmer, M. (1958). *Experentia* **14**, 463–465.
Robbers, J. E., and Floss, H. G. (1968). *Arch. Biochem. Biophys.* **126**, 967–969.
Rosazza, J. P., Kelleher, W. J., and Schwarting, A. E. (1967). *Appl. Microbiol.* **15**, 1270–1283.
Sandoz Pharmaceutical Co. S. A. (1963). Belgian Patent 629,158 (October 21).
Spilsbury, S. F., and Wilkinson, S. (1961). *J. Chem. Soc.*, pp. 2085–2091.
Stauffacher, D., and Tscherter, H. (1964). *Helv. Chim. Acta* **47**, 2186–2194.
Stauffacher, D., Tscherter, H., and Hofmann, A. (1965). *Helv. Chim. Acta* **48**, 1379–1380.
Stoll, A., and Burckhardt, E. (1935). *Compt. Rend.* **200**, 1680–1682.
Stoll, A., and Schlientz, W. (1955). *Helv. Chim. Acta* **38**, 585–594.
Stoll, A., Petrzilka, Th., Rutschmann, J., Hofmann, A., and Günthard, H. H. (1954). *Helv. Chim. Acta* **37**, 2039–2057.
Stoll, A., Brack, A., Hofmann, A., and Kobel, H. (1957). U.S. Patent No. 2,809,920 (October 15).
Taber, W. A., and Vining, L. C. (1957). *Can. J. Microbiol.* **3**, 55–60.
Taber, W. A., and Vining, L. C. (1958). *Can. J. Microbiol.* **4**, 611–626.
Taber, W. A., and Vining, L. C. (1959). *Chem. & Ind.* pp. 1218–1219.
Taylor, E. H., and Ramstad, E. (1960). *Nature* **188**, 494–495.
Thompson, M. R. (1935). *Science* **81**, 636–639.
Tonolo, A. (1966). *Nature* **209**, 1134–1135.
van Urk, H. W. (1929). *Pharm. Weekblad* **66**, 473–481.
Vining, L. C., and Taber, W. A. (1959). *Can. J. Microbiol.* **5**, 441–451.
Voigt, R. (1959). *Pharmazie* **14**, 607–617.
Weygand, F., and Floss, H. G. (1963). *Angew. Chem. Intern. Ed. Engl.* **2**, 243–247.
Weygand, F., Floss, H. G., Mothes, U., Gröger, D., and Mothes, K. (1964). *Z. Naturforsch.* **19b**, 202–210.
Yamano, T., Kishino, K., Yamatodani, S., and Abe. M. (1962). *Ann. Rept. Takeda Res. Lab.* **21**, 95–101.
Yamatodani, S. (1960). *J. Agr. Chem. Soc. Japan* **34**, 584–589.

The Microbiology of the Hen's Egg[1]

R. G. BOARD

*School of Biological Sciences,
Bath University of Technology,
Bath, Somerset, England*

I.	Introduction	245
II.	The Egg	246
	A. Structure	246
	B. Strength	251
	C. Defense	252
III.	The Course of Infection	261
	A. Infection	261
	B. Penetration	263
	C. Colonization	268
	Addendum	274
	References	274

I. Introduction

Systematic study of this subject had its genesis in an argument between Donné and Pasteur. The former maintained that vigorous shaking was all that was needed to induce the addling of an egg. Pasteur (see Gayon, 1873) did not always succeed in producing rots by this method but, when successful, he noted large numbers of bacteria in the decomposing yolk and white. He was uncertain of their origin but favored the view that they came from the oviduct. Even when the arguments occasioned by the concept of spontaneous generation had lapsed, research in egg microbiology was dominated by two questions: Why do eggs rot? How, during the distribution of those intended for human consumption, can this be prevented? Perhaps these have been the main cause of the *ad hoc* nature of the research and for the piecemeal growth of our knowledge—witness, also, this essay in which strings of references have had to be given because of the absence of incisive studies.

Moreover the strong applied bias may be a reason why the data have had such a limited impact in general biology. Thus one can perceive a widening of the gulf between pure and applied research and a loss of rapport between disciplines that have an interest in the problem. This is to be regretted since those engaged in molecular biology, for example, might find that the whole egg could

---

[1] Dedicated to Dr. T. Gibson on the occasion of his seventieth birthday.

provide a model system for testing the biological implications of hypotheses based on detailed study of one of its components. Similarly, such information could lead egg technologists to a fuller appreciation of the antimicrobial defense of the egg and, perhaps, suggest ways in which the biological properties of a component could be exploited or its action aped in the food and allied industries. Likewise, one would imagine that the fate of the developing embryo would be of as much concern to the zoologist as to persons employed in the hatchery industry, yet this surmise receives little support from the literature (Beer, 1967; Lack, 1968). Such considerations influenced the organization of this essay and an attempt has been made to place the data in a biological rather than an applied setting.

II. The Egg

A. Structure

The egg (Fig. 1) is a complex physicochemical system in which enzyme-mediated energy transfer and chemical transformations are limited, in the main, to the cells of the blastoderm (Brooks and Taylor, 1955; Shenstone, 1968). With eggs intended for human consumption, their activities are minimized by low-temperature storage. This delays also the rate of deterioration of the physicochemical systems and loss of the structural integrity of the main components of an egg. The principal changes occurring in a stored egg are summarized in Table I. Of these, the breakdown of the albuminous sac (Brooks and Hale, 1959) and the stretching and weakening of the vitelline membrane (Fromm, 1967) are, from a commercial viewpoint, the chief causes of the loss of "quality" (Wells, 1968) and prolonged storage is required before there are demonstrable changes in the chemical composition of eggs (Evans *et al.*, 1950). During the formation of an egg, a genetic code along with the materials (Table II) needed for its interpretation to the chick stage are included within the shell and its two membranes (Fig. 1).

Intimate association of the hen and eggs ends at oviposition—there is nothing comparable to the uterine immunity discussed by Brambell (1958). The prenatal stages of development occur in an environment separated from that of the parent by the shell. In many ways the egg can be viewed as an ecosystem with a need only for exchange of respiratory gases, a source of heat, and regular movement. This stage of development has been studied in great detail, as witnessed by the monograph of Romanoff (1967), and when the bioenergetics are

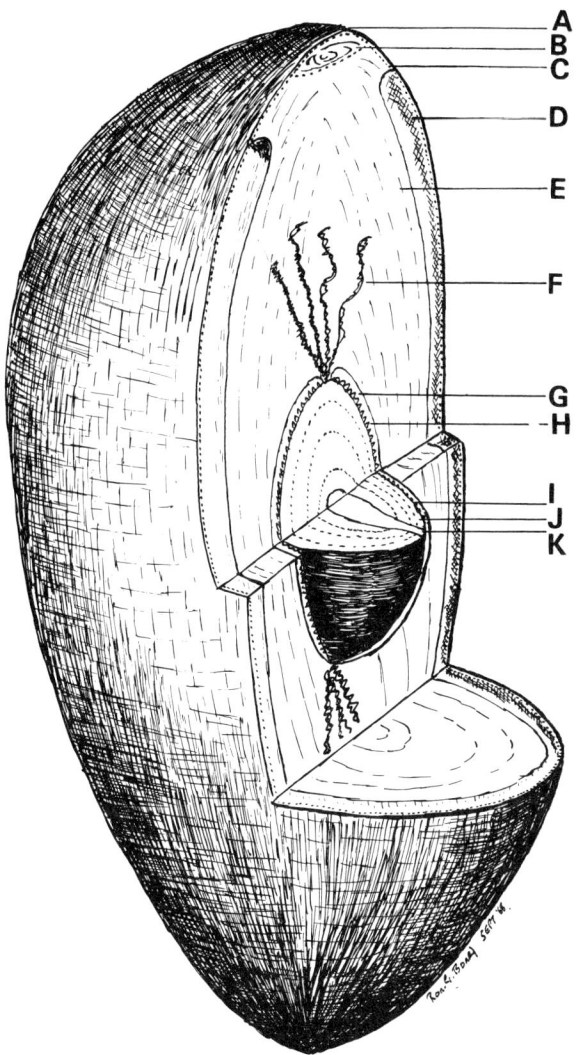

FIG. 1. An artist's impression of the hen's egg as shown by cutaway sections. A, Shell; B, air cell; C, inner shell membranes; D, outer thin white; E, albuminous sac; F, chalazae; G, inner thin white; H, chalaziferous membrane; I, vitelline membrane; J, yolk; K, latebra.

represented graphically, it is found not surprising that the pattern is similar to that which summarizes the changes occurring in a pure culture of a bacterium (Dean and Hinshelwood, 1966). There would

TABLE I
CHANGES OCCURRING IN STORED EGGS

Change	Cause	Effect	Environmental factors influencing	References
1. Shrinkage of cuticle	(i) Shrinkage of the vesicles (ii) Diminution of intervesicular air space	Increased porosity of shell?	In egg cleaning, temperature of water, type of detergent	Simons and Wiertz (1966)
2. Formation of air cell	Differential contraction of the shell and egg contents	Inner shell membrane pulled away from the outer shell membrane	Ambient temperature	Romanoff and Romanoff (1949)
3. Alkaline drift in albumen	Diffusion of CO_2	Breakdown of the buffer system of albumen	Temperature and composition of atmosphere of store	Cotterill et al. (1958)
4. Increase in density of albumen	(i) Evaporation	Increase in volume of air cell	Temperature, relative humidity, shell thickness	Wells (1968)
	(ii) Absorption of water by yolk	Decrease in viscosity of yolk Stretching and weakening of vitelline membrane	Temperature and differences in osmotic pressure between yolk and white	Feeney et al. (1956) Smith (1934)

5. Deterioration of albuminous sac	Unknown in detail: interaction of lysozyme and ovomucin considered important (Cotterill and Winter 1954a,b, 1955)	(i) Fracture of fibers connecting the sac to the shell membranes at the poles of the egg (ii) Contraction of albuminous sac around yolk (iii) Movement of inner thin white to outside of sac (iv) Contraction of the chalazae (v) Loss of gelation	(a) Temperature (b) Heritable factors (c) CO_2 content of storage atmosphere (d) Treatment of egg prior to storage – viz., protection by thermostabilization (pasteurization)	Brooks and Hale (1959) Baker and Stadelman (1958) Baker (1960) Rutherford and Murray (1963) Murray and Rutherford (1963) Funk (1943)
6. Increased buoyancy of yolk	Combination of 4 (i) and (ii)	Yolk moves away from center and comes to rest against shell membranes due to 5 (i), (ii), (iv), and (v)	5a, b, c, and d	Romanoff and Romanoff (1949)

TABLE II
Composition of the Hen's Egg[a]

Distribution of Egg Contents

	Amount (%)	Grams
Total	100.0	51.6
Water	73.6	38.0
Solids	26.4	13.6
Organic matter	25.6	13.2
Proteins	12.8	6.6
Lipids	11.8	6.1
Carbohydrates	1.0	0.5
Inorganic matter	0.8	0.4

Composition of Albumen

	Amount (%)
Total	100.0
Water	87.9
Solids	12.2
Organic matter	11.6
Proteins	10.6
Lipids	Trace
Carbohydrates	0.9
Inorganic matter	0.6

Protein Composition of Albumen

Component	Percent (approx.) of egg white solids
Ovalbumin	54
Conalbumin	13
Ovomucoid	11
Lysozyme	3.5
Ovomycin	1.5
Flavoprotein-apoprotein	0.8
Ovoinhibitor	0.1
Avidin	0.05
Unidentified proteins (mainly globulins)	8

Composition of Egg Shell

	Amount (%)
Total	100.0
Water	1.6
Solids	98.4
Organic matter	3.3
Proteins	3.3
Lipids	0.03
Inorganic matter	95.1

Composition of Yolk

	Amount (%)
Total	100.0
Water	48.
Solids	5.1
Organic matter	5.0
Proteins	16.6
Lipids	32.6
Carbohydrates	1.0
Inorganic matter	1.1

[a] Compiled from data given by Romanoff and Romanoff (1949); Brooks and Taylor (1955); Parkinson (1966); Shenstone (1968).

appear to be, therefore, an *a priori* need for such a limited universe as an egg to be endowed with means whereby the embryonic cells are (*i*) protected from physical damage due to sudden and violent changes in the surrounding environment and (*ii*) maintained as a "pure culture." When the relevant facts are assembled, this thesis can be supported.

B. Strength

When considering protection from physical damage, various roles can be proposed for the structural components of the hen's egg (Table III). Of these, the shell provides an example of a structure whereby natural selection seems to have favored a compromise. It has to provide the prebrooding and prenatal stages with protection against damage by impact or crushing as well as allowing the exchange of respiratory gases (6.6 gm., O_2; 7.6 gm., $CO_2/21$ days; Romanoff, 1967). To allow this exchange the shell contains $7.0–17 \times 10^3$ pores (Tyler, 1953), the diameters of which are in the range 9–35 μ (Romanoff and Romanoff, 1949; Tyler, 1956). When examining their distribution, Tyler (1955) noted that the arrangement lay somewhere be-

TABLE III
THE ROLE OF COMPONENTS OF THE EGG IN THE DEFENSE OF THE EMBRYO[a]

Component	Role
1. (a) Cuticle	Barrier to microbial invasion (a, b, and c)
(b) Shell	Control of rate of evaporation (a, b, and c)
(c) Shell membranes	Regulation of gaseous exchange (a and b)
	Protection against crushing (b and c)
2. Air cell	Control of pressure within egg
3. Albumen	Cushioning against damage due to sudden movement of the egg
	Reservoir of water
	Lag against violent fluctuations in temperature
	Impediment to microbial movement
	Control of rate and extent of microbial growth
	Passive immunity in young chick?
4. Chalaziferous and vitelline membranes	Physical, and to a less extent, chemical isolation of the yolk from the white
5. Yolk	Exchange of material between the yolk and white is controlled by the diffusion gradient existing beneath the vitelline membrane

[a] The roles of the components listed above have been postulated from data given in Romanoff and Romanoff (1949).

tween randomness and complete uniformity. He believed that this spacing maintained a balance between the need for gaseous exchange and the requirement for the shell to resist fracture or crushing. These latter attributes are also of commercial importance (Baskett et al., 1937; Brooks and Hale, 1959; Hale, 1950; Brooks, 1960a; Simons et al., 1966) but, in spite of our detailed knowledge of the structure (Simons and Wiertz, 1963) and chemical composition (Simkiss, 1961, 1968), the actual basis of the shell's strength awaits definition (Tyler, 1968).

C. DEFENSE

What manufacturer would be so foolhardy as to attempt to pack a perishable commodity (Table II) in a semipermeable membrane within a fragile, porous shell and have it distributed and marketed under uncontrolled conditions? Yet this is what the hen appears to have done for many thousands of years. It is thus reasonable to assume that, through selection, some form of defense has evolved whereby the blastoderm and its supply of nutrients are protected from microbial attack during the prebrooding period and, whereby, the early stages of embryo development occur without interference from microbes be they bacteria, fungi, protozoa, worms, or maggots. When discussing this topic Tokin (1959) implied that immunological properties of the egg, particularly of the albumen, provides a passive immunity. Yet, to all intents and purposes, the albumen is free from antibodies. This is not altogether surprising since, as will be seen later (Table VII), the albumen is liable to invasion by organisms with which the hen is unlikely to have had intimate or recent contact. Consequently its chances of containing sufficient antibodies to repel a specific organism are remote. Antibodies are included in the yolk (Frazer et al., 1934; Brandley et al., 1946), and the titer reflects the concentration in the plasma of the hen (Patterson et al., 1962). As these are absorbed from the yolk by the young chick, they could provide a means – a passive immunity – whereby the latter is afforded protection against pathogens that have recently afflicted the mother and are thus liable to be present on or in the hen. Phagocytic cells, even if they did occur in eggs, would seem to be incapable of providing an efficient defense since there would be no method by which they could be transported to the site of infection and, once there, they would lack the complementary action of antibodies.

As the two classic forms of immunity do not and, possibly, could not function efficiently, another means of defense has to be sought. The early studies of the hen's egg indicated that the albumen kills or prevents the growth of a wide range of microorganisms, whereas, the yolk or a mixture of yolk and white does not (for references, see Haines, 1939). This has been confirmed in recent years (Brown et al., 1966a; Zagaevsky and Lutikova, 1944; Ayres and Taylor, 1956). This implies that the structural integrity of the yolk and white are important in the egg's defense. Likewise it has been demonstrated that the fracturing of the shell increases an egg's susceptibility to microbial degradation (Brown et al., 1966c). In the past 20 years, intensive study of the properties of the proteins of the albumen has allowed a tentative but as yet incomplete description of the means whereby the young embryo is maintained as a "pure culture." This work has been summarized in Table IV, and it will be noted that the albumen can be viewed as a medium that is unsuitable for microbial growth. Of the factors listed, lysozyme, conalbumin, and the alkaline reaction appear to be of primary importance.

1. Lysozyme

In 1909, Laschtschenko noted that *Bacillus subtilis* was lyzed when added to egg white; however, this did not occur when the white had been held at 65°–70°C. for 30 minutes. The enzymatic nature of this phenomenon was recognized by Fleming (1922) and he gave the name lysozyme to the lytic agent. Today this enzyme is considered to be a N-acetylhexosaminidase and, along with related enzymes from other sources, it is classified as a muramidase (Salton, 1964). The lysozyme of egg white has been studied to the extent that the amino acid sequence of the protein is now known and, that, from the data concerning the configuration of the macromolecule, it is possible to interpret its catalytic activity in stereochemical terms (Symposium, 1967). In the past 20 years also, this enzyme has been extensively used in studies concerned with the chemical composition and organization of the bacterial cell wall (Salton, 1964; Weidel and Pelzer, 1964; Martin, 1963; Rogers, 1965; Rogers and Perkins, 1968).

The cytoplasmic membrane of the eubacterial cell is contained within a rigid macromolecular net, the basal structure (Work, 1957), murein layer (Martin, 1963), or murein sacculus (Weidal and Pelzer, 1964). In some ways, the structure can be considered to have a role similar to that of the egg shell. It gives the cell a characteristic shape and protection against damage that could ensue from sudden changes in osmotic pressure. The net is formed from a heteropolymer, trivially referred to as mucopeptide (Mandelstam and Rogers, 1959), glyco-

TABLE IV
POSTULATED COMPONENTS OF THE ANTIMICROBIAL DEFENSE OF THE HEN'S EGG

Component	Action	Contribution to defense[a]	References
1. Lysozyme	Lysis of cell walls of eubacteria Flocculation of bacterial cells Hydrolysis of β1–4 glycosidic bonds	A	Laschtschenko (1909); Fleming (1922); Friedberger and Hoder (1932); Berger and Weiser (1957)
2. Conalbumin (interaction with 5; possible interaction with 3)	Chelation of iron, copper, and zinc	A	Schade and Caroline (1944); Alderton et al. (1946)
3. Riboflavin (possible interaction with 2)	Chelation of cations	B	Feeney and Nagy (1952); Ramsey and Wilson (1957)
4. Glucose (possible interaction with 2 and 3)	Repression of respiratory capacity of facultative anaerobes?	B	Schade et al. (1968)
5. pH 9.6 (interaction with 2)	Maximizes chelating potential of conalbumin Provides an unsuitable environment for many organisms	A	Schade and Caroline (1944); Garibaldi (1960)

6. Low content of nonprotein nitrogen (possible interaction with 2, 4, and 5)	Elective for nutritionally nonfastidious organisms	C	Haines (1939); Ducay et al. (1960)
7. Avidin	Combination with biotin	B	Eakin et al. (1940); Wolley and Longsworth (1942)
8. Apoprotein	Combination with riboflavin	B	Rhodes et al. (1959)
9. Ovoinhibitor	Inhibition of fungal proteases	C	Matsushima (1958)
10. Ovomucoid	Inhibition of trypsin	C	Delezenne and Pozerski (1903); Feeney et al. (1963)
11. Uncharacterized proteins			
A	Inhibition of trypsin and chemotrypsin	C	Rhodes et al. (1960)
B	Combination with vitamin B_6	C	Evans et al. (1951)
C	Chelation of calcium	C	Abels (1936)
D	Inhibition of ficin and papain	C	Fossum and Whitaker (1968)

^aRole has been demonstrated in intact eggs (A) or with albumen *in vitro* (B) or postulated on basis of information in literature (C).

peptide (Strominger, 1962), glycosaminopeptide (Salton, 1964) etc., which consists of an axial filament formed from N-acetylglucosamine and N-acetylmuramic acid (linkage β1-4 and β1-6 glycosidic bonds). Side chains consist of peptides formed from some or all of the following: glutamic acid, alanine, lysine, 2,6-diaminopimelic acid, and glycine (Rogers and Perkins, 1968). The β1-4 bonds are attacked by lysozyme (Berger and Weiser, 1957; Salton and Ghuysen, 1960).

In his studies, Fleming (1922) selected the organism *Micrococcus lysodeikticus*, because of its great sensitivity to lysozyme. It is noteworthy that, on a dry weight basis, the heteropolymer contributes upward of 90% of the cell wall of this organism (Perkins and Rogers, 1959). In consequence the marked sensitivity to lysozyme is associated with the exposed position of many of the β1-4 glycosidic bonds. A similar situation is found in *Bacillus megaterium;* depolymerization of the cell wall causes this rod shaped organism to assume a spherical form (a protoplast) that is quickly lyzed unless suspended in an isotonic solution (Weibull, 1953a,b). The organisms noted to date give a positive reaction in Gram's staining method. Although the cell walls of such organisms are rich in the heteropolymer, they show varying resistance to lysozyme. In some instances, as with *Listeria monocytogenes*, resistance appears to be associated with lipids masking the murein layer (Ghosh and Murray, 1967), while with others (e.g., resistant strains of *M. lysodeikticus* and naturally occurring strains of *Staphylococcus aureus*) the substitution of N-acetyl by O-acetyl (Brumfitt, 1959; Park and Griffith, 1964) or N-propionyl residues (Hara and Matsushima, 1967) prevent lysozyme from making effective contact with its substrate.

The murein sacculus makes a relatively small contribution (as low as 5–10%) to the cell walls of gram-negative bacteria (Rogers and Perkins, 1968) and it is buried beneath laminae of lipoproteins and lipopolysaccharides (Kellenberger and Ryter, 1958; Burge and Draper, 1967a,b,c). How these layers are joined is not known, but ionic and hydrogen bonding have been suggested (Gray and Wilkinson, 1965a,b; Rogers, 1965; Wolin, 1966; Birdsell and Cota-Robles, 1967; Asbell and Eagon, 1966; Nermut and Murray, 1967; Weinbaum et al., 1967; Cox and Asbell, 1968). With the possible exception of very young cells (Birdsell and Cota-Robles, 1967), the outer layers of the cell wall provides a permeability barrier whereby lysozyme is prevented from reaching its substrate. Sensitivity to the enzyme can be imposed by treatments that disrupt the surface architecture of the cell wall. This can be achieved by incubation under alkaline conditions (Zinder and Arndt, 1956; Vos, 1964; Birdsell and Cota-Robles, 1967), freezing and thawing (Kohn, 1960), or treatment with alkaline solutions of ethylenediaminetetraacetic acid (Repaske, 1956, 1958; Vos, 1967). Lysis expresses itself by a decline in the opacity of a suspension of treated cells, and this can occur without the prior formation of the spherical (spheroplast) form. Thus, Vos (1964) and Asbell and Eagon (1966) have noted that treated cells retain their normal morphology even though they are extremely sensitive to slight changes in osmotic pressure. This suggests (Carson and Eagon, 1966; Weinbaum and Markham, 1966) that the outer layers contribute to the overall strength of the cell wall of gram-negative bacteria and that this fact will have to be taken into account when the action of lysozyme within the egg is being examined.

2. Conalbumin (Ovotransferrin)

This protein was first isolated from eggs by Osborne and Cambell (1900). It is a glycoprotein with properties similar to transferrin (siderophilin), the β_1-globulin of serum (Marshall and Deutsch, 1951; Karminiski and Durieux, 1956; Aisen et al., 1966); their chief difference is in their sialic acid content of the carbohydrate moiety (Williams, 1962). The proteins have molecular weights in the range 70–90 × 10^3 (Laurell and Ingleman, 1947; Bain and Deutsch, 1948; Warner and Weber, 1951) and they act as chelating agents (Alderton et al., 1946; Fraenkel-Conrat and Feeney, 1950; Schade et al., 1949). The binding of iron, copper, and zinc is through ionic bonds and binding constants are high; the estimates for the pK_1 and pK_2 at pH 7.4 being 27.7 and 30.3, respectively (Davis et al., 1962). The iron within the complex is trivalent (Ehrenberg and Laurell, 1955) as is that occurring in eggs (Halkett et al., 1958). The conalbumin-iron complex is more resistant, than is conalbumin alone, to enzymic digestion, thermal denaturation etc. (Fuller and Briggs, 1956; Azari and Feeney, 1958, 1961). This feature has been exploited in methods devised for the isolation and purification of the protein (Azari and Baugh, 1967), the sterilization of plasma (Keller and Pennell, 1958), and in a method for the pasteurization of egg albumen (Cunningham and Lineweaver, 1965).

In the majority of commercial strains of laying hens, the transferrin locus may be homozygous (Tf_a/Tf_a) or heterozygous (Tf_a/Tf_b) whereas the wild type is apparently homozygous for the allele Tf_b (Morton *et al.*, 1965). Such differences at the chromosome level are reflected in the rate of migration and homogeneity or otherwise of the conalbumin "band" during electrophoresis (Lush, 1961; Ogden *et al.*, 1961; Baker and Manwell, 1962; Baker, 1968). The liver appears to be the most likely site of transferrin synthesis and it has been demonstrated that amino acids can be included in conalbumin by cells and cell debris derived from the oviduct (Mandeles and Ducay, 1961; Williams, 1962; Carey, 1966).

Varieties of conalbumin are widely distributed in the eggs of different species of birds (Clark *et al.*, 1963; Baker, 1968). In the domestic hen it accounts for 10% of the total egg white solids (Longsworth *et al.*, 1940; Alderton *et al.*, 1946); it is uniformly distributed throughout the white (Feeney *et al.*, 1952) and its sequestering power is not detectably reduced by short-term storage (Feeney *et al.*, 1952) unless iron migration from the yolk (Schaible *et al.*, 1946) is expedited through the diffusion gradient at the periphery of the yolk, being modified by physical (Hale, 1950) or chemical methods (Phelps *et al.*, 1965). The bacteriostatic action of conalbumin and its dependance upon a relatively low hydrogen-ion concentration was noted by Schade and Caroline (1944) and confirmed by Brooks (1960b) and Garibaldi (1960). When pure preparations of conalbumin or transferrin are used to inhibit microbial growth, it has been established that they must be present in stoichiometric excess of the iron found in a medium by chemical analysis (Fraenkel-Conrat and Feeney, 1950; Feeney and Nagy, 1952). Inhibition expresses itself by an increase in the lag phase of growth and a decreased rate of multiplication once growth has begun (Theodore and Schade, 1965a), the actual rate and extent of growth being a function of the percentage iron−saturation of the ligand (Schade, 1958, 1963). Feeney and his collaborators have noted that different organisms show different degrees of inhibition (micrococci are more sensitive than *Bacillus* spp. and the latter are more sensitive than gram-negative bacteria).

Indirect evidence that bacteriostasis stems from a disturbed iron metabolism came with the observations (Feeney and Nagy, 1952; Garibaldi, 1967) that conalbumin enhances the production of pigment by pseudomonads. This is a typical response of these organisms when growing in media having a reduced iron content (King *et al.*, 1948; Totter and Moseley, 1953; Paton, 1959). As pseudomonads and physiologically related organisms depend on a cytochrome-containing electron transport system, it would be of interest to know how this system operates when the organisms are exposed to conalbumin. It has been suggested that the organisms scavenge the free Fe^{3+} that is always present through dissociation of the complex (Fraenkel-Conrat and Feeney, 1950; Feeney, 1951; Feeney and Nagy, 1952), or that they themselves form a chelate (Garibaldi and Neilands, 1956; Garibaldi, 1960) which, through competitition, causes iron to be released from the conalbumin complex in much the same way as is brought about by 8-hydroxyquinoline or citric acid (Feeney and Nagy, 1952).

A more detailed picture of the bacteriostatic action of conalbumin has come from studies concerned with the growth and metabolic activities of *Staphylococcus aureus*. With media containing known amounts of iron, it has been shown that the initiation, rate, and extent of growth are related directly to the concentration of ionic iron (Theodore and Schade, 1965a) or, when the medium contains either conalbumin or transferrin, that the growth rate is a function of the percentage of iron-saturation of the ligand (Schade, 1960, 1963; Theodore and Schade, 1965a). Cells harvested from a carbohydrate-free medium containing optimal amounts of iron are capable of complete oxidation of glucose, pyruvate, acetate, formate, and Krebs cycle intermediates (Theodore and Schade, 1965b; Schade *et al.*, 1968). These findings are in accord with those of Strasters and Winkler (1963), who demonstrated that *S. aureus* has a pentose and a TCA cycle; the latter is of primary importance when the organisms grow aerobically in nutrient broth. The presence of glucose in a medium containing readily available iron represses the synthesis of catalase and reduces the cells' capacity to oxidize the substrates noted previously (Theodore and Schade, 1965b); at the same time it enhances the production of dehydrogenases for glyceraldehyde 3-phosphate and lactic acid (Strasters and Winkler, 1963). Moreover, the exclusion of molecular oxygen from cells growing in a complex medium results in a reduction in the synthesis of cytochromes (Strasters and Winkler, 1963; Jacobs and Conti, 1965; Conti *et al.*, 1968), and there are reasons to believe that the concentration of these substances play a role in controlling the synthesis of lactate dehydrogenase (Garrard and Lascelles, 1968). An analogous repression of oxidative capacity follows the addition of glucose to a culture of *Escherichia coli* (Gray *et al.*, 1966a,b). It is noteworthy that Theodore and Schade (1965b) caused modifications of this type by deliberately minimizing the concentration of iron in a carbohydrate-free medium. They found, moreover, that the presence of glucose in such a medium further reduced the oxidative capacity of the organisms. Although such cells must rely upon glycolysis, they appear to differ from anaerobically grown cells in that the end products of fermentation include pyruvate, acetate, and acetoin whereas, normally, lactic acid predominates (Gardner and Lascelles, 1962; Collins and Lascelles, 1962). Thus the picture emerging from the studies of Schade and his collaborators indicates that inadequate levels of iron causes a facultative anaerobe such as *S. aureus* to change from respiration to glycolysis, and, under the most exaggerated conditions, this could be expected to reflect itself in a reduced amount of growth since the latter is a relatively inefficient method for ATP synthesis (Bauchop and Elsden, 1960). To date, however, the data do not permit a complete interpretation and this will not be possible until more is known concerning the control mechanisms in microorganisms. Moreover it would be of interest to know, in view of the low levels (0.14–0.54 μmole/ml.) of amino acid in the albumen of the egg (Ducay *et al.*, 1960), whether or not organisms become nutritionally more demanding if iron starvation were to impair the efficiency of an amphibolic system (Davis, 1961), such as the TCA cycle.

As the white is primarily a reservoir of water during early embryo development (Romanoff and Romanoff, 1949) its other components appear to be concerned, along with the shell, in isolating the major food reserve and the young cells from microorganisms on the mother or in the nest. In other words, they contribute to the isolation of the ecosystem in which early prenatal development occurs. Later they are absorbed by the embryo (Baker, 1968) – does this result in an enhancement of the passive immunity of the young chick? Of the components listed in Table IV, available evidence restricts serious discussion to lysozyme and conalbumin. It is noteworthy that the former is widely distributed in the tissues and secretions of the animal body (Fleming, 1922) and that it is a normal constituent of the phagocytic cell (Hirsch, 1965). Recent studies have suggested that conalbumin and related substances may be an important component of the antimicrobial defense of man and animals (Summers and Hasenclever, 1964; Sword, 1966; Caroline *et al.*, 1964). It is interesting to note that the egg appears to be endowed with those components of the animals' antimicrobial defense system which do not require a vascular system, nervous or hormonal control, and which can function over a range of temperatures. Moreover, do these components constitute the earliest form of immunity in animals? This query is posed because of the claims that lysozyme may be a primitive enzyme (Manwell, 1967) and because of the observation that conalbumin-like substances are plentiful in body fluids of animals in which it is difficult to elicit antibody production (Good *et al.*, 1967; Manwell, 1963).

III. The Course of Infection

The majority of early investigations on the rotting process tend to be muddled since the workers failed to appreciate that addling was merely the culmination of a sequence of events. This was recognized by Gillespie and Scott (1950). They defined three phases: (1) infection and microbial penetration of the shell, (2) colonization of the shell membranes, and (3) contamination of the albumen. This concept still provides a valid approach in applied research, but it does suffer from the fact that emphasis is placed on a particular stage, which detracts from an overall view of the process. A desirable level of integration can be achieved when the egg is considered as an ecosystem and the commercially important stages noted above are considered within a framework provided by the concepts of ecology. Thus the term *association* can be used for the characteristic flora of rotten eggs and its genesis under practical conditions can be considered to be determined by (1) the *infection* of the shell and shell membranes, (2) conditions obtaining in the egg (*intrinsic* factors), (3) conditions external to the egg (*extrinsic* factors), and (4) the properties of the organisms making up the association (*implicit* factors).

A. Infection

Previous reviewers (Brooks and Taylor, 1955; Board, 1968) have deduced that microorganisms are absent from the majority of eggs laid by healthy hens. Even when contamination does occur, the types of organisms in the egg at oviposition differ markedly from those in rotten eggs or those which fail to hatch (Harry, 1963a). Moreover their presence in the yolk sac at the time of hatching is of no practical importance since it appears that they would be unlikely to become established in the young chick (Fuller and Jayne-Williams, 1968). The shell can be infected when passing through the vent (Stuart and McNally, 1943) but the chief contamination results from the shell's contact with dirty surfaces (Rosser, 1942; Board *et al.*, 1964; Harry, 1963b). The extent of contamination is indicated by the data given in Table V. As would be expected, the shell harbors a heterogeneous population (Table VI) in which gram-positive organisms are dominant—does this reflect their resistance to drying? When the major components of the flora are considered, it seems reasonable to deduce that dust, soil, and feces are the major depots of the common contaminants. Infection is largely confined to the surface of the shell of nest clean eggs (Dr. K. Büchli, personal communication), and less than

TABLE V
LEVELS OF MICROBIAL CONTAMINATION OF THE SHELL OF THE HEN'S EGG

Source of supply	No. eggs examined	Number of micro-organisms/shell		Grade and/or treatment of eggs*	Reported by
		Mean	Range		
Farm	36	9.5×10^3	—[a]	Handled with gloved hands	Rosser (1942)
Batteries	25	2.2×10^4	$2.5 \times 10^3 – 8.1 \times 10^4$	—	Harry (1963b)
Experimental farm	—[a]	6.3×10^4	$1.0 \times 10^6 – 1.0 \times 10^6$	—	Forsythe et al. (1953)
Packing station	72	7.0×10^4	—	—	Rosser (1942)
Farm	36	1.5×10^5	—	—	Rosser (1942)
Shops and farms	130	1.3×10^5	—	—	Haines (1938)
Packing station	—	2.2×10^5	$3.0 \times 10^2 – 1.0 \times 10^7$	Clean, Grade A	Board et al. (1964)
Deep litter	25	3.5×10^5	$6.2 \times 10^3 – 2.4 \times 10^6$	—	Harry (1963b)
Packing station	77	9.7×10^5	$1.0 \times 10^3 – 1.9 \times 10^7$	Lightly soiled, Grade B	Board et al. (1964)
Deep litter	96	3.1×10^6	$5.0 \times 10^2 – 1.0 \times 10^7$	Clean and lightly soiled	Board and Wilson (1965)

[a] —, Details not given in original report.

1% of such eggs rot during storage (Brooks and Taylor, 1955). The organisms on the shell do not multiply unless eggs are held under humid conditions (Sharp and Stewart, 1936; Haines, 1938; Forsythe *et al.*, 1953; Board *et al.*, 1964); with pathogens, such as salmonellae, death can occur (Mellor and Banwart, 1965) unless the eggs are held under moist, chilled conditions or the organisms are included in dried mud or fecal material (Wolk *et al.*, 1950; Lancaster and Crabb, 1953; Cotterill and Gardner, 1957; Magwood, 1964b; Rizk *et al.*, 1966b). Thus under normal conditions the shell is contaminated at its surface with organisms which tend to remain quiescent.

When the composition of the flora of the shell (Table VI) is compared with that of rotten eggs (Table VII), it is notable that the incidence of gram-positive organisms in the former is contrary to that in the latter. Thus, potential rot-producing bacteria make only a small contribution to the initial contamination of the shell, even though such organisms (e.g., *Pseudomonas*) may be widely disseminated in the hen's environment (Gordon and Tucker, 1954). It would seem, therefore, that modern methods of poultry husbandry do not impose conditions elective for gram-negative organisms. This situation does not obtain in the hatchery where undesirable microorganisms can be selected unless the hatchery is designed so that cross-contamination is minimized and a high level of hygiene maintained (Magwood, 1964a,b; Magwood and Marr, 1964; Nichols *et al.*, 1967).

B. Penetration

It has long been recognized that the shell imposes a barrier to microbial invasion of the egg (Haines, 1939; Williams and Whittemore, 1967) but support for this concept came from the rather nebulous observations that fracture of the shell results in heavy contamination of the albumen and a high incidence of rotting (McNally, 1953; Miller and Crawford, 1953; Brown *et al.*, 1966c). Earlier in this discussion it was noted that the shell is perforated with $7-17 \times 10^3$ pores having an average diameter of 9–35 μ, a size that could not be expected to hinder the movement of bacterial cells. Although it has been generally accepted that the cuticle contributes to the shell's resistance to invasion (Haines, 1939; Romanoff and Romanoff, 1949; Brooks and Taylor, 1955), direct evidence has had to await the studies of electron microscopists (Masshoff and Stolpmann, 1961; Simons and Wiertz, 1963, 1965, 1966). Their observations have been summarized in Fig. 2. The cuticle is an organized structure that clothes the surface of the shell. Its surface is irregularly fissured, except when traversing the

TABLE VI
Type of Microorganisms Recovered from the Shell of the Hen's Egg

Incidence (%) of organisms recovered from the shell of the hen's egg

Type of organism	Shops and farms[a]	Egg breaking plants[b]			Packing station[c]		
		Clean	Lightly soiled	Heavily soiled	Clean	Lightly soiled	Cracked, etc.
Streptococcus	—	8	5	—	—	—	—
Staphylococcus	5	30	—	—	9	5	11
Micrococcus	18	23	20	—	37	52	42
Sarcina	2	20	—	—	—	—	—
Arthrobacter	—	—	—	—	5	13	10
Bacillus	30	—	13	5	5	2.5	—
Pseudomonas	6	—	—	—	22.5	12.5	24
Achromobacter	19	—	—	—	1.5	2	1
Alcaligenes	—	—	—	—	—	2	—
Flavobacterium	3	—	—	—	—	—	—
Cytophaga	—	—	—	—	—	1	—
Coli-aerogenes	5	19	7	5	10.5	7.5	4
Aeromonas	—	—	—	—	1	—	2
Proteus	1	—	20	20	—	—	—
Serratia	—	—	20	50	—	—	—
Molds	7	—	10	20	—	—	—
Unclassified	—	—	—	—	12[d]	5[d]	6[d]
No. organisms studied	100	NR[e]	NR[e]	NR[e]	130	164	126

[a] Haines (1938).
[b] Zagaevsky and Lutikova (1944).
[c] Board et al. (1964).
[d] Aerobic gram-negative bacteria.
[e] Numbers not given in report.

TABLE VII
Types of Organisms Recovered from Contents of Eggs

Type of egg	Coli-aerogenes	Proteus	Aeromonas	Pseudomonas	Alcaligenes	Achromobacter	Gram+ve bacteria
Rotten eggs							
Haines (1938)	+	+	−	+	+	+	−
Alford et al. (1950)	+	+	−	+	+	+	−
Florian and Trussell (1957)	+	+	+	+	+	+	±
Board (1965b)	+	+	+	+	+	+	±
Board and Board (1968)	+	+	+	+	+	+	±
Tainted eggs							
Richard and Mohler (1950)	+	−	−	+	+	+	±
Incubator rejects							
Board and Board (unpublished observations)[b]	±	−	−	−	−	−	Micrococci and staphylococci
Harry (1957)[c]	+	+	−	−	−	−	Bacillus, micrococcus, streptococcus

[a] +, Isolated on many occasions; ±, isolated occasionally; −, not isolated.
[b] Results obtained from an investigation of >1000 eggs in which only limited embryological development had occurred; rotten eggs or "dead-in-shells" not included.
[c] Organisms derived from the yolk of "dead-in-shells" or chicks which died shortly after hatching.

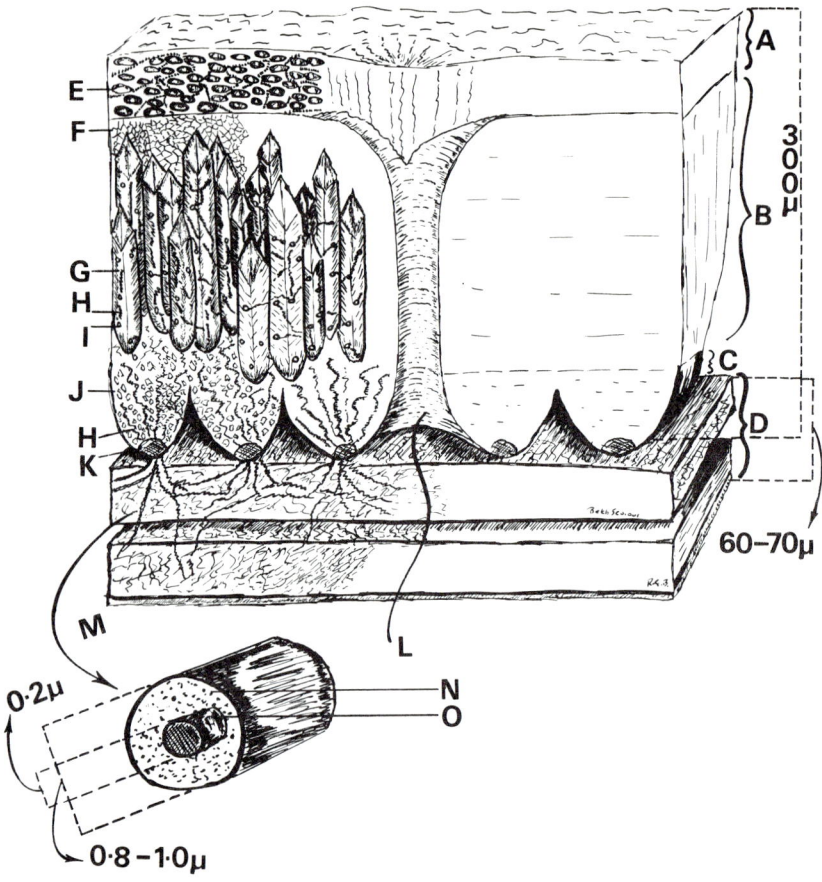

FIG. 2. An artist's impression of the organization of the egg shell as shown by a radial section through a pore. A, Cuticle; B, spongy layer; C, mammillary layer; D, shell membranes; E, vesicles and fibers; F, amorphous crystalline materials; G, columns of calcite crystal; H, fibers; I, vesicles; J, amorphous crystalline material; K, protein plug; L, pore canal; M, large fibers; N, mantle (composition uncertain); O, keratin core.

campanulate orifice of the pores where, presumably due to the stresses arising from drying, the fissures have a radial arrangement. The cuticle is fairly resistant to water or detergents and to gentle rubbing with a cloth (Simons and Wiertz, 1966). It is less resistant to abrasives, and wiping eggs with sandpaper or wire wool causes an increase in (1) the rate of evaporation (Marshall and Cruickshank, 1938; Tyler, 1945), (2) the shell's permeability to dyes (Fromm and Munroe, 1960), and (3) the incidence of rotting (Haines, 1938; Haines and Moran,

1940; Brown *et al.*, 1966a). When egg shells are breached under farm conditions, it is suspected that only ten or so pores provide portals for the entry of microorganisms (Bryant and Sharp, 1934; Orel, 1959). Likewise only a few pores are invaded when eggs are challenged by large numbers of organisms under laboratory conditions (Paton and Ayres, 1964; Board and Board, 1967). How these pores differ from the norm is not known, but the author suspects that they may be exceptionally wide and not capped with cuticle.

Water appears to be an essential agent for microbial invasion of the shell. Thus humid conditions promote the growth of molds at the surface of the shell (Sharp and Stewart, 1936); hyphae penetrate the pores (Weston and Halnan, 1927), and bacteria can be lodged in the shell membranes (Zagaevsky and Lutikova, 1944). When eggs are submerged in water of the same temperature, capillarity causes a flooding of the pore canals (Haines and Moran, 1940). These workers deduced also that water is sucked into the pores when a warm egg contracts in cold water. In view of the observation that an egg, from which a small piece of shell has been removed, is ruptured when placed in water (Lifshitz *et al.*, 1964) it is conceivable that movement of water, and hence microorganisms, through the pores may be accentuated by osmotic forces acting across the semipermeable shell membranes. The mere wiping of a shell with a cloth or brush moistened with a bacterial culture can cause contamination of the egg's contents (Stuart and McNally, 1943; Wilson, 1945; Buxton and Gordon, 1947; Lancaster and Crabb, 1953). It is not clear whether this is due to water being drawn into pores by capillarity or to it being forced (pumped ?) into the orifice of the pores. When any of the above methods have been applied under laboratory or field conditions, it has been found that there is a heavy contamination of the shell membranes and a high incidence of rotting when the eggs are stored (Haines, 1938; Haines and Moran, 1940; Gillespie *et al.*, 1950a). Because the incidence of rotting is highest when eggs are caused to contract in water, this phenomenon has been the subject of many investigations. It has been established that the following factors influence the extent of microbial penetration of the shell: (1) the temperature differential between the egg and the bacterial suspension, the incidence being directly proportional to the difference in temperature over the range $6°-21°C$. (Lorenz *et al.*, 1952; Brant and Starr, 1962), (2) the number of organisms in the suspension (Stokes *et al.*, 1956; Hartung and Stadelman, 1963), (3) the period of immersion—an osmotic effect ?—(Hartung and Stadelman, 1962), and (4) the thickness of the shell, thin shells offering less resistance than thick ones (Orel, 1959). From this evidence, it can be inferred that the shell is poorly adapted to function under wet conditions. This is unfortunate since washing is the most convenient method of cleaning eggs for market. Conflicting results came from early studies concerned with the storage behavior of washed eggs, but they did establish that certain designs of egg washing machines tended to increase the incidence of rotting (Gillespie *et al.*, 1950b,c; Trussell *et al.*, 1955a,b; Knowles, 1957a,b). Moreover the use of a disinfectant, although it reduces the level of contamination of the shell (Winter *et al.*, 1952, 1955), does not guarantee freedom from spoilage during the storage of treated eggs (Gillespie *et al.*, 1950c; Sauter *et al.*, 1962; Sauter, 1966). In certain instances the use of a selective disinfectant can cause a relative increase in the number of gram-negative bacteria (Gillespie *et al.*, 1950c), whereas, in others, the organisms penetrate the pores and are protected against inimical agents by the shell membranes (Cotterill and Hartman, 1956; Elliott and Romoser, 1957; Schmidt and Stadelman, 1957; Bean and McLaury, 1959; Rizk *et al.*, 1966a). To overcome these difficulties, many investigations have been concerned with pasteurization of shell eggs; it has been established that deterioration due to microbial action is of rare occurrence in eggs which have been held briefly at temperatures of 135.5°–158°F. (Goresline *et al.*, 1950; Funk, 1943, 1948; Romanoff and Romanoff, 1944; Murphy and Sutton, 1947; Salton *et al.*, 1951; Funk *et al.*, 1954; Scott and Vickery, 1954; Knowles, 1956). In recent years, experience gained through study of microbial invasion of the shell together with that reviewed immediately above has lead to marked improvements in the design and operation of egg washing machines. Of particular importance is the requirement for the wash water to be maintained at a temperature higher than that of the egg. A differential of 20°F. has been recommended by Brant *et al.* (1966) and 20°C. by Büchli (1967). With the latter, the water used for washing and *rinsing* the eggs is held at 40°C., a temperature which may well have a pasteurizing effect. Moreover when alkaline detergents (pH 10) are used at this temperature, the wash water and the machine are kept relatively free of microorganisms (Büchli, 1967) thus permitting the recirculation of water with obvious savings in operating costs.

C. Colonization

It can be assumed that any agent that causes microbial penetration of the shell will result in a *heterogeneous* population being introduced on or near to the shell membranes; organisms have to be lodged within 30–35 μ of the membranes before they meet water activity (aw) (0.98) suitable for growth (Gillespie and Scott, 1950). Lysozyme-sensitive organisms will be destroyed by the enzyme present in the shell membranes (Korotkova, 1957). As such organisms make only a slight contribution to the microflora of the shell (Board, 1968), this component of the egg's defense would appear to be of relatively little significance at this stage in the infection process. When the shell membranes are suspended in a solution of mineral salts, the level of available nutrients supports appreciable growth of the common rot-producing bacteria (Elliott and Brant, 1957; Board, 1965a; Stokes and Osborne, 1956; Garabaldi and Stokes, 1958). Likewise there is significant growth when organisms are placed on the surface of the shell membranes *in ovo* (Board, 1968). When, however, they become enmeshed in the membranes the amount of growth is slight (Brooks, 1960b; Board, 1964). This has been attributed (Board, 1968) to the albumen influencing the environment within the structures. The response of organisms to these conditions has been the subject of limited study only (Board, 1968; J. Roger Saxon and R. G. Board, unpublished observations). These have indicated that the early phase of infection of the membranes is characterized by a selection of organisms. In practice this means that the gram-negative fraction of the flora increases at the expense of the gram-positive one. The actual rate of change in these fractions is determined by temperature and this influences also the composition of the gram-negative fraction. For example, there is evidence that coliforms do not develop in eggs held at room temperature, but do so with storage at 37°C. These initial observations tend toward explaining the previously noted change from a predominantly gram-positive flora of the shell to a gram-negative one in the rotten eggs.

In eggs contaminated under farm or laboratory conditions and held at ambient temperatures, it is notable that, following invasion of the shell, the infection remains confined to the shell membranes for upward of 15–20 days (Zagaevsky and Lutikova, 1944; Gillespie and Scott, 1950; Bigland and Papas, 1953; Miller and Crawford, 1953; Elliott, 1954; Stokes *et al.*, 1956; Orel, 1959; Fromm and Munroe, 1960; Garibaldi and Bayne, 1960; Rizk *et al.*, 1966b; Büchli, 1967). Is it merely a coincidence that this period is slightly longer than that needed by the wild hen to lay a "clutch" of 12 eggs (Harland, 1927)? There has been a tendency to ascribe (Kraft *et al.*, 1958) this lag in the infection process to the membranes imposing a barrier to microbial movement. Such an interpretation has been prompted, no doubt, by the repeated observation that the shell membranes can be made to act as bacterial filters (Walden et al., 1956; Garibaldi and Stokes, 1958). Under test conditions, however, the membranes are subjected to challenges different from those occurring under practical ones. For example, Haines and Moran (1940) and Garibaldi and Stokes (1958) applied pressure to the outer surface of eggs, the contents of which had been replaced with suspensions of bacteria. It is probable that under these conditions the shell membranes were compressed and forced up against the internal orifice of the pores, thereby increasing resistance to bacterial movement. The results of such studies also show poor agreement with the observations of persons who have used whole eggs. Thus, Bean and McLaury (1959) recovered organisms from the inner surface of the shell membranes immediately following the contraction of a warm egg in a cold suspension of bacteria; Board and Ayres (1965) noted the rapid penetration of the inner shell membrane *in ovo*. This membrane is considered to offer the greatest resistance to bacterial invasion (Lifshitz et al., 1964). Likewise the observations that the mantle surrounding the keratin fibers (Fig. 2) can be removed by bacteria (Brown et al., 1965) does not necessarily imply that they "digest" their way through the membranes. The latter had been in contact with bacteria for 15 days and it is probable that the modification of the mantles occurred when bacteria began to grow in the albumen—a period during which the physiological activity of the organisms is likely to be at a maximum. Moreover, bacteria that release nitrogenous substances from shell membranes *in vitro* do not penetrate the membranes at a rate faster than that of those which have no demonstrable action on these structures (Garibaldi and Stokes, 1958; Board, 1965a). Many factors appear to influence the rate and extent of microbial growth in the shell membranes. With *Serratia marcescens*, for example, no growth occurred when infected eggs were held at 10°C., and it was quickly killed when eggs were stored at 37°C. (Board and Ayres, 1965). At this temperature, coliforms do not die (Ellenor G. H. Wilson and R. G. Board, unpublished observations). When eggs contaminated with serratiae were held at room temperature, slight growth took place in the 2–4 days following inoculation, a response that is typical of the majority of rot-producing bacteria (Board, 1964). With storage at 0°–10°C., psychrotrophic bacteria such as *Pseudomonas fluorescens* grow in the membrane without obvious hindrance (Board and Ayres, 1965). Their growth can be prevented by the storage of infected eggs in high concentrations of CO_2 (Sally Beastall and R. G. Board, unpublished observations). In contrast, traces of iron at the site of infection of the shell membranes promote extensive microbial growth (Board et al., 1968). The same response is observed when the chelating potential of the conalbumin of the albumen is satisfied (Board, 1964). Under commercial conditions it has been demonstrated that contamination, either deliberate or natural, of wash water with iron can result in an increase in the rate and extent of rotting of stored eggs (Garibaldi and Bayne, 1960, 1962a,b).

This evidence indicates that the fate of contaminants of the shell membranes is determined by an interplay of factors intrinsic (lysozyme and conalbumin) and extrinsic (temperature, gaseous environment, and presence or absence of additional iron) to the egg with those implicit to the organisms themselves. It would appear that this phase of the infection process should be studied in greater detail because a better understanding of their interplay might suggest methods that would ensure the safety not only of eggs intended for human consumption but also those used by the hatching industry.

The growth lag noted previously is apparent when relatively large numbers of organisms are placed directly in the albumen (Brown et al., 1966b). Moreover, this lag is not appreciably influenced even when large numbers of organisms are placed on the shell membranes. Thus both Brooks (1960b) and Board (1964) noted a lag of ca. 13 days, although the former used an inoculum 100–1000 times greater than the latter. Moreover, Brooks noted that the albumen was contaminated shortly after the seeding of the shell membranes. There is much indirect evidence that the early contaminants of the albumen fail to multiply and that they may even be killed. This feature has not been studied in detail but the available evidence suggests that the fate of an organism is determined by the temperature of incubation. Thus, it is noteworthy that Sharp and Whitaker (1927) achieved rapid killing by placing young cells in albumen (*in vitro*) at 37°C., whereas Garibaldi (1960), who used essentially the same range of organisms, observed only a slight decline in the numbers of viable organisms when 18-hour cultures were used to seed albumen. With infected, whole eggs held at 10°C., Board and Ayres (1965) noted a progressive build-up in the contaminants of the albumen. Although the evidence is fragmentary, it is suspected that the antimicrobial defense of the albumen decreases in efficiency as the temperature moves away from that of the hen. This feature is worthy of further exploration since the results might indicate means whereby the albumen of commerce could be freed of microorganisms without impairing its functional properties.

The cardinal importance of the albumen in the egg's defense was recognized by Sharp and Whitaker (1927), and they concluded that actual rotting did not occur until organisms made contact with the yolk. This mode of induction was suggested also by Board and Ayres (1965) to account for sudden macroscopic changes in eggs the albumen of which became progressively contaminated during storage at 10°C. These observations suggest that the viscosity of the albumen and the gelatinous nature of the albuminous sac are important components of the egg's defense since they impede microbial movement. This accounts, no doubt, for the observations (Gillespie and Scott, 1950) that the initial contaminants of the albumen occur in clumps, a phenomenon that is particularly notable early in the infection of eggs with fluorescent pseudomonads. The observations of Brooks (1960b), Board (1964), and Board and Ayres (1965) suggest that when eggs are held at room temperature or above, addling begins with the growth of organisms that have reached the junction of the yolk and shell membranes. In other words, the antimicrobial defense system of the albumen is short-circuited by the yolk touching the membranes. Is it for this reason that selection favored hens that turned their eggs during incubation? Regular turning maintains tension in the chalazae, thereby ensuring that the yolk is retained in a central position. Little is known concerning the fate of the initial contaminants of the albumen. Earlier in the discussion it was inferred that they remain quiescent. Such a view receives support from the observations that there are no demonstrable changes in the hydrogen ion, glucose (Board, 1964; Board et al., 1968), or amino acid (Dr. D. J. Stewart, personal communication) content of the albumen until there is visible evidence of infection. At this time there is active multiplication of organisms in the albumen and population of 10^9 organisms/ml. albumen are achieved.

With eggs contaminated under commercial conditions, it is usual for a mixed population of gram-negative organisms to be present in the contents when rotting occurs (Board, 1965b; Board and Board, 1968). This suggests that the selective influences present in the shell membranes occur in an accentuated form in the albumen also. Following the investigations of Miles and Halnan (1937) it became customary to seed fresh eggs with cultures obtained from rots. The organism that produced changes similar to those observed in the original rot was referred to as the rot-producer, whereas, the others were considered to be adventitious contaminants (Haines, 1938) or secondary invaders (Florian and Trussell, 1957). It is noteworthy that the course of infection in eggs inoculated with the former is essentially the same as that of the latter (Board, 1964). The main difference between the two groups resides in the metabolic attributes of the rot-producing organisms. They possess one or more of the following properties (Table VIII): (1) production of a pigment, (2) digestion of protein with or without H_2S production, and (3) an ability to attack lecithin. On occasion an egg can be infected with organisms which do not have these attributes, and such an egg appears normal in a cursory examination but it can be tainted (Richard and Mohler, 1950). This indicates that the terms "adventitious contaminant" or "secondary invaders" are of

TABLE VIII
METABOLIC ATTRIBUTES OF COMMON CONTAMINANTS OF ADDLED EGGS AND THEIR ROLE IN THE ROTTING PROCESS[a]

Attribute	Action	Organisms[b]
Nonwater soluble pigment	Discoloration of shell membrane at site of infection, occasionally in the white and on the surface of the yolk	Cytophaga, Flavobacterium, Serratia
Water-soluble pigment	Discoloration of white	Pseudomonas aeruginosa, P. putida, and P. fluorescens
Proteolysis	Digestion of white and yolk	Proteus, Aeromonas
Production of H_2S	Blackening of the yolk	Proteus, Aeromonas
Lecithinase	Breakdown of yolk emulsion	cloacae
Slime production	Increase in viscosity of the albumen	cloacae (some strains only)
Odor production	Characteristic odor emitted by infected eggs	Pseudomonas maltophilia
None of the above	No macroscopic changes even when eggs harbor 10^8 organisms/ml. albumen	Alcaligenes, Salmonella, Citrobacter

[a] Table compiled from data given by Haines (1939), Platt and Anderson (1939), Florian and Trussell (1957), and Board and Board (1968).

[b] Some of the organisms have several of the metabolic attributes but it is usual for the changes in an egg to be dominated by the activity of one attribute only.

little relevence and it would appear they could be replaced by the terms used in ecology. Thus, the populations of rotten eggs can be considered to consist of three fractions: (1) the dominant (rot- or taint-producing) organisms, (2) the associates (those organisms which fail to produce significant changes in the egg), and (3) the incidentals (i.e., gram-positive bacteria such as streptococci and micrococci which are occasionally present in low numbers only).

In this discussion of the course of the infection process, emphasis has been given to the integrated workings of the shell, the shell membranes, and the albumen in the egg's defense. When the egg is considered as an ecosystem it can be argued that the albumen plays the principal role in isolating the blastoderm and its food supply from the environment of the hen. In other words, it provides an aqueous phase in which embryo development can occur without direct competition from other organisms. As yet we have only an incomplete picture of the role of the majority of components of the albumen. This would appear to be a subject worthy of further investigation especially since the "loss" due to low hatchability costs the British poultry industry more than £1 million per annum (Mr. A. E. Beer, personal communication).

Addendum

Can we now, with the cogency borne of hindsight, offer an explanation of the real basis of the argument twixt Donné and Pasteur? The author suggests that the passion of dogma may have caused Donné to shake violently his eggs thereby rupturing the yolk membranes and negating the antimicrobial defense whereas Pasteur, with sagacious gentleness, maintained the integrity of the egg.

Acknowledgments

The author wishes to thank the British Egg Marketing Board for financial assistance over many years, Mrs. Marilyn J. Dowdell for compiling the references, Mrs. Beth Seviour for preparing figures and tables, and Dr. Derek Shrimpton for constructive criticism of the manuscript.

References

Abels, J. C. (1936). *J. Am. Chem. Soc.* **58**, 2609–2610.
Aisen, P., Leibman, A., and Reich, H. A. (1966). *J. Biol. Chem.* **241**, 1666–1671.
Alderton, G., Ward, W. H., and Fevold, H. L. (1946). *Arch. Biochem.* **11**, 9–13.
Alford, L. R., Holmes, N. E., Scott, W. J., and Vickery, J. R. (1950). *Australian J. Appl. Sci.* **I**, 208–214.
Asbell, M. A., and Eagon, R. G. (1966). *Biochem. Biophys. Res. Commun.* **22**, 664–671.
Ayres, J. C., and Taylor, B. (1956). *Appl. Microbiol.* **4**, 355–359.
Azari, P. R., and Baugh, R. F. (1967). *Arch. Biochem. Biophys.* **118**, 138–144.

Azari, P. R., and Feeney, R. E. (1958). *J. Biol. Chem.* **232,** 293-302.
Azari, P. R., and Feeney, R. E. (1961). *Arch. Biochem. Biophys.* **92,** 44-52.
Bain, J. A., and Deutsch, H. F. (1948). *J. Biol. Chem.* **172,** 547-555.
Baker, C. M. A. (1960). *Brit. Poultry Sci.* **1,** 3-16.
Baker, C. M. A. (1968). In "Egg Quantity: A Study of the Hen's Egg" (T. C. Carter, ed.), p. 67. Oliver & Boyd, Edinburgh.
Baker, C. M. A., and Manwell, C. (1962). *Brit. Poultry Sci.* **3,** 161-174.
Baker, R. C., and Stadelman, W. J. (1958). *Poultry Sci.* **37,** 558-564.
Baskett, R. G., Dryden, W. H., and Hale, R. W. (1937). *J. Min. Agr. Northern Ireland* **5,** 132-142.
Bauchop, T., and Elsden, S. R. (1960). *J. Gen. Microbiol.* **23,** 457-469.
Bean, K. C., and McLaury, D. W. (1959). *Poultry Sci.* **38,** 693-698.
Beer, A. E. (1967). *Poultry Rev.* **6,** 70-75.
Berger, L. P., and Weiser, R. S. (1957). *Biochim. Biophys. Acta* **26,** 517-521.
Bigland, C. H., and Papas, G. (1953). *J. Comp. Med.* **17,** 105-109.
Birdsell, D. C., and Cota-Robles, E. H. (1967). *J. Bacteriol.* **93,** 427-437.
Board, P. A., and Board, R. G. (1967). *Lab. Practice* **16,** 471-482.
Board, P. A., and Board, R. G. (1968). *Brit. Poultry Sci.* **9,** 111-120.
Board, P. A., Hendon, L. P., and Board, R. G. (1968). *Brit. Poultry Sci.* **9,** 211-215.
Board, R. G. (1964). *J. Appl. Bacteriol.* **27,** 350-364.
Board, R. G. (1965a). *J. Appl. Bacteriol.* **28,** 197-205.
Board, R. G. (1965b). *J. Appl. Bacteriol.* **28,** 437-453.
Board, R. G. (1968). In "Egg Quality: A Study of the Hen's Egg" (T. C. Carter, ed.), p. 133. Oliver & Boyd, Edinburgh.
Board, R. G., and Ayres, J. C. (1965). *Appl. Microbiol.* **13,** 358-364.
Board, R. G., and Wilson, E. G. H. (1965). *Edinburgh School Agr. Exptl. Work,* 89-90.
Board, R. G., Ayres, J. C., Kraft, A. A., and Forsythe, R. H. (1964). *Poultry Sci.* **43,** 584-594.
Brambell, F. W. R. (1958). *Biol. Rev.* **33,** 488-531.
Brandley, C. A., Moser, H. E., and Jungherr, E. L. (1946). *Poultry Sci.* **25,** 397-398.
Brant, A. W., and Starr, P. B. (1962). *Poultry Sci.* **41,** 1468-1473.
Brant, A. W., Starr, P. B., and Hamaan, J. A. (1966). *Poultry Tribune,* p. 65.
Brooks, J. (1960a). In "Texture in Foods." Soc. Chem. Inc., London.
Brooks, J. (1960b). *J. Appl. Bacteriol.* **23,** 499-509.
Brooks, J., and Hale, H. P. (1959). *Biochim. Biophys. Acta* **32,** 237-250.
Brooks, J., and Taylor, D. J. (1955). Eggs and Egg Products. *Rept. Food Invest. Bd. No.* **60.** HMSO, London.
Brown, W. E., Baker, R. C., and Naylor, H. B. (1965). *Poultry Sci.* **44,** 1323-1327.
Brown, W. E., Baker, R. C., and Naylor, H. B. (1966a). *Poultry Sci.* **45,** 276-279.
Brown, W. E., Baker, R. C., and Naylor, H. B. (1966b). *Poultry Sci.* **45,** 279-283.
Brown, W. E., Baker, R. C., and Naylor, H. B. (1966c). *Poultry Sci.* **45,** 284-287.
Brumfitt, W. (1959). *Brit. J. Exptl. Pathol.* **40,** 441.
Bryant, R. L., and Sharp, P. L. (1934). *J. Agr. Res.* **48,** 67-89.
Büchli, K. (1967). *Inst. Pluimveeteelt 'Het Spelderholt'—Beekbergen Mededel. No.* **145.**
Burge, R. E., and Draper, J. C. (1967a). *J. Mol. Biol.* **28,** 173-187.
Burge, R. E., and Draper, J. C. (1967b). *J. Mol. Biol.* **28,** 189-204.
Burge, R. E., and Draper, J. C. (1967c). *J. Mol. Biol.* **28,** 205-210.
Buxton, A., and Gordon, R. F. (1947). *J. Hyg.* **45,** 265-281.

Carey, N. H. (1966). *In* "Physiology of the Domestic Fowl" (C. Horton-Smith and E. C. Amoroso, eds.), p. 125–132. Oliver and Boyd, Edinburgh.
Caroline, L., Taschdjian, C. L., Kozinn, P. J., and Schade, A. L. (1964). *J. Invest. Dermatol.* **42**, 415–419.
Carson, K. J., and Eagon, R. G. (1966). *Can. J. Microbiol.* **12**, 105–108.
Clark, J. R., Osuga, D. T., and Feeney, R. E. (1963). *J. Biol. Chem.* **238**, 3621–3631.
Collins, F. M., and Lascelles, J. (1962). *J. Gen. Microbiol.* **29**, 531–535.
Conti, S. F., Jacobs, N. J., and Gray, C. T. (1968). *J. Bacteriol.* **96**, 554–556.
Cotterill, O. J., and Gardner, F. (1957). *Poultry Sci.* **36**, 196–206.
Cotterill, O. J., and Hartman, P. (1956). *Poultry Sci.* **35**, 733–735.
Cotterill, O. J., and Winter, A. R. (1954a). *Poultry Sci.* **33**, 607–611.
Cotterill, O. J., and Winter, A. R. (1954b). *Poultry Sci.* **33**, 679–686.
Cotterill, O. J., and Winter, A. R. (1955). *Poultry Sci.* **34**, 679–686.
Cotterill, O. J., Gardner, F. A., Funk, E. M., and Cunningham, F. E. (1958). *Poultry Sci.* **37**, 479–483.
Cox, S. T., Jr., and Asbell, R. G. (1968). *Can. J. Microbiol.* **14**, 913–922.
Cunningham, F. E., and Lineweaver, H. (1965). *Food Technol.* **19**, 136–141.
Davis, B. D. (1961). *Cold Spring Harbor Symp. Quant. Biol.* **26**, 1–10.
Davis, B., Saltman, P., and Benson, S. (1962). *Biochem. Biophys. Res. Commun.* **8**, 56.
Dean, A. C. R., and Hinshelwood, C. (1966). "Growth Function and Regulation in Bacterial Cells." Clarendon Press, Oxford.
Delezenne, C., and Pozerski, E. (1903). *Compt. Rend. Soc. Biol.* **55**, 935.
Ducay, E. D., Kline, L., and Mandeles, S. (1960). *Poultry Sci.* **39**, 831–835.
Eakin, R. E., Snell, E. E., and Williams, R. J. (1940). *J. Biol. Chem.* **136**, 801–802.
Ehrenberg, A., and Laurell, C. B. (1955). *Acta Chem. Scand.* **9**, 68–72.
Elliott, R. P. (1954). *Appl. Microbiol.* **2**, 158–163.
Elliott, R. P., and Brant, A. W. (1957). *Food Res.* **22**, 241–250.
Elliott, R. P., and Romoser, G. L. (1957). *Poultry Sci.* **36**, 365–375.
Evans, R. J., Davidson, J. A., and Butts, H. A. (1950). *Poultry Sci.* **29**, 104–108.
Evans, R. J., Butts, H. A., and Davidson, J. A. (1951). *Poultry Sci.* **30**, 515–519.
Feeney, R. E. (1951). *Arch. Biochem. Biophys.* **34**, 196–208.
Feeney, R. E., and Nagy, D. A. (1952). *J. Bacteriol.* **64**, 629–643.
Feeney, R. E., Ducay, E. D., Silva, R. S., and MacDonnell, C. R. (1952). *Poultry Sci.* **31**, 639–647.
Feeney, R. E., Weaver, J. M., Jones, J. R., and Rhodes, M. B. (1956). *Poultry Sci.* **35**, 1061–1066.
Feeney, R. E., Stevens, F. C., and Osuga, D. T. (1963). *J. Biol. Chem.* **238**, 1415–1418.
Fleming, A. (1922). *Proc. Roy. Soc.* **B93**, 306–317.
Florian, M. L. E., and Trussell, P. C. (1957). *Food Technol.* **11**, 56–60.
Forsythe, R. H., Ayres, J. C., and Radlo, J. L. (1953). *Food. Technol.* **7**, 49–56.
Fossum, K., and Whitaker, J. R. (1968). *Arch. Biochem. Biophys.* **125**, 367–375.
Fraenkel-Conrat, H. L., and Feeney, R. E. (1950). *Arch. Biochem.* **29**, 101–113.
Frazer, D. T., Jukes, T. H., Branion, H. D., and Halpen, K. C. (1934). *J. Immunol.* **26**, 437–446.
Friedberger, E., and Hoder, F. (1932). *Z. Immunitaetforsch.* **74**, 429–447.
Fromm, D. (1967). *J. Food Sci.* **32**, 1–5.
Fromm, D., and Monroe, R. J. (1960). *Food Technol.* **14**, 401–403.
Fuller, R. A., and Briggs, D. R. (1956). *J. Am. Chem. Soc.* **78**, 5253–5257.
Fuller, R., and Jayne Williams, D. J. (1968). *Brit. Poultry Sci.* **9**, 159–163.

Funk, E. M. (1943). *Res. Bull. Missouri Agr. Exptl. Stat. No.* **364**.
Funk, E. M. (1948). *Res. Bull. Missouri Agr. Exptl. Stat. No.* **426**.
Funk, E. M., Forward, J., and Lorah, M. (1954). *Res. Bull. Missouri Agr. Exptl. Stat. No.* **550**.
Gardner, J. F., and Lascelles, J. (1962). *J. Gen. Microbiol.* **29**, 157–164.
Garibaldi, J. A. (1960). *Food Res.* **25**, 337–344.
Garibaldi, J. A. (1967). *J. Bacteriol.* **94**, 1296–1299.
Garibaldi, J. A., and Bayne, H. G. (1960). *Poultry Sci.* **39**, 1517–1520.
Garibaldi, J. A., and Bayne, H. G. (1962a). *J. Food Sci.* **27**, 57–59.
Garibaldi, J. A., and Bayne, H. G. (1962b). *Poultry Sci.* **41**, 850–853.
Garibaldi, J. A., and Neilands, J. B. (1956). *Nature* **177**, 526–527.
Garibaldi, J. A., and Stokes, J. L. (1958). *Food Res.* **23**, 283–290.
Garrard, W., and Lascelles, J. (1968). *J. Bacteriol.* **95**, 152–156.
Gayon, M. V. (1873). *Compt. Rend.* **76**, 232–233.
Ghosh, B. K., and Murray, R. G. E. (1967). *J. Bacteriol.* **93**, 411–426.
Gillespie, J. H., and Scott, W. J. (1950). *Australian J. Appl. Sci.* **1**, 514–530.
Gillespie, J. H., Scott, W. J., and Vickery, J. R. (1950a). *Australian J. Appl. Sci.* **1**, 215–223.
Gillespie, J. H., Scott, W. J., and Vickery, J. R. (1950b). *Australian J. Appl. Sci.* **1**, 313–329.
Gillespie, J. H., Salton, M. R. J., and Scott, W. J. (1950c). *Australian J. Appl. Sci.* **1**, 531–538.
Good, R. A., Finstad, J., Gewurz, H., Cooper, M. D., and Pollara, B. (1967). *Am. J. Diseases Children* **114**, 477–497.
Gordon, R. F., and Tucker, J. F. (1954). *Contrib.* 109 *10th World's Poult. Congr.* Edinburgh, Department of Agriculture, Scotland.
Goresline, H. E., Moser, R. E., and Hayes, K. M. (1950). *Food Technol.* **4**, 426–430.
Gray, C. T., Wimpenney, J. W. T., Hughes, D. E., and Mossman, M. R. (1966a). *Biochim. Biophys. Acta* **117**, 22–32.
Gray, C. T., Wimpenney, J. W. T., and Mossman, M. R. (1966b). *Biochim. Biophys. Acta* **117**, 33–41.
Gray, G. W., and Wilkinson, S. G. (1965a). *J. Appl. Bacteriol.* **28**, 153–164.
Gray, G. W., and Wilkinson, S. G. (1965b). *J. Gen. Microbiol.* **39**, 385–389.
Haines, R. B. (1938). *J. Hyg.* **38**, 338–355.
Haines, R. B. (1939). *Rept. Food Invest. Bd. No.* **47**. HMSO, London.
Haines, R. B., and Moran, T. (1940). *J. Hyg.* **40**, 453–461.
Hale, H. P. (1950). *J. Sci. Food Agr.* **1**, 46–48.
Halkett, J. A. E., Peters, T., and Ross, J. F. (1958). *J. Biol. Chem.* **231**, 187–199.
Hara, S., and Matsushima, Y. (1967). *J. Biochem.* **62**, 118–125.
Harland, H. C. (1927). *J. Genet.* **18**, 55–62.
Harry, E. G. (1957). *Vet. Record* **69**, 1433–1439.
Harry, E. G. (1963a). *Brit. Poultry Sci.* **4**, 63–70.
Harry, E. G. (1963b). *Brit. Poultry Sci.* **4**, 91–100.
Hartung, T. E., and Stadelman, W. J. (1962). *Poultry Sci.* **41**, 1590–1591.
Hartung, T. E., and Stadelman, W. J. (1963). *Poultry Sci.* **42**, 147–150.
Hirsch, J. G. (1965). *Ann. Rev. Microbiol.* **19**, 339–350.
Jacobs, N. J., and Conti, S. F. (1965). *J. Bacteriol.* **89**, 675–679.
Karminiski, M., and Durieux, J. (1956). *Exptl. Cell Res.* **10**, 590–618.
Kellenberger, E., and Ryter, A. (1958). *J. Biophys. Biochem. Cytol.* **4**, 323–326.

Keller, W., and Pennell, R. B. (1958). *J. Lab. Clin. Med.* **53**, 638–645.
King, J. V., Campbell, J. J. R., and Eagles, B. A. (1948). *Can. J. Res.* **26**, 514–519.
Knowles, N. R. (1956). *J. Appl. Bacteriol.* **19**, 293–300.
Knowles, N. R. (1957a). *Res. Exptl. Record Min. Agr. Northern Ireland* **6**, 100–107.
Knowles, N. R. (1957b). *Res. Exptl. Record Min. Agr. Northern Ireland* **6**, 108–117.
Kohn, A. (1960). *J. Bacteriol.* **79**, 697–706.
Kraft, A. A., Elliott, L. E., and Brant, A. W. (1958). *Poultry Sci.* **37**, 238–240.
Korotkova, G. P. (1957). *Zh. Obshch. Biol.* **18**, 275–287.
Lack, D. (1968). "Ecological Adaptions for Breeding in Birds." Methuen, London.
Lancaster, J. E., and Crabb, W. E. (1953). *Brit. Vet. J.* **109**, 139–148.
Laurell, C. B., and Ingleman, B. (1947). *Acta Chem. Scand.* **1**, 770.
Laschtschenko, P. (1909). *Z. Hyg. Infektionskrankh.* **64**, 419–427.
Lifshitz, A., Baker, R. C., and Naylor, H. B. (1964). *J. Food Sci.* **29**, 94–99.
Longsworth, L. G., Cannan, R. K., and MacInnes, D. A. (1940). *J. Am. Chem. Soc.* **62**, 2580–2590.
Lorenz, F. W., Starr, P. B., Starr, M. P., and Ogasawara, F. X. (1952). *Food Res.* **17**, 351–360.
Lush, I. E. (1961). *Nature* **189**, 981–984.
McNally, E. H. (1953). *Poultry Sci.* **32**, 915.
Magwood, S. E. (1964a). *Poultry Sci.* **43**, 441–449.
Magwood, S. E. (1964b). *Poultry Sci.* **43**, 1567–1572.
Magwood, S. E., and Marr, H. (1964). *Poultry Sci.* **43**, 1558–1566.
Mandeles, S., and Ducay, E. D. (1961). *J. Biol. Chem.* **237**, 3196–3199.
Mandelstam, J., and Rogers, H. J. (1959). *Biochem. J.* **72**, 654.
Manwell, C. (1963). *In* "The Biology of Myxine," (A. Brodal, and R. Fange, eds.), p. 372–455. Universitetsforloget, Oslo.
Manwell, C. (1967). *Comp. Biochem. Physiol.* **23**, 383–406.
Marshall, M. E., and Deutsch, H. F. (1951). *J. Biol. Chem.* **189**, 1–9.
Marshall, W., and Cruikshank, D. B. (1938). *J. Agr. Sci.* **28**, 24–42.
Martin, H. H. (1963). *J. Theoret. Biol.* **5**, 1–34.
Massoff, W., and Stolphann, H. J. (1961). *Z. Zellforsch.* **55**, 818–832.
Matsushima, K. (1958). *Science* **127**, 1178–1179.
Mellor, D. B., and Banwart, G. J. (1965). *Poultry Sci.* **44**, 1244–1248.
Miles, A. A., and Halnan, E. T. (1937). *J. Hyg.* **37**, 79–97.
Miller, W. A., and Crawford, L. B. (1953). *Poultry Sci.* **32**, 303–309.
Morton, J. R., Gilmour, D. G., McDermid, K. M., and Ogden, A. L. (1965). *Genetics* **51**, 97–107.
Murphy, T. W., and Sutton, W. S. (1947). *Agr. Gaz. N. S. Wales Misc. Publ. No.* **3317**.
Murray, M. W., and Rutherford, P. P. (1963). *Poultry Sci.* **42**, 505–508.
Nermut, M. V., and Murray, R. G. E. (1967). *J. Bacteriol.* **93**, 1949–1965.
Nichols, A. A., Leaver, C. W., and Paynes, J. J. (1967). *Brit. Poultry Sci.* **8**, 297–310.
Ogden, A. L. (1961). *Animal Breeding Abstr.* **29**, 127–138.
Ogden, A. L., Morton, J. R., Gilmour, D. G., and McDermid, E. M. (1962). *Nature* **195**, 1026–1028.
Orel, V. (1959). *Poultry Sci.* **38**, 8–12.
Osborne, T. B., and Cambell, F. (1900). *J. Am. Chem. Soc.* **22**, 422–450.
Park, J. T., and Griffith, M. (1964). *Proc. 3rd Intern. Symp. Fleming's Lysozyme*, Session 1a, p. 23.

Parkinson, T. L. (1966). *J. Sci. Food Agr.* **17**, 101–111.
Paton, A. M. (1959). *Nature* **184**, 1254–1255.
Paton, A. M., and Ayres, J. C. (1964). *Nature* **204**, 803–804.
Patterson, R., Youngner, J. S., Weigle, W. O., and Dixon, F. J. (1962). *J. Immunol.* **89**, 272–278.
Perkins, H. R., and Rogers, H. J. (1959). *Biochem. J.* **72**, 647–654.
Phelps, R. A., Shenstone, F. S., Kemmerer, A. R., and Evans, R. J. (1965). *Poultry Sci.* **44**, 388–394.
Platt, A. E., and Anderson, C. F. (1939). *J. Dept. Agr. S. Australia* **42**, 1040–1041.
Ramsey, H. H., and Wilson, T. E. (1957). *Antonie van Leeuwenhoek J. Microbial. Serol.* **23**, 226.
Repaske, R. (1956). *Biochim. Biophys. Acta* **22**, 189–191.
Repaske, R. (1958). *Biochim. Biophys. Acta* **30**, 225–232.
Rhodes, M. B., Bennett, N., and Feeney, R. E. (1959). *J. Biol. Chem.* **234**, 2054–2060.
Rhodes, M. E., Bennett, N., and Feeney, R. E. (1960). *J. Biol. Chem.* **235**, 1040–1041.
Richard, O., and Mohler, A. (1950). *Mitt. Gebiete Lebensm. Hyg.* **41**, 168–180.
Rizk, S. S., Ayres, J. C., and Kraft, A. A. (1966a). *Poultry Sci.* **45**, 764–769.
Rizk, S. S., Ayres, J. C., and Kraft, A. A. (1966b). *Poultry Sci.* **45**, 825–829.
Rogers, H. J. (1965). *In* "Function and Structure in Micro-organisms" (M. R. Pollock and M. H. Richmond, eds.), Symp. Soc. Gen. Microbiol., No. 15, pp. 186–219. Cambridge Univ. Press, Cambridge.
Rogers, H. J., and Perkins, H. R. (1968). "Cell Wall Membranes." Spon, London.
Romanoff, A. L. (1967). "Biochemistry of the Avian Embryo." Wiley (Interscience), New York.
Romanoff, A. L., and Romanoff, A. J. (1944). *Food Res.* **9**, 358–366.
Romanoff, A. L., and Romanoff, A. J. (1949). "The Avian Egg." Wiley, New York.
Rosser, F. T. (1942). *Can. J. Res.* **20D**, 291–296.
Rutherford, P. P., and Murray, M. W. (1963). *Poultry Sci.* **42**, 499–505.
Salton, M. R. J. (1964). "The Bacterial Cell Wall." Elsevier, Amsterdam.
Salton, M. R. J., and Ghuysen, J. M. (1960). *Biochim. Biophys. Acta* **45**, 355–363.
Salton, M. R. J., Scott, W. J., and Vickery, J. R. (1951). *Australian J. Appl. Sci.* **2**, 205–222.
Sauter, E. A. (1966). *Poultry Sci.* **45**, 131–135.
Sauter, E. A., Petersen, C. F., and Lampman, C. E. (1962). *Poultry Sci.* **41**, 468–473.
Schade, A. L. (1958). *Abstr. Proc. 4th Intern. Congr. Biochem. Vienna*, p. 127, 10–42.
Schade, A. L. (1960). *In* "Protides of the Biological Fluids" (H. Peeters, ed, p. 261–263. Elsevier, London.
Schade, A. L. (1963). *Biochem. Z.* **338**, 140–148.
Schade, A. L., and Caroline, L. (1944). *Science* **100**, 14–15.
Schade, A. L., Reinhart, R. W., and Levy, H. (1949). *Arch. Biochem.* **20**, 170–172.
Schade, A. L., Myers, N. H., and Reinhart, R. W. (1968). *J. Gen. Microbiol.* **52**, 253–260.
Schaible, F. J., Bandemer, S. L., and Davidson, J. A. (1946). *Poultry Sci.* **25**, 440–445.
Schmidt, F. J., and Stadelman, W. J. (1957). *Poultry Sci.* **36**, 1023–1026.
Scott, W. J., and Vickery, J. R. (1954). *Australian J. Appl. Sci.* **5**, 89–102.
Sharp, P. F., and Stewart, G. F. (1936). *Mem. Cornell Univ. Agr. Expt. Stat. No.* **91**.
Sharp, P. F., and Whitaker, R. (1927). *J. Bacteriol.* **14**, 17–46.
Shenstone, F. S. (1968). *In* "Egg Quality: A Study of the Hen's Egg" (T. C. Carter, ed.). Oliver & Boyd, Edinburgh.

Simkiss, K. (1961). *Biol. Rev.* **36**, 321–367.
Simkiss, K. (1968). *In* "Egg Quality: A Study of the Hen's Egg" (T. C. Carter, ed.), p. 3. Oliver & Boyd, Edinburgh.
Simons, P. C. M., and Wiertz, G. (1963). *Z. Zellforsch. Mikroskop. Anat. Histochem.* **59**, 555–567.
Simons, P. C. M., and Wiertz, G. (1965). *Brit. Poultry Sci.* **6**, 283–286.
Simons, P. C. M., and Wiertz, G. (1966). *Poultry Sci.* **45**, 1153–1162.
Simons, P. C. M., Tyler, C., and Thomas, H. P. (1966). *Brit. Poultry Sci.* **7**, 309–314.
Smith, A. J. M. (1934). *J. Exptl. Biol.* **11**, 228–242.
Stokes, J. L., and Osborne, W. W. (1956). *Food Res.* **21**, 264–269.
Stokes, J. L., Osborne, W. W., and Bayne, H. G. (1956). *Food Res.* **21**, 510–518.
Strasters, K. C., and Winkler, K. C. (1963). *J. Gen. Microbiol.* **33**, 213–229.
Strominger, J. L. (1962). *In* "The Bacteria," (I. C. Gunsalus and R. Y. Stanier, eds.), Vol. 3, p. 413. Academic Press, New York.
Stuart, L. S., and McNally, E. H. (1943). *U.S. Egg Poultry Mag.* **49**, 28–31, 45–47.
Summers, D. F., and Hasenclever, H. F. (1964). *J. Bacteriol.* **87**, 1–7.
Sword, C. P. (1966). *J. Bacteriol.* **92**, 536–542.
Symposium, (1967). *Proc. Roy. Soc. (London)* **B167**.
Theodore, T. S., and Schade, A. L. (1965a). *J. Gen. Microbiol.* **39**, 75–83.
Theodore, T. S., and Schade, A. L. (1965b). *J. Gen. Microbiol.* **40**, 385–396.
Tokin, B. P. (1959). *Proc. 15th Intern. Congr. Zool. London.*
Totter, J. R., and Moseley, F. T. (1953). *J. Bacteriol.* **65**, 45–47.
Trussell, P. C., Fulton, C. D., and Cameron, C. I. (1955a). *Food Technol.* **9**, 130–134.
Trussell, P. C., Triggs, R. E., and Green, B. A. (1955b). *Food Technol.* **9**, 134–137.
Tyler, C. (1945). *J. Agr. Sci. Camb.* **36**, 111–116.
Tyler, C. (1953). *J. Sci. Food Agr.* **4**, 266–272.
Tyler, C. (1955). *J. Sci. Food Agr.* **6**, 170–176.
Tyler, C. (1956). *J. Sci. Food Agr.* **7**, 483–493.
Tyler, C. (1968). *Brit. Poultry Sci.* **9**, 143–158.
Vos, J. C. (1964). *J. Gen. Microbiol.* **35**, 313–317.
Vos, J. C. (1967). *J. Gen. Microbiol.* **48**, 391–400.
Walden, C. C., Allen, I. V. F., and Trussell, P. C. (1956). *Poultry Sci.* **35**, 1190–1196.
Warner, R. C., and Weber, I. (1951). *J. Biol. Chem.* **191**, 173–180.
Weibull, C. (1953a). *J. Bacteriol.* **66**, 688–695.
Weibull, C. (1953b). *J. Bacteriol.* **66**, 696–702.
Weidel, W., and Pelzer, H. (1964). *Advan. Enzymol.* **26**, 193–232.
Weinbaum, G., and Markham, R. (1966). *Biochim. Biophys. Acta* **124**, 207–209.
Weibaum, G., Rich, R., and Fischman, D. A. (1967). *J. Bacteriol.* **93**, 1693–1698.
Wells, R. G. (1968). *In* "Egg Quality: A Study of the Hen's Egg" (T. C. Carter, ed.), p. 207. Oliver & Boyd, Edinburgh.
Weston, W. A. R. D., and Halnan, E. T. (1927). *Poultry Sci.* **6**, 251–258.
Williams, J. (1962). *Biochem. J.* **83**, 355–364.
Williams, J. E., and Whittemore, A. D. (1967). *Avian Diseases* **11**, 467–490.
Wilson, J. E. (1945). *Vet. Record* **57**, 411.
Winter, A. R., Burhart, B., and Wettling, C. (1952). *Bull. Ohio Agr. Exp. Stat. No.* **710**.
Winter, A. R., Burhart, B., Clements, P., and MacDonald, L. (1955). *Bull. Ohio Agr. Expt. Stat. No.* **762**.
Wolin, M. J. (1966). *J. Bacteriol.* **91**, 1781–1786.

Wolk, J., McNally, E. H., and Spicknell, N. H. (1950). *Food Technol.* **4,** 316–318.
Woolley, D. W., and Longsworth, L. G. (1942). *J. Biol. Chem.* **142,** 285–290.
Work, E. (1957). *Nature* **179,** 841–847.
Zagaevsky, J. S., and Lutikova, P. O. (1944). *U.S. Egg Poultry Mag.* **50,** 17–20, 43–46, 75–77, 89–90, 121–123.
Zinder, N. D., and Arndt, W. F. (1956). *Proc. Natl. Acad. Sci. U.S.* **42,** 586–590.

Training for the Biochemical Industries

I. L. HEPNER

Process Biochemistry, London, England

With the widsom of hindsight the 1950's could be described as the *chemical decade*, and the 1960's as the *electronic decade*. Should the present demand for new biological products continue at the present rate, then the 1970's might well become the biological decade. In other words, the next 11 years could witness the full exploitation in industry of biological (or biochemical) techniques and processes. However this desirable state of affairs may be postponed indefinitely if our system of training biologists—specifically biochemists and microbiologists—is not ameliorated.

Historians of the 20th century will claim that one major factor responsible for the explosive growth of the western world's chemical industry in the 1950's was the existence of a large pool of chemical engineering brainpower trained by the founder fathers of this relatively young technology during the third and fourth decades of this century. These chemical engineers learned how to apply the physical principles of heat transfer, fluid mechanics, and hydrodynamics to the design and operation of chemical process plant. By analogy it can be claimed that the design and operation of biochemical process plant, for the production of enzymes, antibiotics, and microbial protein, will become the task of microbiological engineers or industrial microbiologists of the future. This immediately raises the next problem: who will train these microbiologists?

Until only a few years ago, the biological sciences as taught at British universities were peripheral—or rather satellitic—to medicine and agriculture. In their preclinical years medical and veterinary students were taught bacteriology, biochemistry, physiology, and anatomy. Agricultural students were taught the elements of plant genetics, animal husbandry, and bacteriology. "Independent" departments of botany, zoology, biochemistry, and microbiology within science faculties of universities were generally small. The saddest aspect of this scene was really the lack of communication between biologists working within medical faculties and those within science or agricultural faculties. A good illustration is the pioneering work of medical biochemists during the 1930's in elucidating the function of metabolic cycles in the human system, which was largely ignored by plant biochemists until many years later. As a result of this fragmentation the best students selected the physical sciences and medicine — biology was always considered a soft option. With the growth of molecular biology after the war, this subject, once described by Chargaff as "biochemistry without a license," became the glamour girl of the biological sciences — aided and abetted in no small measure by the large crop of Nobel prizes awarded to molecular biologists in recent years. This was in many ways similar to the attraction that nuclear physics had for physicists and chemical physics for chemists. Young scientists had been conditioned by their professors to believe that the most valuable research involves unraveling the structure of atoms, molecules, or biological macromolecules. That microbiology or biochemistry applied to industrial problems could be of equal scientific worth was maintained only by a handful of professors in Europe and the United States who found faint echos of response from industry. Of course in Japan the picture was entirely different. There it has been deliberate policy since 1945 to concentrate all the country's biological brainpower into industrial microbiology.

Those industries wherein biochemistry or microbiology is involved I have defined as the biochemical industries. There are five major biochemical industries: brewing, distilling and winemaking, food and dairy processing, pharmaceutical and enzyme manufacture, organic chemicals manufacture, and sewage and effluent purification. Some processes in the biochemical industries are as old as civilization, e.g., brewing, breadmaking, cheesemaking; others date back to the beginning of this century, e.g., biological sewage purification and citric acid manufacture, and a small proportion are ultracontemporary, e.g., enzyme manufacture. Apart from the biochemical industries, microbial processes can play a beneficient or deleterious role in other industries. For example, microbial leaching of copper from weak mineral concentrations is now being seriously investigated in the United States, Canada, and Australia and could be the key to releasing new sources of expensive metals. On the other hand, biodeterioration (or microbial spoilage) of fuel oil, cement, paper, textiles, or metallic pipes causes annual losses running into millions of dollars. Consequently, if microbiologists and biochemists harnessed their skills to the biochemical industries as well as to some other industries such as mining or metallurgy, they could catalyze tremendous industrial innovations and at the same time pursue a satisfactory scientific career.

The industries classified as biochemical industries are really product-orientated. The brewing industry is concerned with beer, the dairy industry with milk and cheese, and the baking industry with bread. From a sales and marketing point of view this is perfectly valid, as profitability can only be increased by identifying industry with product. On the other hand, in order to improve and update some of the traditional processes or to pioneer new processes for enzymes and nucleotides, it is necessary for the industry to become process-orientated. Here the microbiologist, with understanding of chemical engineering principles, could play an invaluable role.

Once these industries are regarded from a process point of view one can define four major process operations common to them all. These are: fermentation, separation, preservation, and maturation. In brewing beer all four of these operations are involved, as in the case of wine, spirit, and cheese manufacture. In the production of enzymes or organic acids by fermentation only three of these operations are involved. Each of these operations is based on the principles of biochemistry, microbiology, and chemical engineering. Only by understanding the subtle balance between these biological and physical aspects of each operation, is it possible to optimize them for maximum efficiency. A cardinal example of this is the development of continuous fermentation applied to brewing. This was developed after many years of patient collaborative research between chemical engineers and microbiologists, and is now being widely—and profitably—adopted throughout the British brewing industry. Possibly if the same research efforts had been devoted to fermentation processes for organic acids these might not have been superseded so rapidly by synthetic chemical processes.

A strong plea should be made for devoting more research attention to the second process operation, separation. This operation comprises several of the classical chemical engineering unit operations such as leaching, liquid/liquid extraction, ion exchange, chromatography, and their application in harvesting microbial metabolites or biomass from fermentation liquors. Many potentially exciting fermentation processes are still on the drawing board simply because no one has yet managed to develop large-scale harvesting techniques for handling very dilute liquors.

The third process operation, preservation, has already resulted in the long-term storage of food products using such techniques as high-temperature short-time sterilization or radiation. Here it is necessary to harness the skills of microbiologists to determine the degree of sterility necessary for the preservation of a food or feedstuff or in preserving a human organ *in vitro* needed for a transplant. Equally important is the prevention of biodeterioration in noncomestible products. Surely the preservation processes that will kill or inactivate microorganisms in food products could be modified to kill or inactivate microorganisms in fuel oil and cement.

The least understood of these operations is the fourth, maturation. This operation is time-intensive and involves intricate biochemical changes that proceed in wine, spirit, dairy, and meat products. What would not distillers give if whisky would be matured in one year instead of three? The industrial rewards of accelerated maturation are obvious. The scientific challenge is considerably more daunting than for any other of the three operations.

Nevertheless, the greatest challenge that industrial microbiologists will face during the next decade will be in the development of large-scale processes for the production of enzymes, nucleotides, and other novel biologicals. Here the effects brought about by degree of scale become significant. While the biochemist today may be able to produce a few grams of an enzyme in the laboratory, he can not readily scale-up this process to produce tonnage quantities. (The acceptance of enzyme-containing detergents might well mean that tonnage enzyme plants will soon become integrated with detergent complexes.)

Only the petroleum industry has appreciated the validity of these biochemical process operations. In this industry large-scale continuous processes have been traditionally used, and when pioneering the production of microbial protein grown on petroleum fractions several years ago they did not consider anything but a continuous fermentation followed by continuous harvesting and purification. Microbiologists would be well advised to appreciate this point.

To prepare for the potentially large demand of microbiologists and biochemists in industry it will be imperative to readjust university courses in these subjects to the requirements of the age. University authorities should be encouraged to expand their departments of biochemistry and microbiology at the expense of botany and zoology — the biological relics of the Darwinian age — and every effort should be made to prevent wastage by needless duplicating of biochemistry departments in medical and agricultural schools. Inside these departments of biochemistry and microbiology there should be active nuclei of chemical engineers engaged in teaching and research. In Table I, the basic requirements for a modern course in industrial microbiology are outlined.

TABLE I
Basic Science Requirements for a Course in Industrial Microbiology

A four-year degree course[a]			
Year 1	Year 2	Year 3	Year 4[b]
Biology	Biochemistry	Biochemical engineering	Biochemical processes
Chemistry	Microbiology	Fermentor design and control	Food manufacture
Physics	Mathematics	Aeration and agitation	Brewing and distilling
Mathematics	Unit operations	Industrial sterilization	Sewage and effluent treatment
	Chemical reaction kinetics	Preservation of biochemical products	Production of biochemicals (antibiotics, vaccines, enzymes)
		Biochemistry (particularly bio-organic chemistry and biotransformations)	
		Microbiology	

[a] **Further Education:** Provision should be made in the universities teaching industrial microbiology to run further education courses for older biochemists and microbiologists. These courses, lasting a summer vacation, should try to instill the elements of chemical engineering which are useful and necessary to those engaged in the biochemical industries. The following subjects should be taught: unit operations, chemical reaction kinetics, heat transfer, and the chemical engineering basis of all major biochemical processes.

[b] As part of this degree examination, the student should have to design a specific biochemical process plant under the direction of a professor.

Of course, the reverse pattern is well known: establishment of biochemical engineering units within chemical engineering departments for postgraduate training of chemical engineers in the elements of microbiology and biochemistry. However, biochemical engineers so trained tend to be a minority elite and it would surely make more sense to produce a greater number of industrial microbiologists than a relatively small number of biochemical engineers. Perhaps an alternative to this scheme might be the establishment of several institutes of industrial microbiology linked to universities, where postgraduate training is carried out hand in hand with research and development. The staff in these institutes would comprise biologists of all varieties as well as engineers, chemists, and physicists. A culture collection would form a linchpin of such an institute and its major departments would comprise biological activities such as screening and taxonomy, biodeterioration, chemical engineering, analysis and statistics, and solid and liquid fermentations. The four biochemical process operations outlined previously would feature prominently within each of these departments. There is little doubt that Japanese success in industrial microbiology owes a lot to the existence of such institutes. We must be clearly aware at this stage that industrial microbiology could become the key to illimitable human progress. Only by radically redesigning our educational and training program can the biological decade become a reality and not remain a dream.

AUTHOR INDEX

Numbers in parentheses are reference numbers and indicate that an author's work is referred to, although his name is not cited in the text. Numbers in italics show the page on which the complete reference is listed.

A

Abe, M., 218, 221, 233, 240, *242, 244*
Abels, J. C., 255, *274*
Abraham, E. P., 18(2), 19(20), 20, 21(3, 38, 147), 35(3), 36(3, 147), 46(2), 61(174), 62, 63(174), *68, 69, 74*
Abramson, E., 139, *155*
Acklin, W., 240, *243*
Acred, P., 37(5), 46(6), 49, 57(7), 58(7), *68*
Acton, H. W., 114, *131*
Adamcova, Z., 177(71), *209*
Agurell, S. L., 219, 221, 237, 240, *242*
Ahrens, G., 139, *153*
Aiba, S., 163(2), 164(2), *207*
Aisen, P., 257, *274*
Aiso, K., 107, 120, *131*
Akers, G. K., 180(77), 181(77, 78), 183(78), *209*
Akiba, F., 90, *100*
Albertsson, P. A., 174(48), *208*
Alburn, H. E., 46(8), *68*
Alderton, G., 254, 257, 258, *274*
Alford, L. R., 265, *274*
Allen, I. V. F., 270, *280*
Allison, M. J., 55(58), *70*
Altgelt, K. H., 174, 177(45), 178(45), *208*
Ambler, C. M., 164(8), *207*
Amici, A. M., 220, 233, 234, 235, *242*
Ammerlaan, A. C. F., 203(124), 205(124), *210*
Anderson, C. F., 273, *279*
Anderson, D. M. W., 174(51), 180(51), *208*
Anderson, D. W., 169(31), 170(31), *208*
Anderson, E. S., 94, *99*
Anderson, F. S., 55(9), *68*
Anderson, J. A., 241, *243*
Andrews, P., 174(53), *208*
Arai, T., 107, 120, 121, *131*
Aran, A. P., 137, 142, 144, *157*
Arcamone, F., 221, 222, 224, 225, 226, 227, 228, 230, 231, 232, *242*

Archard, H. O., 142, *154*
Arigoni, D., 240, *243*
Arima, K., 81, *99*
Arndt, W. F., 257, *281*
Asbell, M. A., 256, 257, *274*
Asbell, R. G., 256, *276*
Ashbrook, A. W., 171(37), *208*
Auskaps, A. M., 151, *153*
Averill, H. M., 140, *153*
Ayliffe, G. A. J., 39(10), *68*
Ayres, J. C., 253, 261, 262, 263, 264, 267, 268, 269, *274, 275, 276, 279*
Azari, P. R., 257, *275*

B

Babel, R. B., 24(89), *71*
Bach, J. A., 61(11), 63(11), 64(11), *68*
Backus, E. J., 122, *132*
Baer, P. N., 150, *153*
Baes, C. F., Jr., 171(40), *208*
Bain, J. A., 257, *275*
Baitsch, H., 143, *154*
Baker, C. M. A., 249, 258, 260, *275*
Baker, G. E., 113, 120, *132*
Baker, R. C., 249, 253, 263, 267, 270, 271, *275, 278*
Baldacci, E., 102, 104, 113, *131*
Ballio, A., 20, 65(12), *68*
Bandemer, S. L., 258, *279*
Banghart, S., 143, 144, 145, 151, *154*
Bangs, S. E., 166(16), *207*
Banwart, G. J., 263, *278*
Barber, M., 20(15), 34(16), 40(16), 41(16), 44(16), 47(32), 51(32), 53(14), *68, 69*
Barclay, W. R., 64(119), *72*
Barger, G., 211, 213, *242*
Barndin, R. L., 24(17), 25(17), *68*
Barnes, M. W., 55(80), *71*
Barrett, F. F., 92, *99*
Bartels, C. R., 168(30), 169, *208*
Baskett, R. G., 252, *275*
Batchelor, F. R., 17, 22, 23(19, 168), 34

AUTHOR INDEX

(167), 55(167), 62, 63(200), 65(20), 66(22), 68, 69, 73, 74, 189(92), 209
Bauchop, T., 260, 275
Bauer, A. W., 21(23), 23(122), 69, 72
Baugh, R. F., 257, 274
Baxter, R. M., 238, 242
Bayne, H. G., 267, 269, 271, 277, 280
Bean, K. C., 268, 270, 275
Bechhold, H., 194, 210
Becker, B., 123, 131
Beer, A. E., 246, 275
Behrens, O. K., 19, 20(24, 27), 21, 65(170), 69, 74
Bennett, J. V., 95, 100
Bennett, N., 255, 279
Bennich, H., 188, 209
Benoit, M., 19(141), 73
Benson, S., 257, 276
Berg, G., 168, 208
Berger, L. P., 254, 256, 275
Berman, K. S., 142, 151, 154
Berryman, G. H., 65(29), 69
Beveridge, J., 141, 155
Bhattacharji, S., 238, 242
Bibby, B. G., 135, 136, 138, 140, 144, 147, 150, 153, 154, 156, 157
Bigland, C. H., 269, 275
Billmeyer, F. W., Jr., 185, 186(86), 209
Binkley, S. B., 24(55), 70
Birch, A. J., 238, 242
Bird, A. E., 49(30), 69
Birdsell, D. C., 256, 257, 275
Biro, B. E., 19(87), 71
Bisset, K. A., 111, 131
Bixler, H. J., 191(102), 193(102), 195(102), 200(102), 202(102), 209
Black, J., 114, 133
Blake, C. A., Jr., 171(40), 208
Blayney, J. R., 135, 137, 138, 140, 142, 156
Board, P. A., 265, 267, 271, 272, 273, 275
Board, R. G., 261, 262, 263, 264, 265, 267, 269, 270, 271, 272, 273, 275
Böhmer, M., 220, 244
Boger, W. P., 20(31), 65 (159), 69, 73
Bolhofer, W. A., 112, 131
Bond, J. M., 44(32), 47(32), 51(32), 69
Bondi, A., 45(60), 70
Bonino, C., 221, 222, 224, 225, 226, 227, 228, 230, 231, 232, 242

Bonns, W. W., 222, 242
Bonow, B. E., 139, 143, 155
Bortnick, L., 150, 156
Bowen, W. H., 136, 142, 144, 147, 153
Boyles, W. A., 166(21), 207
Brack, A., 216, 238, 242, 243, 244
Bradley, S. G., 123, 132
Brady, L. R., 212, 213, 242
Brain, E. G., 52(33), 53(33), 69
Brambell, F. W. R., 246, 275
Brandl, E., 20(34), 21, 42, 69
Brandley, C. A., 252, 275
Brandtzaeg, P., 152, 153, 154
Branion, H. D., 252, 276
Brant, A. W., 267, 268, 269, 275, 276, 278
Braude, A. I., 91, 99
Braun, W., 83, 99
Breed, R. S., 150, 154
Bretz, H. W., 166(17), 207
Brian, P. L. F., 204(125), 210
Bricker, H. M., 144, 157
Briggs, D. R., 257, 276
Brodie, J. L., 45(186), 46(99, 171), 71, 74
Brodsky, B. C., 110, 119, 125, 127, 128, 130, 132
Brooks, J., 246, 249, 250, 252, 258, 261, 263, 269, 271, 272, 275
Brown, D. M., 34(129), 37(5), 46(6), 49(6), 57(7), 58(8), 68, 72
Brown, E. V., 21(177), 74
Brown, K. B., 171(36, 39, 40), 172(39), 208
Brown, W. E., 253, 263, 267, 270, 271, 275
Brownlie, L. E., 95, 99
Bruce, M. A., 150, 156
Bruce, W. F., 21(201), 74
Brumfitt, W., 57(36), 69, 256, 275
Brunius, E., 139, 155
Brunner, R., 215, 243
Bryant, R. L., 267, 275
Buchanan, R. E., 120, 131
Büchel, K. G., 233, 244
Büchli, K., 268, 269, 275
Bugie, E., 109, 133
Bulger, R. J., 45(186), 74
Bunn, P. A., 19(141), 45(37), 69, 73
Buono, N., 61(11), 63(11), 64(11), 68
Buonocore, M. G., 138, 146, 154
Burckhardt, E., 219, 244
Burge, R. E., 256, 275
Burhart, B., 268, 280

Burley, D. M., *74*
Burton, H. S., 21(38), *69*
Butler, K., 48, *69*
Butterworth, D., 23(168), *74*
Butts, H. A., 246, 255, *276*
Buxton, A., 267, *275*

C

Cain, C. K., 3, *16*
Cameron, C. I., 268, *280*
Cameron-Wood, J., 23(168), 34(167), 55(167), *73, 74*
Campbell, F., 257, *278*
Campbell, J. J. R., 258, *278*
Cannan, R. K., 258, *278*
Canniff, T. L., 138, *155*
Carere, R. P., 95, *99*
Carey, N. H., 258, *276*
Carlsson, J., 143, 147, *154*
Carmack, C. C., 28(93), 34(94), 43(94), 44(93), 52(93), 53(93), 54(93), *71*
Caroline, L., 254, 258, 260, *276, 279*
Carpenter, F. H., 24(40), *69*
Carroll, C. J., 3, *16*
Carroll, R. A., 141, *157*
Carson, K. J., 257, *276*
Carter, H. E., 19(26), *69*
Casey, J. I., 92, *99*
Cassel, E. A., 166(19), 167(19), *207*
Castagnoli, N., 222, 241, *242, 243*
Cavallito, C. J., 25(41), *69*
Cavender, F., 241, *243*
Cerrai, E., 171(41), *208*
Chabbert, Y. A., 55(42), 56(43), *69*
Chaikovskaya, S. M., 63(44), *69*
Chain, E., 4, 14, *16*, 18(2), 19(2), 20(12), 23(19, 168), 34(167), 46(12), 55(167), 65(12), *68, 69, 73, 74*, 221, 222, 224, 225, 226, 227, 228, 230, 231, 232, *242*
Charney, J., 112, *131*
Charney, W., 114, *133*
Chase, S. W., 135, 136, 138, *157*
Cheney, L. C., 17(155), 23(153), 24(89, 154), 25(75), 26(153), 52(97), 54(97), *71, 73*
Chev, K. K., 65(170), *74*
Chiang, Ming-Chien, 19(25), *69*
Chisholm, D., 61(11), 63(11), 64(11), 65, 68, *70*

Chow, A. W., 25(47), *70*
Christensen, B. G., 57(133), *72*
Cignarella, G., 26(48), *70*
Claesen, M., 26(76), 30(209), *70, 71, 75*
Claridge, C. A., 23(50), *70*
Clark, D. E., 46(8), *68*
Clark, J. R., 258, *276*
Clark, R., 146, *154*
Clarke, H. T., 20(51), *70*
Clements, P., 268, *280*
Clopper, P. W., 135, 136, 138, *157*
Coghill, R. D., 18(52, 53, 144), *70, 73*
Cohen, S., 66(116), *72*, 91, *99*
Cole, M., 23(168), 39(54), *70, 74*
Coleman, C. F., 171(39), 172(39), *208*
Collings, C. K., 145, *154*
Collins, F. M., 260, *276*
Colmer, A. R., 108, *131*
Comaschi, G. F., 102, 104, *131*
Comby, L., 110, *132*
Coniglio, C. F., 114, *133*
Conner, H. D., 145, *154*
Conti, S. F., 259, *276, 277*
Cooper, D. E., 24(55), *70*
Cooper, M. D., 260, *277*
Cope, Z., 13, *16*
Coriell, L. L., 21(56), *70*
Cornblath, M., 143, *154*
Cornick, D. E., 136, 142, *153*
Corse, J., 19(25, 26), 20(27), 21(27), *69*
Cortis-Jones, B., 179(72), *209*
Cota-Robles, E. H., 256, 257, *275*
Cotterill, O. J., 248, 263, 268, *276*
Couch, J. N., 126, *131*
Cowan, S. T., 102, *131*
Cox, M. A., 135, 136, 138, *157*
Cox, S. T., Jr., 256, *276*
Crabb, W. E., 263, 267, *278*
Crane, R. K., 143, *154*
Crast, L. B., 23(153), 24(89, 154), 26(153), *73*
Crawford, L. B., 263, 269, *278*
Cronk, G. A., 19(57), *70*
Crouse, D. J., Jr., 171(36, 39), 172(39), *208*
Crowley, M. C., 135, 136, 138, *157*
Croydon, E. A. P., 66(130), *72*, 189(93), *209*
Cruikshank, D. B., 266, *278*
Crump, S. L., 140, *154*

D

Cueto, E. I., 138, 146, *154*
Culbreth, W., 97, *99*
Cummins, C. S., 126, *131*
Cunningham, F. E., 248, 257, *276*

Dadabo, K., 25(75), *71*
Dahlqvist, A., 143, 144, *154*
Dahlstrom, D. A., 164(7), *207*
Dain, J. A., 144, *156*
Dale, A. C., 150, *156*
Dalton, H. P., 55(58), *70*
Daniels, S. L., 167(25), *207*
Datta, N., 55(9), *68*, 86, 87, 88, 89, *99*, *100*
Davidson, J. A., 246, 255, 258, *276*, *279*
Davies, A. M., 145, *154*, *155*
Davis, B., 257, *276*
Davis, B. D., 260, *276*
Davis, N. S., 166(16), *207*
Dawes, C., 141, *154*
Dean, A. C. R., 247, *276*
de Coursaz, P. A., 140, 144, 147, 148, *154*
Deindoerfer, F. H., 168(28), *208*
de la Isla, M. de. L., 212, *243*
Delezenne, C., 255, *276*
Der Marderosian, A., 218, *243*
De Somer, P., 30(49, 209), *70*, *75*
Determann, H., 174, 175(43), 176(43), 177(43), 180(43), 183(43), 187(43), *208*
Deuel, H., 175, *209*
Deutsch, H. F., 257, *275*, *278*
Dewdney, J. M., 65(20), 66(21, 22), *69*, 189(92), *209*
De Weck, A. L., 65(59, 151), *70*, *73*
Di Accadia, F., 20(12), 65(12), *68*
Dibbern, H. W., 220, *243*
Dick, D. S., 151, *154*, *156*
Dirksen, T. R., 140, *154*
Dixon, F. J., 252, *279*
Dlouhy, P. E., 164(7), *207*
Doepfner, W., 220, *243*
Doisy, E. A., 3, *16*
Dolan, M. A., 45(60), *70*
Donnan, F. G., 194, *210*
Dorson, W., 205(128), *210*
Doty, D. B., 66(173), *74*
Dowling, H. F., 13, *16*, 19(61), *70*, 84, *99*
Doyle, F. P., 17(18), 22(18), 43(63, 67, 68, 70), 44(70), 52(33, 63), 53(33), 54(63, 69, 71), *68*, *69*, *70*
Draper, J. C., 256, *275*
Dresner, L., 194(112), 199(112), 202(112), 204(125), *210*
Dryden, W. H., 252, *275*
Dubey, G. A., 203(124), 205(124), *210*
DuBois, R., 19(141), *73*
Dudley, H. W., 219, *243*
Ducay, E. D., 255, 258, 260, *276*, *278*
Dunbar, J. B., 136, *154*
Dunn, W., 21(4), *68*
Durieux, J., 257, *277*
Dutrochet, H., 194, *210*
DuVigneaud, V., 22(72), 24(40), *69*, *70*
Dwyer, D. M., 150, *154*

E

Eagles, B. A., 258, *278*
Eagon, R. G., 256, 257, 274, *276*
Eaker, D., 179, *209*
Eakin, R. E., 255, *276*
Edgar, R. S., 84, *100*
Edmonds, E. J., 136, *156*
Edsall, G., 190(99), *209*
Edwards, C. D., 21(201), *74*
Edwards, J. P., 20(27), 21(27), *69*
Edwardsson, S., 143, 144, *154*
Egelberg, J., 143, *154*
Ehrenberg, A., 257, *276*
Eichhorn, H. H., 84, *100*
Eisen, H. N., 65(151, 152), *73*
Eisenstein, B., 66(116), *72*
Ekstrom, B., 35, *70*
Elliott, L. E., 269, *278*
Elliott, R. P., 268, 269, *276*
Elliott, V. B., 78, *99*
Ellwood, P., 167(24), *207*
Elsden, S. R., 260, *275*
Emneus, A., 188(88, 90, 91), 189(90, 91), *209*
Endo, T., 120, *131*
Englander, H. R., 138, 141, 151, *155*
English, A. R., 23(107, 110), 24(111), 31(74), 65(46), *70*, *72*
Erge, D., 240, 241, *243*
Ericsson, Y., 141, *154*
Erikson, D., 103, 106, 109, 113, 123, *131*

Ervin, F., 137, 142, *156*
Essery, J. M., 24(89, 154), 25(75), *71, 73*
Esumi, S., 164(6), *207*
Eustace, G. C., 23(168), *74*
Evans, A. H., 86, *100*
Evans, R. J., 246, 255, 258, *276, 279*
Evans, R. M., 24(17), 25(17), *68*
Evrard, E., 26(76), *71*

F

Falkow, S., 81, 86, 87, 89, *99*
Farrar, W. E., Jr., 63(77), *71*, 78, *99*
Fasold, H., 175(58), *208*
Feeney, R. E., 248, 254, 255, 257, 258, 259, *275, 276, 279*
Fehr, T., 240, *243*
Feinberg, J. G., 66(22), *69*, 189(92), *209*
Feinberg, S. M., 65(78), *71*
Feingold, D. S., 145, *155*
Feretti, A., 221, 222, 224, 225, 226, 227, 228, 230, 231, 232, *242*
Ferlauto, R. J., 45(60), *70*
Ferry, J. D., 199, *210*
Fevold, H. L., 254, 257, 258, *274*
Field, F. W., 21(202), *75*
Finland, M., 19(79), 21(211), 31(139), 40(125), 41(125, 126), 55(80), 56(175), 63(175), 64(175), 66(123), *71, 72, 73, 74, 75*, 92, *99*
Finlayson, M. G., 95, *99*
Finn, R. K., 165(12), *207*
Finn, S. B., 146, *154*
Finstad, J., 260, *277*
Fischer, R., 238, *244*
Fischman, D. A., 256, *280*
Fish, N. A., 95, *99*
Fisher, M. W., 21(4), *68*
Fisher, R. E., 204(125), *210*
Fisher, W. P., 112, *131*
Fitzgerald, D. B., 136, *154*
Fitzgerald, R. J., 136, 137, 138, 141, 142, 144, 147, 150, 151, *153, 154, 155*
Flanigan, C. C., 52(97), 54(97), *71*
Fleming, A., 18(81, 82, 83), *71*, 253, 254, 256, 260, *276*
Fletcher, C. M., 18(2), 19(2), 46(2, 8), *68*
Flodin, P., 174, 175(57), 176, 183(46), 184(83), 185, 186(83), *208, 209*

Florey, H. W., 2, 4, *16*, 18(2, 45, 85), 19(2, 85), 21(105), 36(105), 46(2), 65(84), *68, 69, 71, 72*
Florian, M. L. E., 265, 273, *276*
Floss, H. G., 237, 238, 239, 240, 241, *242, 243, 244*
Foley, G., 138, *156*
Folkers, K., 24(117), *72*
Fornefeld, E. J., 241, *243*
Forsythe, R. H., 261, 262, 263, 264, 275, *276*
Forward, J., 268, *277*
Fosdick, L. S., 135, 136, 138, *157*
Fossum, K., 255, *276*
Foster, G. E., 219, *243*
Foulerton, A., 104, *131*
Fraenkel-Conrat, H. L., 257, 258, 259, *276*
Fraher, M. A., 63(86), *71*
Franklin, T. J., 81, 83, *99*
Fraser, R. R., 24(89, 154), *71, 73*
Frazer, D. T., 252, *276*
Free, A. H., 19(87), *71*
Freeman, R. R., 163(1), 165(1), 166(1), 167(1), 168(1), *207*
Freire, P. S., 140, *153*
Frey, A. J., 215, *244*
Friedberger, E., 254, *276*
Friedlander, H. Z., 190(100), 192(100), 193(100), 194(119), 195(119), 196(119), 205(119), *209*
Froesch, E. R., 143, *154, 156*
Fromm, D., 246, 266, 269, *276*
Frostel, G., 147, *154*
Fuentes, S. F., 212, *243*
Fujii, S., 112, *132*
Fujita, T., 48(112), *72*
Fuller, R., 261, *276*
Fuller, R. A., 257, *276*
Fulton, C. D., 268, *280*
Funk, E. M., 248, 249, 268, *276, 277*
Fusari, S. A., 21(4), *68*

G

Gaby, W. L., 3, *16*
Gaden, E., Jr., 164(5), *207*
Gale, G. O., 82, 83, 84, *99*
Gallup, D. M., 165(14, 15), *207*
Gardell, S., 143, 144, *154*

Gardner, A. D., 4, 16, 18(2, 45), 19(2), 46(2), 68, 69
Gardner, F., 263, 276
Gardner, F. A., 248, 276
Gardner, J. F., 260, 277
Gardner, P., 93, 99
Garibaldi, J. A., 254, 258, 259, 269, 270, 271, 277
Garrard, W., 259, 277
Garrison, L., 20(27), 21(27), 69
Garside, J. S., 83, 85, 99
Garson, W., 21(202), 75
Gaudin, A. M., 166(16, 22), 207, 208
Gayon, M. V., 245, 277
Gazzard, D., 65(20), 69
Geiger, W. B., 109, 133
Gelman, C., 191(104), 210
Gelotte, B., 174, 179(73), 188(90), 189(90), 208, 209
Genghof, D. S., 24(40), 69
Gerhardt, P., 165(14, 15), 207
Gerke, J., 143, 146, 154
Gewurz, H., 260, 277
Ghosh, B. K., 256, 277
Ghuysen, J. M., 256, 279
Gibbons, R. J., 142, 143, 144, 145, 150, 151, 154, 156
Gibson, W. A., 150, 155
Giddings, J. C., 183(82), 184(84), 186, 209
Gillespie, J. H., 261, 267, 268, 269, 272, 277
Gilman, A., 218, 243
Gilmour, D. G., 258, 278
Giovannani, M., 21(35), 69
Glass, D. G., 21(205), 35(205), 36(205), 75
Godfrey, J. C., 24(89, 154, 214), 71, 73, 75
Goebel, W., 241, 243
Goldberg, I. M., 80, 99
Goldsmith, L., 21(205), 35(205), 36(205), 75
Golueke, C. G., 165(13), 167(13), 168(13), 207
Gomez-Revilla, A., 35(73), 70
Good, R. A., 260, 277
Gooder, H., 39(88), 71
Goodman, L. S., 218, 243
Goose, D. H., 146, 155
Gordon, H. A., 135, 137, 138, 140, 142, 156

Gordon, R. E., 127, 128, 132
Gordon, R. F., 83, 85, 99, 263, 267, 275, 277
Goresline, H. E., 268, 277
Gottlieb, D., 80, 99, 110, 119, 127, 128, 132
Gottstein, W. J., 24(89), 25(75), 71
Gourevitch, A., 17(155), 23(50), 28(93), 31(96), 34(92, 94), 40(90), 43(94), 44(93), 47(91), 48, 52(93, 97, 98), 54(97), 55(98), 61(11), 63(11), 64(11), 68, 70, 71, 73
Graf, H., 140, 144, 147, 154
Graham, T., 194, 210
Grahnen, H., 139, 143, 155
Grainger, R. M., 138, 141, 143, 156
Grant, N. H., 46(8), 68
Grau, F. H., 95, 99
Gravenkemper, C. F., 46(99), 71
Gray, C. T., 259, 276, 277
Gray, G. W., 256, 277
Green, B. A., 268, 280
Gregory, G., 141, 155
Grenby, T. H., 140, 155
Grieves, R. B., 166(17, 18, 20), 167(20), 207
Griffith, M., 256, 278
Grisamore, T. L., 150, 156
Grisebach, H., 237, 238, 244
Gröger, D., 222, 223, 224, 228, 231, 232, 237, 238, 239, 240, 241, 243, 244
Gruebbel, A. O., 135, 136, 138, 157
Grunberg, E., 79, 84, 100
Guddal, E., 24, 25(100), 71
Günthard, H. H., 214, 244
Günther, H., 238, 240, 243
Guggenheim, B., 138, 142, 143, 144, 147, 148, 154, 155
Guinee, P. A. M., 95, 99
Gundlach, G., 175(58), 208
Gupta, O. P., 151, 153
Guseva, A. S., 63(44), 69
Gustafson, G., 139, 155
Gustafsson, B. E., 139, 143, 155
Guy, D. B., 205(127), 210

H

Haberman, S., 145, 148, 154, 155

Haines, R. B., 253, 255, 262, 263, 264, 265, 266, 267, 270, 273, 277
Haines, W. J., 19(26), 69
Haldi, J., 139, 155
Hale, C. W., 21(3), 35(3), 36(3), 68
Hale, H. P., 246, 249, 252, 258, 275, 277
Hale, R. W., 252, 275
Halkett, J. A. E., 257, 277
Hall, N. M., 25(47), 70
Hall, R. H., 82, 83, 99
Haller, W., 177, 209
Hallstrand, A., 25(75), 71
Halnan, E. T., 267, 272, 278, 280
Halpen, K. C., 252, 276
Hamaan, J. A., 268, 275
Hamburger, M., 44(213), 51(215), 75
Hamilton-Miller, J. M. T., 25, 54(191), 55(102), 56(190), 61, 63(103), 71, 74
Hamlet, J. C., 24(17), 25(17), 68
Han, F., 92, 99
Handleman, S. L., 147, 155
Hansch, C., 48(104, 112), 49(104), 71, 72
Hanson, J. C., 43(70), 44(70), 54(71), 70
Hara, S., 256, 277
Harden, M. E., 108, 132
Hardwick, J. L., 141, 155
Hardy, D., 53(69), 54(69), 70
Hardy, K., 52(33), 53(33), 69
Hardy, T. L., 46(6), 49(6), 68
Hargie, M. P., 21(4), 68
Harland, H. C., 269, 277
Harley, J. H., 25(41), 69
Harris, H., 126, 131
Harris, P. N., 65(170), 74
Harris, R., 141, 155
Harris, R. S., 140, 156
Harrison, R. W., 135, 137, 138, 140, 142, 156
Harry, E. G., 261, 262, 265, 277
Hart, M. V., 23(168), 74
Hartles, R. L., 146, 155
Hartman, P., 268, 276
Hartung, T. E., 267, 277
Harvey, R. F., 151, 155
Hasenclever, H. F., 260, 280
Havens, G. G., 205(127), 210
Hawes, R. R., 147, 155
Hay, D. I., 146, 154
Hayes, J. A., 122, 132

Hayes, K. M., 268, 277
Hays, E. E., 3, 16
Heatley, N. G., 4, 16, 18(2, 45), 19(2), 21(105), 36(105), 46(2), 68, 69, 72
Hedger, F. H., 21(177), 74
Hehre, E. J., 145, 155
Held, A. J., 148, 155
Helfferich, F., 168(27), 169(27), 208
Hems, B. A., 24(17), 25(17), 68
Henderson, N. D., 21(180), 74
Henderson, O., 166(19), 167(19), 207
Hendon, L. P., 271, 272, 275
Hendricks, F. D., 21(212), 75
Henery-Logan, K. R., 22, 74
Henley, E. J., 193(105), 196(105), 202(105), 210
Herbst, R. N., 60(106), 72
Herzog, H. L., 114, 133
Hess, J. C., 151, 155
Hesseltine, C. W., 123, 132
Hestrin, S., 145, 154, 155
Hill, T. J., 147, 155
Hinshelwood, C., 247, 276
Hirsch, H. L., 19(61), 70
Hirsch, J. C., 260, 277
Hjerten, S., 175(60), 177, 208
Hobbs, D. C., 23(107), 72
Hoder, F., 254, 276
Hoffman, I. D., 149, 155
Hoffman, O. E., 135, 136, 138, 157
Hofmann, A., 214, 215, 216, 220, 221, 238, 242, 243, 244
Hogfeldt, E., 171(38), 208
Hohl, H., 120, 132
Holderby, J. M., 163(4), 207
Holdrege, C. T., 24(89, 154, 214), 52(97), 54(97), 71, 73, 75
Holloway, P. J., 141, 155
Holm, S., 190(98), 209
Holmes, K. K., 55(163), 73
Holmes, L. A., 151, 156
Holmes, N. E., 265, 274
Holt, J. G., 120, 131
Holt, R. J., 56(108), 72
Honda, H., 120, 131
Hopper, M. W., 45(109), 72
Hoppert, C. A., 138, 155
Hoover, J. R. E., 25(47), 45(60), 70

AUTHOR INDEX

Hornemann, U., 238, *242*
Houseley, J. R., 44(158), *73*
Howell, A., Jr., 151, *155*
Huang, H. T., 23(110), 24(111), *72*
Hughes, D. E., 259, *277*
Hütter, R., 123, *132*
Humphrey, A. W., 163(2), 164(2), 168(28), *207, 208*
Hungate, R. E., 111, 123, *132*
Hunt, A. D., 21(56), *70*
Hunt, G. A., 28(93), 34(94), 40(90), 43(94), 44(93), 47(91), 48(91), 52(93, 97), 53(93), 54(93, 97), *71*
Hussey, H. H., 19(61), *70*
Hwang, S. W., 236, *243*

I

Ilavsky, J., 105, *132*
Immel, H., 238, 239, *244*
Ingleman, B., 257, *278*
Irish, D. B., 168(30), 169(30), *208*
Isaeva, N. L., 173(42), *208*
Iwasha, J., 48(112), *72*
Izaki, K., 33(113), *72*, 81, *99*

J

Jablon, J. M., 137, 142, 144, *155, 157*
Jackson, G. G., 55(138), *73*
Jackson, L. J., 60(178), *74*
Jacobs, F. A., 3, *16*
Jacobs, N. J., 259, *276, 277*
Jamison, H. C., 146, 152, *153, 154*
Jamrack, W. D., 171(34), 172(34), *208*
Jansen, A. B. A., 24(17, 114), 25(17, 114), *68, 72*
Jarolmen, H., 90, 91, 97, *99*
Jawetz, E., 63(86), *71*
Jayne-Williams, D. J., 261, *276*
Jeanes, A., 144, *157*
Jenkins, G. N., 141, *154*
Jennings, M. A., 4, *16*, 18(2, 45), 19(2), 46(2), *68, 69*
Jensen, H. L., 103, 104, 105, 106, 110, *132*
Jensen, S., 150, *156*
Joe, E. G., 171(37), *208*
Johansson, M., 240, *242*
John, K., 139, *155*
Johnson, D. A., 24(115), *72*

Johnson, G. W., 185(85), *209*
Johnson, J. D., 166(19), 167(19), *207*
Johnson, J. F., 174, *208*
Johnson, J. R., 20(51), *70*
Johnson, J. S., Jr., 194(112), 199(112), 202(112), *210*
Johnson, M. J., 164(9), *207*
Jones, J. R., 248, *276*
Jones, L. A., 123, *132*
Jones, L. R., 3, *16*
Jones, R. G., 19(25, 26), 20(27), 21(27), *69*, 241, *243*
Jones, W. F., Jr., 55(80), *71*
Jordan, H. V., 138, 142, 143, 150, 151, *154, 155*
Joustra, M. K., 188(88), *209*
Jukes, T. H., 252, *276*
Jungherr, E. L., 252, *275*

K

Kabins, S. A., 66(116), *72*, 91, *99*
Kaczka, E., 24(117), *72*
Kammermeyer, K., 194(117), 195(117), *210*
Kampelmacher, E. H., 94, *99*
Kandel, S. I., 238, *242*
Kaplan, S. I., 173, *208*
Kapsimalis, B., 142, 150, 151, *154*
Karger, B. L., 166(20), 167(20), *207*
Karminiski, M., 257, *277*
Karshan, M., 138, *156*
Kasik, J. E., 61, 64(119), *72*
Kato, K. J., 22, *72*
Katsman, P. A., 3, *16*
Kaufmann, K., 23(122), *72*
Keefer, C. S., 19(156), *73*
Keilich, G., 238, *244*
Kelleher, W. J., 220, 222, 223, 224, 225, 227, 228, 229, 230, 231, 232, 236, *243, 244*
Kellenberger, E., 256, *277*
Keller, W., 257, *278*
Kellett, G. L., 174(54), *208*
Kelley, R. N., 185(85), 186(86), *209*
Kemmerer, A. R., 258, *279*
Kemp, G., 90, 91, 97, *99*
Kempe, L. L., 167(25), *207*
Kempf, B., 26(146), 34(146), *73*
Kendrick, W. B., 122, *132*

AUTHOR INDEX

Kenner, D. D., 120, *132*
Kent, T. H., 78, *99*
Kern, M., 65(152), *73*
Kessner, D. M., 92, *99*
Keyes, P. H., 136, 137, 138, 141, 142, 143, 144, 147, 150, 151, *154, 155*
Kharasch, M. S., 219, *243*
Khokhlov, A. S., 63(44), *69*
Killander, J., 180, 181(75), *209*
Kim, B. K., 229, 230, 231, 232, *243*
Kind, A. C., 46(171), *74*
King, J. V., 258, *278*
King, W. J., 141, *157*
Kirby, W. M. M., 21(23), 45(186), 46(99, 171), *69, 71, 74*
Kishino, K., 218, *244*
Kislak, J. W., 66(123), *72*
Kiser, J. S., 84, *99*
Kitchen, D. K., 66(124), *72*
Kite, O. W., 137, *155*
Kiuchi, K., 81, *99*
Klapper, C. E., 139, *155*
Kleiderer, E. C., 19(26), *69*
Klein, J. O., 40(125), 41(125, 126), *72*
Kleinberg, I., 140, *155*
Kleiner, E. M., 63(44), *69*
Kleinman, G., 168(30), 169(30), *208*
Klemm, L., 91, *100*
Klemt, W., 143, 146, *154*
Kline, G. B., 241, *243*
Kline, L., 255, 260, *276*
Knoettner, P., 142, 151, *154*
Knowles, N. R., 268, *278*
Knox, R., 42, 54(191), 63(103), *71, 72, 74*
Knudsen, E. F., 34(129), 45(128), 57(7), 58(7), 66, 68, 72, 189(93), *209*
Kobel, H., 214, 215, 216, 222, 223, 224, 225, 232, 236, 238, *242*
Kock, R. S., 18(53), *70*
Koenig, M. G., 55(163), *73*
Kohn, A., 257, *278*
Konig, K. G., 138, 139, 140, 142, 143, *155, 156*
Kornfeld, E. C., 241, *243*
Korotkova, G. P., 269, *278*
Korzun, J. N., 168(30), 169(30), *208*
Koster, K. H., 66(131), *72*
Koyama, Y., 120, *131*
Kozinn, P. J., 260, *276*

Kozu, Y., 233, *242*
Kraft, A. A., 261, 262, 263, 264, 268, 269, *275, 278, 279*
Krainsky, A., 113, *132*
Krasilńikov, N. A., 102, 107, 108, 128, *132*
Krasse, B., 137, 139, 140, 142, 143, 144, 147, *154, 155, 156*
Kraus, K. A., 194(112), 199(112), 202(112), *210*
Krause, J. M., 63(77), *71*
Kriss, A. E., 102, 103, 104, 105, 106, 107, 108, *132*
Küster, E., 103, *132*
Kullander, S., 139, *156*
Kunin, C. M., 45(132), 46(132), 47(132), 49(132), 50(132), 51(132), *72*
Kunin, R., 168(29), 175(62), 194(29), *208, 209*
Kuroda, S., 107, 120, 121, *131*
Kusumoto, M., 233, *242*

L

Labhart, A., 143, *154*
Lack, D., 246, *278*
Lamanna, C., 165(11), *207*
Lamb, J. C., III, 166(19), 167(19), *207*
Lampman, C. E., 268, *279*
Lancaster, J. E., 263, 267, *278*
Lanke, L. S., 139, 143, *155*
Larson, L. M., 19(26), *69*
Larson, R. H., 139, 147, *154, 156*
Lascelles, J., 259, 260, *276, 277*
Laschtschenko, P., 254, *278*
Latham, W. C., 190(99), *209*
Lathe, G. H., 175, *209*
Lau, E. F., 169(31), 170(31), *208*
Laubach, G. D., 25(183), *74*
Laurell, C. B., 257, *276, 278*
Laurent, T. C., 180, 181(75), 183(81), *209*
Law, M. L., 139, *155*
Lawr, A. M., 87, 89, *100*
Leach, S. A., 141, *155*
Leanza, W. J., 57(133), *72*
Leaver, C. W., 263, *278*
Lechevalier, H. A., 79, 80, *99*, 102, 103, 104, 123, 129, *131, 132, 133*
Lechevalier, M. P., 79, 80, *99*, 103, 123, *131, 132*

Lederberg, J., 86, *100*
Legault, R. R., 219, *243*
Leibman, A., 257, *274*
Leigh, D. A., 57(36), *69*
Leigh, T., 26(134), *72*
Lein, J., 23(50), 28(93), 31(96), 34(92, 94), 40(90), 43(94), 44(93), 47(91), 48(91), 52(93, 97, 98), 53(93), 54(93, 97), 55(98), 61(95), *70, 71*
LeMinor, L., 55(42), 56(43), *69*
Lemlich, R., 166(20), 167(20), *207*
Lenney, W. S., 142, *156*
Leonards, J. R., 19(87), *71*
Lepper, M. H., 92, *99*
Lessel, E. F., Jr., 120, *131*
Levine, B. B., 65(135), *72*
Levy, H., 257, *279*
Lewis, C. J., 171(35), 173(35), *208*
Lewis, M. J., 94, *99*
Li, N. N., 193(105), 196(105), 202(105), *210*
Lifshitz, A., 267, 270, *278*
Lightbown, J. W., 44(32), 47(32), 51(32), *69*
Lincoln, R. E., 166(21), *207*
Lind, V., 148, *156*
Lindgren, J. E., 240, *242*
Lineweaver, H., 257, *276*
Lingens, F., 241, *243*
Little, M. F., 140, *154*
Littleton, N. W., 136, 148, 151, *156*
Livermore, A. H., 24(40), *69*
Locci, R., 103, 113, *131, 132*
Löe, H., 150, *156*
Long, A. A. W., 43(68), 44(68), 52(33), 53(33), 54(71), *69, 70*
Long, R. B., 193(105), 196(105), 202(105), *210*
Longsworth, L. G., 255, 258, *278, 281*
Lorah, M., 268, *277*
Lorenz, F. W., 267, *278*
Love, J., 65(136), *72*
Loveless, A. R., 212, 213, *244*
Lowery, D. L., 60(178), *74*
Luckey, T. D., 135, 137, 138, 140, *156*
Luedemann, G. M., 110, 114, 119, 125, 127, 128, 130, *132, 133*
Lund, C. G., 66(131), *72*
Lundqvist, C., 139, 143, 146, *155, 156*

Lunn, J. S., 45(37), *69*
Luoma, H., 141, *156*
Lush, I. E., 258, *278*
Lutikova, P. O., 253, 264, 267, 269, *281*
Luttinger, J., 28(93), 34(92), 44(93), 52(93), 53(93), 54(93), *71*
Lwowski, W., 238, *244*

M

McAllister, R. M., 21(56), *70*
McBride, T. J., 24(111), 31(74), *70, 72*
McCabe, W. R., 55(138), *73*
McCall, C. E., 56(175), 63(175), 64(175), *74*
McCarthy, C. G., 31(139), *73*
McCoy, E., 104, 108, *131, 132*
McCullagh, D. R., 19(87), *71*
McDermid, K. M., 258, *278*
McDermott, W., 18(204), 19(141), 20(140), 46(204), 47(204), *73, 75*
MacDonald, J. B., 150, *154, 156*
MacDonald, L., 268, *280*
McDonald, R. E., 135, 136, 138, *157*
MacDonnell, C. R., 258, *276*
McDougall, W. A., 143, *156*
MacGregor, A. B., 140, 143, *156*
McGregor, W. C., 165(12), *207*
McGuire, C., 114, *131*
MacInnes, D. A., 258, *278*
McKee, W. M., 62, 63(142), *73*
McLaren, H. R., 135, 136, 138, *157*
McLaury, D. W., 268, 270, *275*
MacLean, I. H., 18(83), *71*
MacLean, N. A., 65(46), *70*
McNally, E. H., 261, 263, 267, *278, 280, 281*
McNamara, T. F., 84, *99*
McPhail, A. T., 215, *244*
Madsen, K. O., 136, *156*
Magwood, S. E., 263, *278*
Maitland, C. C., 174(52), *208*
Makarova, R. A., 63(44), *69*
Male, C. A., 174(55), *208*
Mallette, M. F., 165(11), *207*
Mallik, K. L., 184(84), *209*
Malone, C, P., 182(79), *209*
Mandel, I. D., 136, *156*
Mandeles, S., 255, 258, 260, *276, 278*

AUTHOR INDEX

Mandelstam, J., 39(162), 73, 253, 278
Manly, R. S., 144, 156
Mann, M. J., 19(25), 69, 241, 243
Mansford, K. L. R., 46(6), 49(6), 68
Mantle, P. G., 222, 242
Manwell, C., 258, 260, 275, 278
Margreiter, H., 20(34), 21(35), 42, 69
Markham, R., 256, 257, 280
Markowitz, M., 205(128), 210
Marquez, J. A., 114, 133
Marr, H., 263, 278
Marshall, A. C., 49(30), 69
Marshall, M. E., 257, 278
Marshall, W., 266, 278
Marthaler, T. M., 143, 156
Martin, H. H., 253, 278
Martin, J. E., Jr., 21(202), 75
Martin, L. F., 177(70), 209
Martin, W. J., 64(137), 73, 146, 156
Mary, N. Y., 223, 224, 225, 227, 229, 236, 244
Massoff, W., 263, 278
Mastropietro-Cancellieri, M. F., 20(12), 65(12), 68
Mateles, R. I., 168(28), 202(123), 205(123), 208, 210
Matsuhashi, M., 33(113), 72
Matsushima, K., 255, 278
Matsushima, Y., 256, 277
Maxted, W. R., 39(88), 71
Maynell, G. G., 89, 100
Meads, M., 19(79), 71
Mehta, M. D., 52(33), 53(33), 69
Mellor, D. B., 263, 278
Merten, U., 193(106), 194(106), 199(106), 202(106), 210
Messersmith, R. E., 97, 99
Meyers, R. J., 175(62), 209
Meynell, E., 86, 87, 88, 89, 99, 100
Meynell, G. G., 87, 99
Michaels, A. S., 191(102, 103), 193(102), 194(116), 195(102), 199(116), 200(102, 116), 201(116), 202(102, 116), 203(116), 209
Michelon, L. E., 220, 244
Michelsen, C. B., 190(99), 209
Mickelsen, O., 140, 156
Mihm, J. M., 127, 132

Miles, A. A., 272, 278
Milicich, S., 45(37), 69
Miller, D., 52(33), 53(33), 69
Miller, G., 236, 244
Miller, H. C., 21(197), 74
Miller, W., 135, 140, 156
Miller, W. A., 263, 269, 278
Millis, N. F., 163(2), 164(2), 207
Mills, J. R., 147, 155
Millward, E., 146, 156
Minghetti, A., 220, 221, 222, 233, 234, 242
Minor, W. F., 23(153), 24(89, 154), 26(153), 52(97), 54(97), 71, 73
Mintz, M. S., 194(113), 210
Mitchell, D. F., 151, 156
Mitsuhashi, S., 55(143), 73
Mizrahi, A., 236, 244
Moggio, W. A., 163(4), 207
Mohler, A., 265, 273, 279
Moir, J. C., 219, 243
Mollberg, R., 35(73), 70
Moller, P., 136, 154
Monet, G. P., 194(114), 195(114), 196(114), 197(114), 210
Monroe, R. J., 266, 269, 276
Monty, K. J., 182(80), 209
Moore, J. C., 174, 175(61), 177(45), 178(45), 208
Moore, J. G., 171(39), 172(39), 208
Moore, J. W., 64, 74
Moran, T., 266, 267, 270, 277
Morch, P., 24(100), 25(100), 71
Morgan, J. G., 44(213), 75
Morpurgo, G., 20(12), 65(12), 68
Morquer, R., 110, 132
Morris, C. J. O. R., 174, 208
Morris, P., 174, 208
Morrison, D. E., 241, 243
Morse, H. N., 194, 210
Morton, J. R., 258, 278
Mosbach, R., 175(60), 208
Moseley, F. T., 258, 280
Moser, R. E., 268, 277
Moses, A. J., 34(94), 43(94), 71
Moses, H. E., 252, 275
Mossman, M. R., 259, 277
Mothes, K., 237, 238, 239, 240, 243, 244
Mothes, U., 238, 239, 240, 242, 243, 244

Moyer, A. J., 18(144), 73
Mühlemann, H. R., 138, 142, 155
Muggleton, P. W., 64(148), 73
Muhler, J. C., 141, 155
Muir, R. D., 3, 16
Mular, A. L., 166(22), 207
Munden, J. E., 169(32), 173(32), 208
Murao, S., 22, 23, 74
Murphy, T. W., 268, 278
Murray, E. G. D., 150, 154
Murray, M. W., 249, 278, 279
Murray, R. G. E., 256, 277, 278
Myers, N. H., 254, 259, 279

N

Nagy, D. A., 254, 258, 259, 276
Nakamura, A., 89, 100
Nakamura, H., 165(10), 168(10), 207
Nakaya, R., 89, 100
Naumann, D. E., 19(57), 70
Naumann, P., 26(146), 34(146), 73
Nayler, J. H. C., 17(18), 22(18), 43(63, 67, 68, 70), 44(68, 70), 52(33, 63), 53(33, 69), 54(63, 69, 71), 68, 69, 70
Naylor, H. B., 253, 263, 267, 270, 275, 278
Neilands, J. B., 259, 277
Neill, J. M., 145, 155
Nermut, M. V., 256, 278
Neter, E., 92, 99
Neubert, M. E., 24(89, 154), 71, 73
Neukom, H., 175, 209
Newman, M., 147, 155
Newman, R. D., 197, 210
Newton, G. G. F., 21(3, 4, 147), 35(3), 36(3, 147), 68, 73
Newton, W. L., 150, 153
Nichols, A. A., 263, 278
Nizel, A. E., 140, 156
Novick, R., 53(14), 68
Noyes, H. J., 135, 136, 138, 157
Nyström, S., 148, 156

O

Obrink, B., 183, 209
O'Callaghan, C. H., 64(148), 73
O'Connor, R. F., 166(22), 207
O'Dell, N. M., 63(77), 71
Oden, E. M., 114, 119, 133

Ørskov, J., 102, 103, 104, 106, 129, 132
Ogasawara, F. X., 267, 278
Ogden, A. L., 258, 278
Ogi, K., 107, 120, 131
Oginsky, E. L., 83, 100
Okamoto, S., 81, 100
Okany, A., 238, 242
Olson, B. H., 21(4), 68
Olsson, I., 143, 144, 154
O'Malley, J., 146, 157
O'Malley, J. E., 142, 156
Onderka, D., 238, 242
Orel, V., 267, 268, 269, 278
Orland, F. J., 135, 137, 138, 140, 142, 156
Orr-Ewing, J., 4, 16, 18(45), 69
Orth, H. D., 238, 244
Ory, E. M., 19(79), 71
Osborne, T. B., 257, 278
Osborne, W. W., 267, 269, 280
Osuga, D. T., 255, 258, 276
Oswald, W. J., 165(13), 167(13), 168(13), 207
Ott, H., 215, 244
Ovary, Z., 65(135), 72
Owen, G., 45(172), 74
Owen, R. D., 84, 100

P

Pacifici, L. R., 222, 223, 224, 228, 230, 244
Panetta, C. A., 24(89), 71
Papas, G., 269, 275
Parikh, S. R., 150, 156
Park, J. T., 256, 278
Parker, C. W., 65(149, 150, 151, 152), 73
Parkinson, T. L., 250, 279
Partyka, R. A., 24(154), 73
Patchett, A. A., 57(133), 72
Paton, A. M., 258, 267, 279
Patterson, R., 252, 279
Pattison, D. A., 194(120), 198(120), 210
Paynes, J. J., 263, 278
Pearce, L. E., 88, 100
Pedersen, K. O., 175(59), 208
Pederson-Bjergaard, K., 66(131), 72
Pelzer, H., 253, 280
Pennell, R. B., 257, 278
Pennella, P., 221, 222, 224, 225, 226, 227, 228, 230, 231, 232, 242
Percival, A., 57(36), 69

Perkins, H. R., 253, 256, 279
Permar, D., 135, 136, 138, 157
Perron, Y. G., 23(153), 24(154, 214), 26(153), 73, 75
Perry, D. M., 21(23), 69
Perry, M. I., 21(202), 75
Peters, T., 257, 277
Petersen, C. F., 268, 279
Petrzilka, T., 214, 244
Phelps, R. A., 258, 279
Pierce, J. A., 194, 210
Pifferi, G., 26(48), 70
Pirie, J. M., 194(115), 210
Platt, A. E., 273, 279
Plieninger, H., 238, 239, 244
Pocurull, D., 96, 100
Polevitsky, K. A., 135, 136, 138, 157
Pollara, B., 260, 277
Pollock, M. R., 80, 81, 100
Polson, A., 175(65), 177, 209
Porath, J., 174(48, 50), 175(57), 176, 179, 180, 188, 208, 209
Porter, R. S., 174, 208
Porterfield, J. S., 145, 156
Potter, L. F., 113, 132
Pozerski, E., 255, 276
Prader, A., 143, 154
Preston, E., III, 21(56), 70
Prévot, A. R., 111, 132
Price, K. E., 17, 61(11), 63(11), 64(11), 68, 73
Pridham, T. G., 102, 110, 119, 123, 132
Pursiano, T. A., 34(94), 43(94), 52(98), 55(98), 61(11, 95), 63(11), 64(11), 71
Pyke, R. G., 238, 242

Q

Quensel, C. E., 139, 143, 155

R

Rachmeler, M., 81, 100
Radlo, J. L., 262, 263, 276
Rammelkamp, C. H., 19(156), 73
Ramsey, H. H., 254, 279
Ramstad, E., 221, 237, 238, 240, 242, 244
Randall, W. A., 21(212), 75
Rapp, G., 146, 157
Rasch, C,, 147, 155

Rassbach, H., 233, 244
Regna, P. P., 21(157), 73
Reich, H. A., 257, 274
Reid, D. B. W., 138, 141, 143, 156
Reilly, E., 86, 100
Rein, C. R., 66(124), 72
Reinhart, R. W., 254, 257, 259, 279
Reisner, S. H., 143, 154
Repaske, R., 257, 279
Reyniers, J. A., 135, 137, 138, 140, 142, 156
Reynolds, M. E., 19(141), 73
Rhodehamel, H. J., Jr., 21(197), 74
Rhodes, M. B., 248, 255, 276, 279
Rich, R., 256, 280
Richard, O., 265, 273, 279
Richards, H. C., 44(158), 73
Richards, M., 23(19, 168), 69, 74
Rickles, R. N., 190(100), 192(100), 193(100), 194(118, 119), 195(119), 196(119), 202(118), 205(119), 209
Rieder, H. P., 220, 244
Rigler, R., 183(81), 209
Risman, G., 65(159), 73
Ritcey, G. M., 171(37), 208
Rizk, S. S., 263, 268, 269, 279
Robbers, J. E., 240, 244
Roberts, E. C., 3, 16
Robinson, H. J., 57(203), 75
Robinson, O. P. W., 57(161), 66(130), 72, 73, 189(93), 209
Robinson, R., 20(51), 70
Rochelmeyer, H., 220, 233, 243, 244
Rodriquez, A. E., 212, 243
Roegner, F. R., 96, 100
Rogers, D. E., 55, 73, 92, 100
Rogers, E. F., 57(133), 72
Rogers, H. J., 39(162), 73, 253, 256, 278, 279
Rolinson, G. N., 17(18), 22(18), 23(19, 168), 25, 26(164), 34(129, 167), 45(128), 47(165), 50(165), 55(167), 57(7, 166), 58(7), 68, 69, 72, 73, 74
Rollo, I. M., 74
Romanoff, A. J., 248, 249, 250, 251, 260, 263, 268, 279
Romanoff, A. L., 246, 248, 249, 250, 251, 260, 263, 268, 279
Romoser, G. L., 268, 276

Roots, M., 141, 155
Rosazza, J. P., 227, 229, 244
Rose, C. L., 65(170), 74
Rosebury, T., 138, 156
Rosen, S. L., 142, 156
Rosenblatt, J. E., 46, 74
Rosenman, S. B., 45(172), 74
Rosenstein, S. N., 146, 156
Rosenthal, E., 150, 156
Rosenthal, I. M., 143, 154
Ross, J. F., 257, 277
Rosselet, J. P., 114, 133
Rosser, F. T., 261, 262, 279
Rothschild, P. D., 66(173), 74
Rotman-Kavka, G., 19(61), 70
Rovelstad, G. H., 147, 156
Rowberry, S. A., 138, 141, 151, 155
Rowland, S. P., 177(70), 209
Rownd, R., 89, 100
Rozwadowska-Dowzanko, M., 20(15), 68
Rubin, A. J., 166(19, 20, 23), 167(19, 23), 207, 208
Rubin, M., 139, 156
Russell, T. J., 24(114), 25(114), 72
Rutherford, P. P., 249, 278, 279
Ruthven, C. R. J., 175, 209
Rutschmann, J., 214, 222, 223, 224, 225, 232, 236, 243, 244
Ryter, A., 256, 277

S

Sabath, L. D., 56(175), 61(174), 62, 63(174, 175), 64(174), 74
Sakaguchi, K., 22, 23, 74
Salivar, C. J., 21(177), 74
Saltman, P., 257, 276
Salton, M. R. J., 253, 256, 268, 277, 279
Salzman, T. C., 91, 100
Samuelsson, E. G., 190(98), 209
Sanders, A. G., 4, 16
Saslaw, M. S., 137, 142, 144, 157
Saunders, A. G., 18(45), 69
Sauter, E. A., 268, 279
Sawyer, S., 150, 154
Saz, A. K., 60, 74
Schade, A. L., 254, 257, 258, 259, 260, 276, 279, 280
Schaffer, L. H., 194(113), 210

Schaible, F. J., 258, 279
Schamschula, R. G., 141, 155
Schenck, J. R., 21(4), 68
Schlientz, W., 215, 244
Schmid, J., 46(179), 74
Schmid, P., 139, 156
Schmid, R., 139, 156
Schmidt, F. J., 268, 279
Schmidt-Nielsen, B., 139, 156
Schnetzer, J. D., 141, 157
Schnitzer, R. J., 79, 84, 100
Scholtan, W., 46(179), 74
Schram, C. J., 146, 154
Schreier, E., 214, 222, 223, 224, 225, 232, 236, 243
Schroeder, H. E., 143, 144, 147, 155
Schroeder, S. A., 95, 100
Schultz, S., 18(204), 46(204), 47(204), 75
Schultz-Haudt, S., 150, 156
Schumacher, J., 44(213), 75
Schuurmans, D. M., 21(4), 68
Schwarting, A. E., 222, 223, 224, 225, 227, 228, 229, 230, 231, 232, 236, 243, 244
Schwimmer, B., 21(180), 74
Scott, W. J., 261, 265, 267, 268, 269, 272, 274, 277, 279
Scotti, T., 102, 104, 131, 220, 233, 234, 235, 242
Sebald, M., 111, 132
Sebba, F., 166(20), 207
Sebek, O., 81, 100
Serlupi-Crescenzi, G., 20(12), 65(12), 68
Sermonti, G., 20(12), 65(12), 68
Seto, T. A., 23(110), 24(111), 72
Shannon, I. L., 139, 150, 155, 156
Shapiro, J., 65(152), 73
Sharp, P. F., 263, 267, 271, 272, 275, 279
Shaw, J. H., 136, 137, 140, 151, 153, 154, 155, 156
Shaw, P. D., 80, 99
Shaw, W. V., 81, 100
Sheehan, J. C., 22, 25(183), 74
Shelley, W. B., 64(184), 74
Shenstone, F. S., 246, 250, 258, 279
Sherwood, T. K., 204(125), 210
Shibata, M., 136, 140, 156
Shidara, I., 106, 107, 120, 131, 132
Shilo, M., 145, 154, 155

Ship, I. I., 140, 156
Shirato, S., 164(6), 207
Shirley, R. L., 64, 74
Shirling, E. B., 127, 128, 132
Shmidov, P. N., 85, 100
Shull, G. M., 23(110), 24(111), 72, 110, 132
Sidell, S., 45(186), 46(99), 71, 74
Siegel, L. M., 182(80). 209
Siemienski, I. C., 91, 99
Silva, R. S., 258, 276
Silvers, S. H., 66(187), 74
Sim, G. A., 215, 244
Simkiss, K., 252, 280
Simmons, E. G., 101, 132
Simon, H., 238, 243
Simons, P. C. M., 248, 252, 263, 266, 280
Sims, J., 147, 155
Sinrod, H. S., 135, 140, 156
Sjoberg, B., 35(73), 70
Slack, G. L., 146, 156
Smith, A. J. M., 248, 280
Smith, D. C. C., 238, 242
Smith, D. H., 56, 74, 90, 93, 99, 100
Smith, H., 43(67), 70, 238, 242
Smith, H. W., 94, 98, 100
Smith, J. T., 54(191), 55(102), 56(189, 190), 61, 63(103), 71, 74
Smith, M. M., 127, 128, 132
Smith, N. R., 150, 154
Snell, E. E., 255, 276
Sobin, B. A., 23(110), 72
Socransky, S. S., 144, 150, 151, 154, 156
Sognnaes, R. F., 137, 155
Sojka, W. J., 94, 100
Sokal, J. E., 92, 99
Somers, G. F., 74
Sonesson, B., 139, 156
Sonoyama, T., 202(123), 205(123), 210
Soper, Q. F., 19(25, 26), 20(27), 21(27), 69
Soulal, M. J., 52(33), 53(33, 69), 54(69), 69, 70
Spalla, C., 102, 104, 131, 220, 233, 234, 235, 242
Spicknell, N. H., 263, 281
Spilsbury, S. F., 218, 244
Spinell, D. M., 147, 154
Spizizen, J. B., 86, 100

Spooner, D. F., 44(158), 73
Spoor, H. J., 66(124), 72
Srb, A. M., 84, 100
Stacy, G. W., 24(40), 69
Stadelman, W. J., 249, 267, 268, 275, 277, 279
Stålfors, A., 140, 157
Stanley, H. R., 138, 142, 154
Starr, M. P., 267, 278
Starr, P. B., 267, 268, 275, 278
Stauffacher, D., 216, 218, 244
Steigbigel, N. H., 56(175), 63(175), 64(175), 74
Steinhauer, B. W., 66(123), 72
Stelling, Em., 139, 155
Stephan, R. M., 135, 136, 138, 140, 144, 156, 157
Stevens, F. C., 255, 276
Stevens, S., 25, 26(164), 34(167), 55(167), 73
Steward, A. R., 48(104), 49(104), 71
Stewart, G. F., 263, 267, 279
Stewart, G. T., 19, 22(193), 23, 35(193), 38, 39(193), 41(193), 45(193), 46(192), 49(193), 56(108), 65, 66(195), 72, 74
Stewart, R. C., 45(60), 70
Stoddart, F., 174(51), 180(51), 208
Stokes, J. L., 267, 269, 270, 277, 280
Stoll, A., 214, 215, 219, 244
Stolpmann, H. J., 263, 278
Stookey, G. K., 141, 157
Stoudt, T. H., 147, 154
Stove, E. R., 43(67, 68), 44(68), 52(33), 53(33, 69), 54(69, 71), 69, 70
Strasters, K. C., 259, 280
Strominger, J. L., 33(113, 196), 72, 74, 80, 100, 256, 280
Stuart, L. S., 261, 267, 280
Sullivan, N. P., 21(197), 74
Summers, D. F., 260, 280
Sus, O., 22(198), 74
Sutherland, R., 39(54, 199), 47(165), 50(165), 57(7, 166), 58(7), 62, 63(200), 68, 70, 73, 74
Sutton, W. S., 268, 278
Suzuki, Y., 81, 100
Sweedler, D. R., 46(99), 71
Sword, C. P., 260, 280
Sylvester, J. C., 19(26), 65(29), 69

Symmes, A. T., 21(197), 74
Szabo, J. L., 21(201), 74

T

Taber, W. A., 219, 220, 233, 238, 244
Taira, T., 112, 132
Takayasu, H., 90, 100
T'ao, L. H., 112, 120, 132
Taschdjian, C. L., 260, 276
Taylor, B., 253, 274
Taylor, D. J., 246, 250, 261, 263, 275
Taylor, E. H., 238, 244
Taylor, J., 94, 100
Tebyakina, A. E., 63(44), 69
Tees, E. C., 66(130), 72, 189(93), 209
Terewaki, Y., 90, 100
Terry, J. M., 139, 156
Terry, P. M., 95, 100
Testa, E., 26(48), 70
Teuscher, G. W., 135, 136, 138, 157
Thayer, J. D., 21(202), 75
Thayer, S. A., 3, 16
Theilade, E., 150, 156
Thelin, H., 35(73), 70
Theodore, T. S., 258, 259, 280
Thiel, J. A., 65(150), 73
Thiele, E. H., 57(203), 75
Thomas, E. W., 66(124), 72
Thomas, G. R., 43(70), 44(70), 52(33), 53(33), 69, 70
Thomas, H. P., 252, 280
Thompson, M. R., 219, 244
Tibbling, P., 188(88), 190(98), 209
Tipper, D. J., 33(196), 74, 80, 100
Tiselius, A., 174, 208
Tognoli, L., 220, 233, 234, 235, 242
Tokin, B. P., 252, 280
Tompsett, R., 18(204), 46, 47(204), 75
Tonolo, A., 221, 222, 224, 225, 226, 227, 228, 230, 231, 232, 233, 241, 242, 243, 244
Tosoni, A. L., 21(205), 35(205), 36(205), 75
Toto, P. D., 146, 150, 156, 157
Totter, J. R., 258, 280
Tramp, M. I., 135, 136, 138, 157
Tresner, H. D., 122, 132
Trevett, M. E., 24(17), 25(17), 68
Trexler, P. C., 135, 137, 138, 140, 142, 156
Treybal, R. E., 170(33), 208

Triggs, R. E., 268, 280
Trubnikova, I. N., 173(42), 208
Trussell, P. C., 265, 268, 270, 273, 276, 280
Tscherter, H., 216, 218, 243, 244
Tsubakimoto, K., 150, 157
Tsuchiya, H. M., 144, 157
Tsutsui, M., 150, 157
Tucker, J. F., 83, 85, 99, 263, 277
Tumilowicz, R., 45(60), 70
Turba, F., 175(58), 208
Turck, M., 55, 62, 63(142), 73, 75
Turner, D. H., 37(5), 68
Tuwiner, S. B., 191(101), 194(101), 196(101), 209
Tweedie, M. C. K., 146, 155
Tybring, L., 24(100), 25(100), 71
Tyler, C., 251, 252, 280
Tyler, V. E., 222, 223, 224, 228, 231, 232, 243

U

Uesseler, H., 241, 243
Ullstrup, A. J., 212, 243
Umbreit, W. W., 83, 100, 104, 108, 132
Unowsky, J., 81, 100
Utsumi, N., 150, 157

V

Van Abeele, F. R., 19(25, 26), 20(27), 21(27), 69
VanArsdel, P. P., Jr., 65(208), 75
Vanderhaeghe, M., 26(76), 30(49, 209), 70, 71, 75
Van Dijck, P. J., 30(49, 209), 70, 75
van Urk, H. W., 220, 244
Vero, L., 221, 222, 224, 225, 226, 227, 228, 230, 231, 232, 242
Vickery, J. R., 265, 267, 268, 274, 277, 279
Vining, L. C., 219, 220, 233, 238, 244
Völkl, A., 238, 239, 244
Voigt, R., 221, 244
Volker, J. F., 136, 139, 146, 154, 155, 157
Vos, J. C., 257, 280

W

Waddington, H. R. J., 43(70), 44(70), 53(69), 54(69), 70

Wade, N. J., 3, *16*
Waehaug, J., 149, *157*
Wagg, B. J., 146, *154*
Wagman, G. H., 114, 119, *133*
Wagner, M., 135, 137, 138, 140, 142, *156*
Waksman, S. A., 2, 10, *16*, 102, 103, 106, 109, 114, 121, 123, *133*
Walden, C. C., 270, *280*
Wallmark, G., 21(211), 40(210), *75*
Walton, J. R., 91, 94, 95, *100*
Warren, G. H., 45(109, 172), *72, 74*
Wang, C. J. K., 111, *133*
Wang, D. I. C., 202(123), 205(123), *210*
Wang, S. L., 166(17, 18), *207*
Ward, W. H., 254, 257, 258, *274*
Warner, R. C., 257, *280*
Watanabe, T., 88, 89, 91, 98, *100*
Waterworth, P. M., 34(16), 40(16), 41(16), 44(32), 47(32), 51(32), *68, 69*
Weaver, J. M., 24(111), *72*, 248, *276*
Webb, F. C., 163(3), 164(3), 177(3), *207*
Webb, G. B., 24(17), 25(17), *68*
Webber, P. A., 138, *155*
Weber, I., 257, *280*
Weber, L. S., 45(172), *74*
Weber, M., 64(119), *72*
Weibaum, G., 256, *280*
Weibull, C., 256, *280*
Weidel, W., 253, *280*
Weigle, W. O., 252, *279*
Weinbaum, G., 256, 257, *280*
Weinstein, M. J., 114, 119, *133*
Weir, R., 65(136), *72*
Weiser, R. S., 254, 256, *275*
Weiss, S., 141, *157*
Welch, H., 21(212), *75*
Wells, I. C., 3, *16*
Wells, R. G., 246, 248, *280*
Welsh, M., 109, *133*
Wendt, H. J., 238, *244*
Wennerholm, G., 148, *156*
Wertheimer, F., 135, 136, 138, *157*
Weston, R. D., 66(21, 22), *69*, 189(92), *209*
Weston, W. A. R. D., 267, *280*
Wettling, C., 268, *280*
Weygand, F., 237, 238, 239, 240, *243, 244*
Wheeler, A. W., 66(21), *69*
White, C. L., 150, *153*
Whitaker, J. B., 255, *276*

Whitaker, R., 271, 272, *279*
Whitehead, C. W., 20(27), 21(27), *69*
Whitehouse, A. C., 44(213), *75*
Whittemore, A. D., 263, *280*
Wiertz, G., 248, 252, 263, 266, *280*
Wiley, A. J., 203(124), 205(124), *210*
Wilke, C. R., 197, *210*
Wilham, C. A., 144, *157*
Wilkinson, S. G., 218, *244*, 256, *277*
Williams, J., 257, 258, *280*
Williams, J. E., 263, *280*
Williams, N. B., 135, 136, 138, *157*
Williams, R. A. D., 141, *155*
Williams, R. J., 255, *276*
Wilson, E. G. H., 262, *275*
Wilson, J. E., 267, *280*
Wilson, M. J., 37(5), *68*
Wilson, T. E., 254, *279*
Wilson, W. W., 20(31), *69*
Wimpenney, J. W. T., 259, *277*
Winberg, E., 64(119), *72*
Winford, T. E., 145, *154*
Winkler, G., 240, *244*
Winkler, K. C., 259, *280*
Winter, A. R., 268, *280*
Winter, J., 238, *242*
Wolf, E. P., 143, *154*
Wolfe, S., 24(214), 31(96), *71, 75*
Wolff, A. E., 136, *154*
Wolff, D. A., 51(215), *75*
Wolin, M. J., 256, *280*
Wolk, J., 263, *281*
Wolley, D. W., 255, *281*
Wolman, B., 145, *155, 157*
Wolman, M., 145, *157*
Wolpert, B., 147, *155*
Wood, J. L., 22(72), *70*
Wood, J. M., 143, 144, *157*
Woodward, R. B., 241, *243*
Work, E., 253, *281*
Wright, D. E., 139, *157*
Wright, M. E., 22(72), *70*
Wybregt, S. H., 143, *154*

X, Y, Z

Yamaguchi, T., 123, 126, *133*
Yamano, T., 218, 233, 242, *244*
Yamatodani, S., 218, 221, 240, *242, 244*
Yau, W. W., 182(79), *209*

Youngner, J. S., 252, *279*
Yurchenko, J. A., 45(109), *72*
Zagaevsky, J. S., 253, 264, 267, 269, *281*
Zamenhof, S., 84, *100*

Zander, H. A., 135, 136, 138, *157*
Zinder, N. D., 86, *100*, 257, *281*
Zinner, D. D., 137, 142, 144, *155*, *157*
Zipkin, I., 139, *156*

SUBJECT INDEX

A

Actinomycetin, 5
Actinomycin, 4
Actinomycin A, 5
Actinomycin D, 11
Albomycin, 6
Allylthiomethyl penicillin, 20
Amicetin, 11
p-Aminobenzyl penicillin, 35
6-Aminopenicillanic acid, 5, 17, 22
  carboxylic acid derivatives of, 33
  disubstituted acetic acid derivatives of, 29, 30, 31
  intrinsic activity of 6-substituted penicillins, 25, 26, 27
  monosubstituted acetic acid derivatives of, 27, 28
  trisubstituted acetic acid derivatives of, 33
Ampicillin, 34, 35
n-Amylpenicillin, 19
Anthramycin, 7
Antibiotic era, 2
Antibiotics
  as natural products, 80
  as secondary metabolites, 79
  produced by bacteria, 2
Antitumor antibiotics, 9
Antiviral antibiotics, 9
Aterrimin, 3
Ayfivin, 3
Azaserine, 11

B

Bacilipins, 3
Bacillin, 3
Bacillomycin, 3
Bacilysin, 3
Bacitracin, 3
Benzyl penicillin, 19
  derivatives of, 38
  salts of, 21
Biochemists, training, 283
Blasticidins, 7

Bulbiformin, 3
Butylthiomethylpenicillin, 20

C

Candicidin, 6
Caries
  antimicrobial agents in treatment, 146, 147
  causes of, 138–144
  dental, 135
  fluoride, 141
  interaction of host, diet, and microbes, 138, 139, 140, 141
  methods of study, 136, 137
  role of sucrose, dextran and levan, 144, 145
Carzinophilin, 11
Cellocidin, 7
Cephalosporin, 5
Cephalosporin N, 21
Chloramphenicol, 6
Chlorellin, 8
Clavacin, 5
Claviformin, 5
Clavine alkaloids, 215
  fermentations, 221
Coli-mycin, 6
Clitocybin, 8

D

Datemycin, 3
Daunomycin, 11
Dental caries, *see* Caries
Diazo-5-oxo-L-norleucine, 10
Dicloxacillin, 45
Diromycin, 11
Drug resistance
  adapted flora, 79
  in agricultural environment, 93
  antibiotic inactivation, 81
  biochemical basis of, 83
  episomal transfer, 87
  evolution, 78
  extrachromosomal, 86
  genetic basis of, 83
  in hospital environment, 91

E

Edeine, 3
Ergot alkaloids
　chanoclavines, 217
　chemistry of, 214
　extraction and analysis, 219
　history of, 211
　mycology of, 212, 213
　pharmacology of, 218
　structures of peptide-type alkaloids, 216
Erythromycins, 6
Eumycin, 3

F

Fluvomycin, 3
Fumagillin,
Fumigacin, 5
Fumigatin, 5
Fungistatin, 3
Fusidin, 5

G

Gentamicin, 7
Globilin, 3
Grisein, 6
Griseofulvin, 5

H

Hen's egg
　changes on storage, 248, 249
　composition of, 250
　course of infection, 261
　　colonization, 269–273
　　penetration, 263–267
　defenses, 252, 254
　　conalbumin, 257–259
　　lysozyme, 253
　strength of, 251
　structure of, 246, 247
N-Heptylpenicillin, 19
p-Hydroxylbenzyl penicillin, 19

I

Industrial fermentations
　compounds of interest, 159–161
　recovery of products, 162
Iturin, 3

K

Kanamycin, 7
Kantellin, 3
Krebiozen, 10

L

β-Lactamase resistance, 17
Levomycetin, 6
Lichenin, 8
Lysergic acid alkaloid fermentations,
　peptide alkaloids, 233
　　biosynthesis of, 236–240
　　organisms producing, 234, 235
　simple lysergic acid derivatives, 222
　　media used in, 225–230
　　organisms producing, 222, 223
　　processes, 224, 225, 231, 232

M

Meserin, 6
Methicillin, 44
Microcins, 112
Micromonospora
　carbonacea, 119
　echinospora, 114
　halophytica, 119
　melanosporea, 113
　narasminoesis, 120
　taxonomy
　　aquatic species, 109
　　Jensen's isolates, 104
　　relationship to actinomycetes, 103
　　Russian species, 107
　　unnatural isolate, 102
Micromonosporin, 6, 7, 109
Mitomycin C, 11
Mycobacillin, 3
Mycosubtilin, 3

N

Nafcillin, 44
Neocidin, 3
Neomycins, 6
Notatin, 4
Nystatin, 6

O

Oxacillin, 45

P

Patulin, 5
Penicidin, 4
Penicin, 22
Penicillin, 4, 13, 15
  biological properties of
    absorbability, 42
    acid stability, 42–44
    gram-negative β-lactamase resistance to, 55–60
    hypersensitivity to, 64, 65
    semisynthetic, 17
      processes for production, 23
    serum binding, 46–50
    staphlococcal β-lactamase resistance to, 51–55
Penicillin B, 4
Penicillin G, 18
  biological activity of derivatives, 24, 25
Penicillin K, 18
Penicillin N, 35
Penicillin V, 17, 21, 22, 44, 45
Penicillin amidase, 23
Penicillin-crustosin, 5
Penicillinase, producers, 20
2-Pentenyl penicillin, 19
Periodontal disease in man and animals, 135
  methods of study, 148, 149
  interactions of microbes, host, and diet, 149, 150
  treatment with antimicrobial agents, 151, 152
Petrin, 3
Phenoxylmethyl penicillin, 20
Polysporin, 8
Purification of biochemicals
  gel filtration, 173–175
    application of, 188, 189
    commercial materials used in, 176, 177
  theory of, 178–183
  liquid ion exchange, 168–171
    commercial sources of, 172
  membrane separations, 190–195
    dialysis, 196, 197
    ultrafiltration, 198–204
  removal of suspended solids, 163, 164
    centrifuge, 164
    dialysis culture, 165
    flocculant, 165
    flotation, 166
    solvent sublation, 167
Puromycin, 11
Pyocyanase, 3
Pyrollnitrin, 7

R

Refuin (anthramycin), 7
Rubomycin, 11

S

Sarkomycin, 10
Semisynthetic penicillins, see Penicillin, semi-synthetic
Streptomycin, 13, 15
Subtenolin, 3
Subtilin, 3
Synnematin B, 21

T

Tetracyclines, 6
Thermomycin, 7
Thermoviridin, 7
Tyrothricin, 4
Trichomycin, 6, 7
Tyrothricin, 3

X, Y, Z

Yeastcidin, 3